Sustainable Mobility and Transport

Sustainable Mobility and Transport

Editors

Tommi Inkinen
Tan Yigitcanlar
Mark Wilson

MDPI • Basel • Beijing • Wuhan • Barcelona • Belgrade • Manchester • Tokyo • Cluj • Tianjin

Editors

Tommi Inkinen
Department of Geography
and Geology
University of Turku
Turku
Finland

Tan Yigitcanlar
School of Architecture
Engineering and Built
Environment
Queensland University of
Technology
Brisbane
Australia

Mark Wilson
School of Planning, Design
and Construction
Michigan State University
East Lansing,Michigan
United States

Editorial Office
MDPI
St. Alban-Anlage 66
4052 Basel, Switzerland

This is a reprint of articles from the Special Issue published online in the open access journal *Sustainability* (ISSN 2071-1050) (available at: www.mdpi.com/journal/sustainability/special_issues/Mobility).

For citation purposes, cite each article independently as indicated on the article page online and as indicated below:

LastName, A.A.; LastName, B.B.; LastName, C.C. Article Title. *Journal Name* **Year**, *Volume Number*, Page Range.

ISBN 978-3-0365-4274-4 (Hbk)
ISBN 978-3-0365-4273-7 (PDF)

Cover image courtesy of Tommi Inkinen

Contents

About the Editors

Tommi Inkinen

Tommi Inkinen has been a Professor of Economic Geography at the University of Turku (UTU, Finland) since 2021. Formerly, he worked as a Research Director and Vice Director of the Brahea Centre at the UTU (2016–2020). He is the current chair of the International Geographical Union's (IGU) Information, Innovation and Technology Commission. He is a chairperson of the University of Turku's strategic multidisciplinary profiling area: 'Maritime Studies'. He is a member of the Academy of Finland's Research Council for Culture and Society (2022–2024). Previously, he worked as the Vice Director of the Department of Geosciences and Geography at the University of Helsinki, and as a Professor of Economic Geography/Human Geography (2008–2016). He was a Visiting Professor at the Queensland University of Technology (QUT, Australia) in 2014.

Tan Yigitcanlar

Tan Yigitcanlar is an eminent Australian researcher with international recognition and impact in the field of urban studies and planning. He is a Professor of Urban Studies and Planning at the School of Architecture and Built Environment, Queensland University of Technology, Brisbane, Australia. Along with this post, he carries out the following positions: Honorary Professor at the School of Technology, Federal University of Santa Catarina, Florianopolis, Brazil; Founding Director of the Australia–Brazil Smart City Research and Practice Network; and Founding Co-Director of QUT City 4.0 Lab. His research findings are disseminated in over 250 articles published in high-impact journals, and 19 key reference books published by the esteemed international publishing houses. According to the 2020 Science-wide Author Databases of Standardised Citation Indicators, amongst urban and regional planning scholars, he is ranked as the #1 most highly cited researcher in Australia, and the #7 most highly cited researcher worldwide. For this achievement, he was recognised as an 'Australian Research Superstar'in the Social Sciences Category at The Australian's 2020 Research Special Report.

Mark Wilson

Mark Wilson is a Professor of Urban and Regional Planning in the School of Planning, Design and Construction at Michigan State University, and Program Director for the PhD in Planning, Design & Construction. His research and teaching interests address urban planning, disruptive technologies, mega-events, and economic development. His current projects include assessing the urban implications of autonomous technologies, planning for industrial parks in Africa and the Middle East, mega-event planning for World Fairs and the Olympics, and the role of innovation, knowledge, and information technology in urban development.

Preface to "Sustainable Mobility and Transport"

Understanding Sustainable Mobility and Transport in the Light of Technological Advancements

This Special Issue is dedicated to sustainable mobility and transport, with a special focus on technological advancements. Global transport systems are significant sources of air, land, and water emissions. A key motivator for this Special Issue was the diversity and complexity of mitigating transport emissions and industry adaptions towards increasingly stricter regulation. Originally, the Special Issue called for papers devoted to all forms of mobility and transports. The papers published in this Special Issue cover a wide range of topics, aiming to increase understanding of the impacts and effects of mobility and transport in working towards sustainability, where most studies place technological innovations at the heart of the matter. The goal of the Special Issue is to present research that focuses, on the one hand, on the challenges and obstacles on a system-level decision making of clean mobility, and on the other, on indirect effects caused by these changes.

All published papers have an international context and contribute to the contemporary debate of sustainable mobility and transport. Papers included in this Special Issue deal with transport studies, logistics, environmental research, transport technologies, and social scientific transport and technology studies. They provide theoretical views as well as empirical cases of the contemporary and future possibilities in clean transport and mobility.

In terms of content, the paper by Moeletsi (2021) explores electric vehicles through resident surveys conducted in South Africa. The results confirm several problems related to electric vehicle adoption, mainly due to their current high prices and availability. The paper proposes several tax and fiscal subsidies policy alternatives in order to make electric cars more feasible. The paper stresses stakeholder collaboration and future initiatives in the development of charging infrastructure. Likewise, the paper by Pipitone et al. (2021) continues electric vehicle studies. This paper points out energy source sustainability as well as environmental stress from the manufacturing process to provide a life cycle assessment. The results indicate a need for systematic scenario building as well as the importance of the vehicle life cycle in the calculation of environmental stress.

The Special Issue includes two papers on computer simulation, modelling, and data driven research. The first one, by Mikušová et al. (2021), explores solutions for transport and mobility sustainability through discrete computer simulation and moves the Special Issue towards simulation models. This paper also applies a selected decision-making methodology and provides insights specifically to meet the needs of sustainable transport. The paper addresses practical problems of urban congestion with applications. The second paper by, Ahmad et al. (2022), considers data-driven deep learning in journals with a specific focus on transport. The paper deals with decision support and governance by applying data resources from newspapers, technology magazines, and science articles from Web of Science. The results stress the importance of sectoral collaboration and the identification of knowledge gaps in reporting on environmental sustainability.

Ruiz-Pérez and Seguí-Pons (2020) move the Special Issue's content to tourism studies and residents transport choices in popular tourism destinations. Their paper addresses the understudied topic of local resident's mobility, and they performed an empirical survey. The results verify the most important variables affecting mobility decisions and transport modes. This is interlinked with the general functionality of public transport. The paper reflects on public transport development plans and highlights the importance of strategic planning in the pursuit of sustainable mobility.

The focus of the Special Issue then moves to concern rail transportation in China (Li et al. 2020). This paper addresses the important topic of regional economic prosperity, and particularly the distinction between cities and rural areas. The authors take the construction of high-speed train networks and their spatial–economic impacts as the focal point. The study utilised statistical nonlinear methods and logit regression models to empirically verify that the convergence effect between Chinese regional types is still relatively weak. Thematically, Lähdeaho and Hilmola (2020) provide a case study of European transport business. The authors looked at the regulation changes that have an impact on international trade and business, focusing on logistics and manufacturing companies. The authors used survey and interview methods to identify and fine-tune three business models that were integrated with circular and sustainable economy goals. The study explains that the majority of companies are not well prepared for rapid external changes, hindering the emergence of optimal business conditions.

A connection between smart cities and sustainable mobility is drawn by Jettanasen et al. (2020); their study looks towards alternative energy sources, particularly piezoelectric material, converting mechanical energy into electricity. This paper presents a pilot device integrated into a bicycle. The paper illustrates that the applied energy harvesting system has the potential to meet the requirements of mobility equipment requiring low power inputs. Cities and urban deliveries are essential in sustainable transport, particularly for the final short-leg deliveries (i.e., door to door). Inkinen and Hämäläinen (2020) continue this in their literature review, and identified a block of studies focusing on sustainable solutions for truck transport. The authors paid particular attention to finding research looking at mitigating emissions from various angles. The review considered different solutions for engines and fuels, resulting into different types of emissions.

The concluding perspective paper by Vilathgamuwa et al. (2022) provides an extensive overview of electric vehicles and its supportive technologies (e.g., infrastructure and power systems). The paper establishes a linkage between urban power systems and system-level problems. The paper presents an integrative novel concept of Mobile-Energy-as-a-Service (MEaaS) for the systemic development and management of urban mobile energy production and consumption for electric vehicles. The concept combines various elements, ranging from measurements to public acceptability and pricing mechanisms.

All ten contributions of this Special Issue study and analyse the extent and diversity of great changes that are needed to support sustainable mobility and transport. All of these papers provide academic as well as practical interpretations of emission control and mitigation, and technological offerings of our time. We hope that the Special Issue provides stimulating ideas and new knowledge of sustainability in transport, with special twist on logistics in general. Based on the published set of papers, we consider that future studies are still needed, particularly on comparative aspects around the world, in order to elaborate the complexity of environmentally friendly and economically sustainable development. As Guest Editors, we thank all of the contributors for their diligent work in revising original drafts. We also thank the large number of reviewers who have dedicated their time and competence to comment on and discuss the drafts in order to improve them. This has increased the overall quality and robustness of this Special Issue. We also thank the editorial staff of *Sustainability* for their smooth and effortless collaboration, resulting in this final product.

References

Ahmad, Istiak, Fahad Alqurashi, Ehab Abozinadah, and Rashid Mehmood. 2022. "Deep Journalism and DeepJournal V1.0: A Data-Driven Deep Learning Approach to Discover Parameters for Transportation"Sustainability 14, no. 9: 5711. https://doi.org/10.3390/su14095711

Inkinen, Tommi, and Esa Hämäläinen. 2020. "Reviewing Truck Logistics: Solutions for Achieving Low Emission Road Freight Transport"Sustainability 12, no. 17: 6714. https://doi.org/10.3390/su12176714

Jettanasen, Chaiyan, Panapong Songsukthawan, and Atthapol Ngaopitakkul. 2020. "Development of Micro-Mobility Based on Piezoelectric Energy Harvesting for Smart City Applications"Sustainability 12, no. 7: 2933. https://doi.org/10.3390/su12072933

Li, Weidong, Xuefang Wang, and Olli-Pekka Hilmola. 2020. "Does High-Speed Railway Influence Convergence of Urban-Rural Income Gap in China?"Sustainability 12, no. 10: 4236. https://doi.org/10.3390/su12104236

Lähdeaho, Oskari, and Olli-Pekka Hilmola. 2020. "Business Models Amid Changes in Regulation and Environment: The Case of Finland–Russia"Sustainability 12, no. 8: 3393. https://doi.org/10.3390/su12083393

Mikušová, Nikoleta, Gabriel Fedorko, Vieroslav Molnár, Martina Hlatká, Rudolf Kampf, and Veronika Sirková. 2021. "Possibility of a Solution of the Sustainability of Transport and Mobility with the Application of Discrete Computer Simulation—A Case Study"Sustainability 13, no. 17: 9816. https://doi.org/10.3390/su13179816

Moeletsi, Mokhele E. 2021. "Future Policy and Technological Advancement Recommendations for Enhanced Adoption of Electric Vehicles in South Africa: A Survey and Review"Sustainability 13, no. 22: 12535. https://doi.org/10.3390/su132212535

Pipitone, Emiliano, Salvatore Caltabellotta, and Leonardo Occhipinti. 2021. "A Life Cycle Environmental Impact Comparison between Traditional, Hybrid, and Electric Vehicles in the European Context"Sustainability 13, no. 19: 10992. https://doi.org/10.3390/su131910992

Ruiz-Pérez, Maurici, and Joana M. Seguí-Pons. 2020. "Transport Mode Choice for Residents in a Tourist Destination: The Long Road to Sustainability (the Case of Mallorca, Spain)"Sustainability 12, no. 22: 9480. https://doi.org/10.3390/su12229480

Vilathgamuwa, Mahinda, Yateendra Mishra, Tan Yigitcanlar, Ashish Bhaskar, and Clevo Wilson. 2022. "Mobile-Energy-as-a-Service (MEaaS): Sustainable Electromobility via Integrated Energy–Transport–Urban Infrastructure"Sustainability 14, no. 5: 2796. https://doi.org/10.3390/su14052796

Tommi Inkinen, Tan Yigitcanlar, and Mark Wilson
Editors

MDPI

Article

Future Policy and Technological Advancement Recommendations for Enhanced Adoption of Electric Vehicles in South Africa: A Survey and Review

Mokhele Edmond Moeletsi [1,2,3]

[1] Agricultural Research Council—Natural Resources and Engineering, Private Bag X79, Pretoria 0001, South Africa; moeletsim@arc.agric.za or mmoeletsi@hotmail.com; Tel.: +27-12-310-2537; Fax: +27-12-323-1157
[2] Business School, Nelson Mandela University, 2nd Avenue Campus, Summerstrand, Port Elizabeth 6001, South Africa
[3] Risk and Vulnerability Science Centre, University of Limpopo, Private Bag X1106, Sovenga, Polokwane 0727, South Africa

Abstract: There are major concerns globally on the increasing population of internal combustion engine (ICE) vehicles and their environmental impact. The initiatives for the advancement of alternative propulsion systems, such as electric motors, have great opportunities, but are marked by a number of challenges that require major changes in policies and serious investment on the technologies in order to make them viable alternative mobility sources around the world. South Africa has struggled a lot in adopting electric vehicles among all the emerging countries. This is mostly attributed to a non-conducive environment for electric vehicle adoption. This study administered a survey consisting of Likert-scale questions in the Gauteng Province to gather information on people's views on some of the major concerns around electric vehicle technology. The survey results demonstrated that Gauteng residents perceive electric vehicle price as the main constraint towards adoption of the technology and introduction of government policy towards addressing this challenge would be helpful. Some of the suggested interventions, such as the rollout of purchasing subsidies and tax rebates, received a high level of satisfaction among the respondents. Future initiatives that tackle issues of charging infrastructure network also received high satisfaction. Thus, there is a need for all stakeholders in the South African automotive industry to improve the enabling environment for the adoption of electric vehicles.

Keywords: electric vehicle policy; electric vehicle incentives; charging infrastructure; green transport strategy; Gauteng province

Citation: Moeletsi, M.E. Future Policy and Technological Advancement Recommendations for Enhanced Adoption of Electric Vehicles in South Africa: A Survey and Review. *Sustainability* **2021**, *13*, 12535. https://doi.org/10.3390/su132212535

Academic Editors: Tommi Inkinen, Tan Yigitcanlar and Mark Wilson

Received: 10 August 2021
Accepted: 27 September 2021
Published: 13 November 2021

1. Introduction

Ownership of vehicles is growing at a faster rate than the human population. The world had 50 million cars in 1950 and the figure rose to 600 million after 50 years; this is expected to increase to 3 billion vehicles by 2050 [1,2]. Without downplaying the importance of vehicles in our lives, their ever increasing population has a negative impact on the environment through increased pollution and emission of greenhouse gases resulting from the combustion of fossil fuels [2]. This has resulted in emissions from the transport sector alone being at around 25% of the total global emissions [3]. In South Africa, transport contributes close to 11% of total emissions with road transport's share being around 91% [4]. It is evident that carbon emissions from the transport sector must be prioritised if the global community is serious about reducing greenhouse gas emissions [5]. At a country level, shifting towards electric vehicles will enable South Africa to attain its national development goals as well as the sustainable development goal (SDG) targets, which include transitioning to a low-carbon economy and providing a sustainable transport system [6].

Substituting internal combustion engine (ICE) vehicles with electric vehicles (EVs) can be a solution for offering urban dwellers a better quality of life with less pollution [7]. The use of EVs has a huge potential of reducing greenhouse gas emissions in the transport sector [7,8]. Thus, it is important for governments around the world to initiate policy measures that support the market penetration of EVs both at district and national level [5]. Electric cars have other advantages as compared to conventional cars, including high energy efficiency, reduced noise and low operating costs [9]. Nevertheless, these socio-economic benefits are accompanied by multiple challenges, such as high purchase prices, low driving range in between recharging, low variety of models, small loading capacity, a need for regular charging, low maximum speed and low acceleration [2,10]. Policy makers should seek ways to understand these challenges and propose strategies that ensure the advancement of the technology while taking into consideration local factors [11].

In South Africa, the adoption of electric vehicles has been extremely slow with less than 1000 battery EVs and plug-in EVs sold since their introduction on the market in 2013 [12,13]. Sales figures of new EVs in 2020 were less than 100, with the best year so far being 2015, when around 120 EVs were sold [14,15]. Preliminary studies in South Africa on consumer perceptions of electric vehicle purchases demonstrated a number of socio-economic and technological barriers that impede the diffusion of EVs in the automotive market [16,17]. The data from the sales figures do not show an increase in the vehicles as in other emerging countries, such as China and India, implying that there is a non-conducive environment for the adoption of EVs in South Africa [12,18–21]. In India alone, close to a million EVs were sold in 2017 [22]. Is this a matter of technology introduced; supporting infrastructure; prices of the vehicles or the type of vehicles? According to Dane [23], South Africa does not have any major policy that can assist the country in shaping the way forward regarding the electric vehicle market and proportion of local industry. It remains to be seen whether the newly adopted green transport strategy will provide the stimulus needed to attract the investment into electric mobility exceeding US$513 billion by 2050 on electric vehicles and US$488 billion by 2030 on hybrid vehicles stated in South Africa's intended contributions to climate change response [4,24].

This study assesses some of the policy recommendations and technological advances that might act as a stimulus to future electric vehicle adoption in South Africa. The research question is as follows: What are the possible impetuses that can be explored to improve the diffusion of electric vehicles in the South African automotive market? The article seeks to augment studies that are already available from other regions in the world with consumer perceptions on EVs from the African continent.

2. Materials and Methods

To assess the policy recommendations that will entice consumers to purchase electric vehicles in the future, a web-based survey was conducted on Survey Monkey from 28 September to 8 November 2018. The study employed a convenience sampling approach in which questionnaires were distributed at first to all the colleagues of the author's workplace, friends, family members and people working in municipalities, universities and other major companies in the Gauteng Province. The initial contacts were encouraged to pass the questionnaire to their network of people. The weblink (https://www.surveymonkey.com/r/electriccars [last accessed on 12 December 2018]) was distributed to the respondents via email, Facebook, short messaging system (SMS) and WhatsApp [16,25]. The study targeted people residing in the Gauteng Province with the potential to purchase a vehicle in the future.

The survey was composed of two parts: general questions and recommendations for policy and improvement questions. In total, 25 of the 26 questions were close-ended multiple choice, which enabled quantitative analysis of the information. Almost all the questions were of a Likert-scale type, mostly asking the degree to which the respondent agrees or disagrees with a certain statement. All these multiple-choice questions were compulsory for all the respondents. The core questions concentrated on both the technological and policy recommendations that were geared towards high consumer acceptance of electric mobility. The two sections present in the questionnaire were as follows [26,27]:

1. General information: Age, gender, race, education, monthly disposable income, region of residence, type of dwelling, property ownership, number of cars owned, type of car owned, type of fuel of the car, engine size of the car, kilometres travelled on a daily basis, awareness that cars emit greenhouse gases that cause climate change, awareness of any full electric vehicle in the South African market and awareness of anybody owning an electric car.
2. Recommendations on policy and adjustments: Chances of buying a full electric vehicle if driving range is increased drastically, chances of buying a full electric vehicle if there will be a charging station at workplace, chances of buying a full electric vehicle if there will be charging station at each of the major fuel filling stations, chances of buying a full electric vehicle if vehicle variety in South African industry is increased, chances of buying a full electric vehicle if there will be government financial incentives in the form of tax rebates, savings in annual car license renewal and toll gate exemptions, chances of buying an electric car if the government mandates a dedicated lane for EVs in major congested roads and any other suggestions that would make them buy an electric vehicle.

To check for the reliability or internal consistency of the questionnaire used to investigate the policy recommendations for South Africa for the adoption of electric vehicles, Cronbach's alpha statistic was used. Cronbach's alpha tests whether the Likert scale type of multiple-choice questions are reliable [28,29]. The equation used is as follows:

$$\propto = \frac{N * \overline{c}}{\overline{v} + (N-1) * \overline{c}}$$

where N is the number of items, $\overline{c}$ is the average covariance between item-pairs and $\overline{v}$ is the average variance.

The results of the Cronbach's alpha showed a value of 0.899, which is above 0.8, indicating that the survey questions for the future recommendation of policy relating to electric vehicle adoption in South Africa show high level of reliability [29].

The responses to the closed-ended survey questions were converted to percentages based on the total number of responses for that particular question that relates to a certain perceived recommendation for policy adjustments in an effort to improve the adoption of EVs in South Africa. The frequency table generated was used as a basis for analysis of the responses from the consumers. The XLSAT statistical software was also used to test whether there were significant differences in the responses against the demographical factors using the chi-square test at 95% and 90% confidence level. The recommendations on the role that government policies can play towards increasing the market share of electric vehicles based on the results of the survey was also deliberated.

3. Results and Discussion

3.1. Basic Information

In total, 402 complete responses were obtained from the survey administered from 24 September to 8 November 2018. All the data were used in the analysis. The respondents were mostly aged between 25 and 44, with over 70% of the total survey population. Less than 2% of the respondents were aged below 18 years and over 65 years. There was a close to 50–50 gender representation, with most people being of African descent (89%). Most of the respondents were highly literate (over 80% post-matric qualification) in contrast to the results of the 2011 census, which stated that over 80% of people living in Gauteng had a matric qualification or lower. This is an indication of the biasness in the sample obtained in the study. Over 50% of the people were living in the Tshwane metropolitan area (Pretoria). In terms of car ownership, the majority of the people had one car (45%), with 26% having no car and 21% having two cars, while less than 10% owned more than two cars. Over 90%

of the respondents were aware of the negative impacts that fumes from ICE vehicles have on the environment.

3.2. *Recommendations on Policy and Advancements of Adoption of Electric Vehicles in South Africa*

Respondents were presented with an opportunity to state their willingness to purchase electric vehicles in the future given certain technological advances and policy changes, and their results are shown in Figure 1.

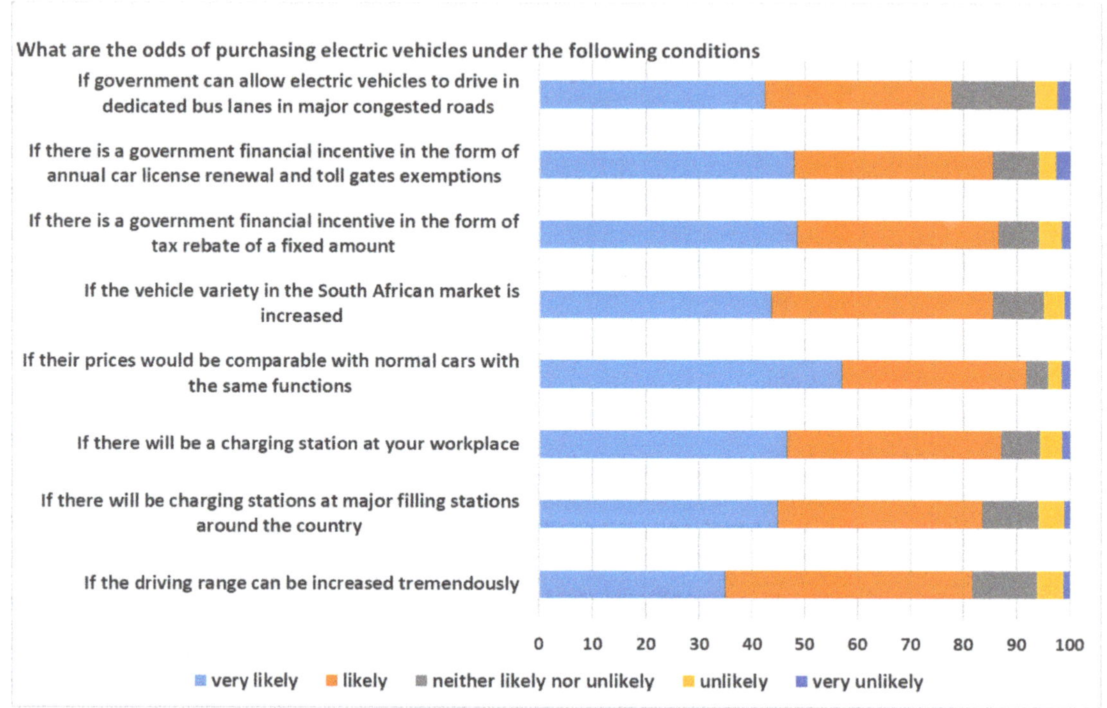

Figure 1. Results of the questionnaire, showing the likelihood of participants to purchase an electric vehicle based on future technological and policy changes in percentages.

Since the driving range of EVs is one of the major thorns in their adoption all over the world [30], respondents are willing (over 80%) to make future purchases of EVs if this problem is solved (Figure 1). The chi-square results showed a significant difference in responses by gender (at 5%), with a considerable percentage of females not eager (7% female vs. 3% male) to purchase EVs even if the driving range is increased (Table 1). In order to fulfil this promise, there has to be major breakthroughs in the development of rechargeable batteries [31]. There are signs of success in this area, as some of the new cars on the market in Europe and North America are proclaimed to have a driving range exceeding 500 km [31]. Based on this information, consumers will have less range anxiety in the future and the number of electric vehicles on South African roads will definitely increase.

Table 1. Chi-square test between demographic factors and factors that tend to improve electric vehicle market penetration.

Factors	a +	b +	c +	d +	e +	f +	g +	h +
Age	0.517	0.955	0.526	0.962	0.538	0.255	0.504	0.503
Gender	0.000 **	0.002 **	0.017 **	0.367	0.141	0.262	0.795	1.000
Education	0.118	0.533	0.603	0.269	0.265	0.034 **	0.165	0.098 *
Region	0.252	0.939	0.911	0.276	0.089 *	0.037 **	0.430	0.049 **
Settlement type	0.013 **	0.791	0.441	0.019 **	0.586	0.357	0.163	0.067
Dwelling type	0.261	0.241	0.562	0.194	0.254	0.255	0.500	0.361
Type of car driven	0.946	0.722	0.450	0.846	0.681	0.024 **	0.027 **	0.095 *
Daily commuting distance	0.717	0.076 *	0.459	0.492	0.006 **	0.389	0.673	0.797
Awareness of climate change	0.205	0.018 **	0.037 **	0.728	0.040 **	0.209	0.068 *	0.737

+: a = driving range can be increased tremendously; b = electric vehicle charging stations are installed at major fuel stations across the country; c = electric vehicle charging stations are installed at workplaces; d = electric vehicle price is reduced and is comparable with internal combustion engine vehicles; e = increased variety of electric vehicles; f = availability of government tax rebates on electric vehicle purchases; g = exceptions for annual car licenses renewals and toll gates; h = government allows electric vehicles to drive on bus lanes on congested roads and during busy times. **: (significant at 5%). *: (significant at 10%).

The poor network of charging stations is currently a deterrent measure that prevents consumers from purchasing EVs [2,32]. The charging points that are present in South Africa are not on the major roads, and this can cause a huge inconvenience to motorists. Even though the majority of the respondents in Gauteng indicate that their adoption of electric vehicles can be enhanced if charging stations are present in major filling stations across the country (Figure 1), there is a significant difference in the responses by gender and awareness to climate change at the 5% significance level (Table 1). A study by Lieven also showed that respondents around the world are dissatisfied by the absence of charging stations on freeways. The presence of this infrastructure at these strategic points would not change the driving patterns of motorists, thus enhancing positive experience on this emerging technology. Thus, there is a need for government to be more involved in the expansion of the charging stations network in South Africa; currently, this expansion has mostly been spearheaded by the private sector, with minimal involvement of the government.

One of the major obstacles that was presented is the lack of convenient charging spots in some countries, including South Africa. It is suggested that placing charging stations at workplaces will resolve this problem. The respondents attest to this notion, with 87% stating that they are likely to buy an EV if there is a policy that encourages employers to install charging stations in their parking lots (Figure 1). However, there is also a significant difference in the responses by gender (Table 1). This idea should be supported by government subsidies for the employers to compensate their electricity usage during charging. An increase in charging spots will also help curb the pre-conceived range anxiety.

It has been documented all over the world that purchase price is the main constraint for purchasing EVs [2,33]. Currently, EVs are more than twice as expensive as ICE vehicles in their class and the drive to be more environmentally friendly can be suppressed by this constraint [34]. Over 90% of the respondents in the Gauteng Province show their willingness to purchase an EV if and only if their prices become comparable to the traditional cars (Figure 1). However, there is a significant difference in responses according to settlement type at a 95% confidence level (Table 1). In order to bridge this price gap, a number of countries in the developed world and emerging economies have introduced lucrative incentives to new EV purchases [35]. Incentives of reducing purchase price are found to be the most effective in the adoption of EVs and increase in market share [11]. This has been a successful intervention by governments in Europe and China, with other countries being encouraged to follow [20,36]. Norway managed to achieve a market share of 18% in 2015 due to the incentives, which have been consistently enhanced since 1990 [11]. It has to be noted that EV adoption needs greater persistence from all stakeholders.

In most areas, electric vehicles are compact cars, and the variety is very low compared with that of ICE cars. The difficulty to obtain an electric SUV or bakkie in South Africa can be challenging to some of the consumers whose travel routes include untarred roads. Even though the respondents show that they are willing to buy an electric vehicle if there can be a variety of classes (Figure 1), there are significant differences in responses according to the daily community distances and awareness of climate change at a 5% significance level (Table 1). In other areas where the technology has been piloted for a number of years, there have been a lot of new classes that have been introduced to the market to stimulate the adoption [37,38]. In recent times, there were also new entrants to the South African market that has increased the variety of EVs.

As stated earlier, governments have a major role to play in ensuring that electric vehicles are adopted by the majority in order to achieve their mitigation obligations. One of the main initiatives has been the introduction of tax rebates of a fixed amount [33,36]. This makes the EVs more affordable and is an indication that governments are also committed to the cause of combating climate change. Around 86% of the people show strong support to the idea (Figure 1), but there are significant differences in responses according to education, region and type of car driven (Table 1). In Norway, EVs are exempted from registration tax, which can exceed 3000 Euros depending on the emission level, engine size and weight of the vehicle [11]. It can be noted that rebates alone will not change the purchasing behaviour of consumers. Other initiatives, such as basic educational programs on TV and print and online campaigns, can ensure that people are well informed and are in a position to make the right choices [39,40].

The introduction of exemptions on annual car registration renewal has the potential to induce more people to consider purchasing electric vehicles [11]. This can be coupled with toll gate exemptions and free charging in malls [36]. The results show that over 85% of the people believe that this incentive can entice them to purchasing EVs, even though there is a significant difference in responses by type of car driven (Table 1). A number of countries, such as Norway, have adopted this policy, and people have responded positively to this initiative [41]. Other incentives in some countries include VAT exemptions, which were introduced to enable the EVs to compete with ICE vehicles on the market [11].

Another intervention will be that of allowing EVs to use dedicated bus lanes in major congested roads [11,33]. This initiative showed the lowest appreciation, with only 77% of the people stating that it might influence them in purchasing an EV (Figure 1) with significant differences according to region (5%), education (10%) and type of car driven (10%). This can be a major incentive to motorists residing in Gauteng, especially in peak hours taking into consideration the density of charging stations. This initiative has worked extremely well in other parts of the world [36].

3.3. *Role of the South African Policies towards EV Adoption*

The importance of advancing the electrification of the transport sector has been highlighted in many studies, as the greenhouse gas emissions from the transport sector continue to rise [42–44]. Research findings by Pautasso et al. [45] and other scientists also showed that adjusting government policies to enable shifting from ICE to electric vehicles has the potential to bear the following benefits over and above environmental considerations: (a) better health conditions for the citizens due to reduced air pollution [46], (b) potential monetary savings in the public health sector due to reduced respiratory diseases caused by air pollution [45], (c) reduction of noise pollution, which will benefit society in a number of ways [47,48] and (d) improved energy security caused by reduced reliance on exported fossil fuels and high volatility attached to crude oil prices [15,49].

There is an extremely low penetration of electric vehicles on the South African market. The main barriers identified include: low percentage of climate-conscious consumers [15], high import duties for vehicles [15], current products not covering an array of consumer needs in South Africa [15] and the high costs of electric vehicles [16]. Most of these barriers can be addressed by drafting and implementation of sound government policies. The lack of policy to address these issues demonstrates to a certain extent a country's lack of commitment in meeting its reduction targets. Consumer educational programmes can be increased to sensitize the communities of the importance of EVs in the fight against climate change. Other factors can be addressed by further developments in the electric vehicle technology and increased drive for current world leaders in vehicle manufacturing to expand their EV production.

Government policies are found to play a critical role towards the adoption of electric vehicles around the world [21,36,41]. In the case of South Africa, the climate change response white paper highlighted the importance of EVs in the fight against climate change by instructing the Department of Transport to initiate a programme that promotes the use of hybrid and electric vehicles [50]. The Gauteng government has drafted a policy on promoting green transport in 2014, but the emphasis was on the decarbonisation of public transport [51]. The green transport strategy is the first main legislature that specifically set the scene for the reduction of greenhouse gas emissions in the transport sector through the promotion of low carbon mobility solutions [4]. The strategy aims at creating and enabling an environment that incentivises the adoption of low or zero tailpipe emission vehicles [4].

The Department of Transport seeks to engage all government institutions in the future to adopt low carbon technologies under the vehicle energy efficiency programme [4]. There is also a drive to develop a policy for new energy vehicles, as outlined in the Auto Green Paper under discussion and public consultations in 2021 [52]. To stimulate the private sector, the Department of Transport together with the Department of Science and Innovation will pave the way for manufacturing incentives for electric vehicles in an effort to increase the local content in future mobility solutions in the country [4]. The government is also proposing to introduce a policy that will scrap all ICE vehicles with a mileage of 400,000 kilometres or more on the roads. This will aid in reducing high-emitting vehicles on the roads and increasing electric vehicle purchases through other complementary strategies envisaged. The other initiative is on the possible introduction of incentives to lower the prices of EVs in an effort to improve competition with ICE vehicles on the market [4]. This strategy has been highly successful in Europe and China, resulting in an enhanced adoption of EVs in those regions [20,33]. In an effort to educate the public, the Department of Transport seeks to initiate green transport awareness campaigns around the country [4].

The introduction of a carbon tax on new vehicle purchases which produce tailpipe emissions is another initiative that tends to promote electric vehicles [53]. Starting from September 2010, all new ICE vehicles that have emission rates exceeding 120 g/km have been subjected to this carbon tax [54]. An amount of 75 Rands for every g/km is charged above the threshold [55]. The upper-limit for double cabs has been set at 175 g/km, with additional g/km being charged at 100 Rands [55]. Vehicles that lack credible emissions data are charged based on their engine size. These efforts are meant to accelerate the introduction of cleaner technologies on South African roads. In the future, the current environmental levy will be reviewed to include other vehicle categories, such as commercial vehicles, which are currently excluded in a drive to improve environmental performance of the country's vehicles [4]. Over and above the ones-off vehicle carbon tax, the government of South Africa is reviewing the introduction of an annual tax on cars based on their emission levels [4]. These future initiatives have the potential to change the electric vehicle environment in South Africa.

As shown in a number of countries where electric vehicles have been successful, government intervention has been key towards providing both financial and non-financial incentives. Early adopters of EVs were mostly lured by the fiscal incentives, but the latest research shows that incentives are no longer the main driver for the market share increase in developed countries, with factors such as an increase in fuel prices and increased public charging stations being the main factors [56]. It is, thus, critical for the South African automotive industry and government to put in place incentives that will increase the diffusion of EVs on the market and break some of the barriers present.

4. Conclusions

Even though the participants of the questionnaire on barriers to adoption of electric vehicles in Gauteng showed positive sentiments, they still indicate that a number of factors need to be changed in order for them to buy an electric vehicle in the future. Policy incentives are considered as one of the most effective measures that can ensure that electric vehicle sales increase. The government has a role to play in ensuring that there are enough incentives for companies that are willing to invest in manufacturing EVs in South Africa. The government should also carefully consider the issue of consumer subsidies for the purchase of the vehicles, as this was found to be the main constraint in this study and other studies across the world. Dedicating resources to research and development in the country will ensure local content that can improve the acceptability of the technology by the consumers. Other considerations for the government are the issuing of exemptions on toll gates and other transport-related levies. In order to improve on range anxiety, the government should consider the issue of expanding the quick charging stations at strategic areas in partnership with the private sector. Currently, in South Africa, vehicle manufacturers are the ones heavily ensuring that charging station density is increased across the country. In countries which have government support for the technology, the market share of EVs has increased tremendously. The government of South Africa has adopted the Green Transport strategy, which will enable the widespread adoption of electric vehicles through a number of proposed initiatives. The implementation of the strategies suggested in this strategy will enhance the chances of South Africa meeting their mitigation obligations. Since this study has some limitations regarding sample size and coverage, further comprehensive studies targeting all segments of the population of South Africa are needed in the future.

Funding: This research was funded by the Agricultural Research Council (South Africa).

Institutional Review Board Statement: Not applicable.

Informed Consent Statement: Informed consent was obtained from all subjects involved in the study.

Data Availability Statement: Not applicable.

Acknowledgments: Authors thank all the colleagues from the Agricultural Research Council who assisted in the collection of data. Thomas Fyfield is thanked for editing this manuscript.

Conflicts of Interest: The authors declare no conflict of interest.

References

1. Samara, S. Electric Cars: Perception and Knowledge of the New Generation Towards Electric Cars. Master's Thesis, School of Economics, Erasmus University Rotterdam, Rotterdam, The Netherlands, 2016.
2. Gärling, A.; Thøgersen, J. Marketing of electric vehicles. *Bus. Strateg. Environ.* **2001**, *10*, 53–65. [CrossRef]
3. Zhang, R.; Fujimori, S. The role of transport electrification in global climate change mitigation scenarios. *Environ. Res. Lett.* **2020**, *15*, 034019. [CrossRef]
4. Department of Transport. *Draft Green Transport Strategy (2017–2050) (Government Gazette No. 41064, 25 August 2017)*; Department of Transport: Pretoria, South Africa, 2017.
5. Ajanovic, A.; Haas, R. Dissemination of electric vehicles in urban areas: Major factors for success. *Energy* **2016**, *115*, 1451–1458. [CrossRef]
6. Stats SA. Sustainable Development Goals: Country Report 2019. Statistics South Africa. 2019. Available online: http://www.statssa.gov.za/MDG/SDGs_Country_Report_2019_South_Africa.pdf (accessed on 25 September 2021).
7. Daina, N.; Sivakumar, A.; Polak, J.W. Modelling electric vehicles use: A survey on the methods. *Renew. Sustain. Energy Rev.* **2017**, *68*, 447–460. [CrossRef]
8. Woo, J.R.; Choi, H.; Ahn, J. Well-to-wheel analysis of greenhouse gas emissions for electric vehicles based on electricity generation mix: A global perspective. *Transp. Res. Part D Transp. Environ.* **2017**, *51*, 340–350. [CrossRef]
9. Thell, L.; Ly, S.; Sundin, H. Electric Cars, the Climate Impact of Electric Cars, Focusing on Carbon Dioxide Equivalent Emissions. 2012. Available online: https://www.diva-portal.org/smash/get/diva2:529571/FULLTEXT01.pdf (accessed on 18 December 2018).
10. Jabeen, F.; Olaru, D.; Smith, B. Acceptability of Electric Vehicles: Findings from a Driver Survey. 2012. Available online: https://therevproject.com/publications/uwa/C2012-ATRF-Acceptability%20Of%20Electric%20Vehicles:%20Findings%20From%20A%20Driver%20Survey-Jabeen%20Braunl%20Et%20Al-ref.pdf (accessed on 26 December 2018).
11. Figenbaum, E. Perspectives on Norway's supercharged electric vehicle policy. *Environ. Innov. Soc. Transit.* **2017**, *25*, 14–34. [CrossRef]
12. Wheels24.co.za Electric Vehicles in SA: Uptake of EVs Slow but Steady. 2018. Available online: https://www.wheels24.co.za/News/Gear_and_Tech/electric-vehicles-in-sa-uptake-of-evs-slow-but-steady-20180116 (accessed on 16 November 2018).

13. Venter, I. Electrified Vehicle Sales in South Africa Remain Dismal, Says uYilo. Engineering News. 2021. Available online: https://www.engineeringnews.co.za/article/electrified-vehicle-sales-in-south-africa-remain-dismal-says-uyilo-2021-04-12 (accessed on 4 September 2021).
14. Daniel, L. Just 92 Electric Vehicles Were Sold in SA Last Year—Here's How Govt Plans to Boost the Market. Business Insider SA. 2021. Available online: https://www.businessinsider.co.za/governments-plan-to-sell-more-electric-vehicles-in-south-africa-2021-5 (accessed on 4 September 2021).
15. Raw, B.; Jack, R. Electric Vehicles 2020: Market Intelligence Report. 2020. Available online: https://www.greencape.co.za/assets/Uploads/ELECTRIC_VEHICLES_MARKET_INTELLIGENCE_REPORT_25_3_20_WEB.pdf (accessed on 12 September 2021).
16. Moeletsi, M.E. Socio-Economic Barriers to Adoption of Electric Vehicles in South Africa: Case Study of the Gauteng Province. *World Electr. Veh. J.* **2021**, *12*, 167. [CrossRef]
17. Kwame, B.M. Unpacking the Technical and Perception Barriers To Electric Vehicle Uptake in South Africa. Master's Thesis, Stellenbosch University, Stellenbosch, South Africa, 2019.
18. Lai, I.K.W.; Liu, Y.; Sun, X.; Zhang, H.; Xu, W. Factors Influencing the Behavioural Intention towards Full Electric Vehicles: An Empirical Study in Macau. *Sustainability* **2015**, *7*, 12564–12585. [CrossRef]
19. Butler, N. Why the Future of Electric Cars Lies in China. Financial Times. 2018. Available online: https://www.ft.com/content/1c31817e-b5a4-11e8-b3ef-799c8613f4a1 (accessed on 31 December 2018).
20. Zhang, X.; Xie, J.; Rao, R.; Liang, Y. Policy incentives for the adoption of electric vehicles across countries. *Sustainability* **2014**, *6*, 8056–8078. [CrossRef]
21. Wang, S.; Wang, J.; Li, J.; Wang, J.; Liang, L. Policy implications for promoting the adoption of electric vehicles: Do consumer's knowledge, perceived risk and financial incentive policy matter? *Transp. Res. Part A Policy Pract.* **2018**, *117*, 58–69. [CrossRef]
22. Globaldata. Energy Electric Vehicle Market Gaining Momentum in India. 2018. Available online: https://www.power-technology.com/comment/electric-vehicle-market-gaining-momentum-india/ (accessed on 2 January 2019).
23. Dane, A. The Potential of Electric Vehicles to Contribute to South Africa's Greenhouse Gas Emissions Targets and Other Developmental Objectives: How Appropriate Is the Investment in Electric Vehicles as a NAMA? Cape Town. 2013. Available online: https://media.africaportal.org/documents/13-Dane-Electric_vehicles.pdf (accessed on 28 December 2018).
24. Department of Environmental Affairs. South Africa's Intended Nationally Determined Contribution (INDC). 2015. Available online: https://www4.unfccc.int/sites/ndcstaging/PublishedDocuments/South%20Africa%20First/South%20Africa.pdf (accessed on 12 December 2018).
25. Moeletsi, M.E. Electric Cars: Their Carbon Implications and Adoption in South Africa. Master's Thesis, Business School, Nelson Mandela University, Port Elizabeth, South Africa, 2019.
26. Krupa, J.S.; Rizzo, D.M.; Eppstein, M.J.; Lanute, D.B.; Gaalema, D.E.; Lakkaraju, K.; Warrender, C.E. Analysis of a Consumer Survey on Plug-In Hybrid Electric Vehicles. *Transp. Res. Part A Policy Pract.* **2014**, *64*, 14–31. [CrossRef]
27. Jabeen, F. The Adoption of Electric Vehicles: Behavioural and Technological Factors. Ph.D. Thesis, Businnes School, The University of Western Australia, Perth, Australia, 2016.
28. StatisticsHowTo. Cronbach's Alpha: Simple Definition, Use and Interpretation. 2014. Available online: https://www.statisticshowto.com/cronbachs-alpha-spss/ (accessed on 12 November 2018).
29. Tavakol, M.; Dennick, R. Making sense of Cronbach's alpha. *Int. J. Med. Educ.* **2011**, *2*, 53–55. [CrossRef]
30. Tsang, F.; Pedersen, J.S.; Wooding, S.; Potoglou, D. *Bringing the Electric Vehicle to the Mass Market a Review of Barriers, Facilitators and Policy Interventions*; Working Papers; RAND Corporation: Santa Monica, CA, USA, 2012; pp. 1–77.
31. Bessenbach, N.; Wallrap, S. Why Do Consumers Resist Buying Electric Vehicles? 2013. Available online: http://studenttheses.cbs.dk/bitstream/handle/10417/4329/nadine_bessenbach_og_sebastian_wallrapp.pdf (accessed on 15 December 2018).
32. Bonges, H.A.; Lusk, A.C. Addressing electric vehicle (EV) sales and range anxiety through parking layout, policy and regulation. *Transp. Res. Part A Policy Pract.* **2016**, *83*, 63–73. [CrossRef]
33. Langbroek, J.H.M.; Franklin, J.P.; Susilo, Y.O. The effect of policy incentives on electric vehicle adoption. *Energy Policy* **2016**, *94*, 94–103. [CrossRef]
34. Ginsberg, J.M.; Bloom, P.N. Choosing the Right Green Marketing Strategy. *MIT Sloan Manag. Rev.* **2004**, *46*, 79–84.
35. Rezvani, Z.; Jansson, J.; Bodin, J. Advances in consumer electric vehicle adoption research: A review and research agenda. *Transp. Res. Part D Transp. Environ.* **2015**, *34*, 122–136. [CrossRef]
36. Lieven, T. Policy measures to promote electric mobility—A global perspective. *Transp. Res. Part A Policy Pract.* **2015**, *82*, 78–93. [CrossRef]
37. PluginIndia. Electric Vehicles in India. 2018. Available online: http://www.plugindia.com/vehicles.html (accessed on 2 January 2019).
38. ITDP. China Tackles Climate Change with Electric Buses. 2018. Available online: https://www.itdp.org/2018/09/11/electric-buses-china/ (accessed on 2 January 2019).
39. Jin, L.; Slowik, P. Literature Review of Electric Vehicle Consumer Awareness and Outreach Activities. 2017. Available online: https://theicct.org/publications/literature-review-electric-vehicle-consumer-awareness-and-outreach (accessed on 30 December 2018).
40. Argue, C.; Hou, E. Emotive—Promoting the Electric Vehicle Experience. *World Electr. Veh. J.* **2016**, *8*, 685–690. [CrossRef]
41. Contestabile, M.; Alajaji, M. Will Current Electric Vehicle Policy Lead to Cost-Effective Electrification of Passenger Car Transport? *Green Energy Technol.* **2018**, *110*, 75–99. [CrossRef]

42. Umar, M.; Ji, X.; Kirikkaleli, D.; Alola, A.A. The imperativeness of environmental quality in the United States transportation sector amidst biomass-fossil energy consumption and growth. *J. Clean. Prod.* **2021**, *285*, 124863. [CrossRef]
43. Song, Y.; Zhang, M.; Shan, C. Research on the decoupling trend and mitigation potential of CO2 emissions from China's transport sector. *Energy* **2019**, *183*, 837–843. [CrossRef]
44. Raza, M.Y.; Lin, B. Decoupling and mitigation potential analysis of CO_2 emissions from Pakistan's transport sector. *Sci. Total Environ.* **2020**, *730*, 139000. [CrossRef] [PubMed]
45. Pautasso, E.; Osella, M.; Caroleo, B. Addressing the Sustainability Issue in Smart Cities: A Comprehensive Model for Evaluating the Impacts of Electric Vehicle Diffusion. *Systems* **2019**, *7*, 29. [CrossRef]
46. Wu, J.; Liao, H.; Wang, J.W.; Chen, T. The role of environmental concern in the public acceptance of autonomous electric vehicles: A survey from China. *Transp. Res. Part F Traffic Psychol. Behav.* **2019**, *60*, 37–46. [CrossRef]
47. Pardo-Ferreira, M.d.C.; Rubio-Romero, J.C.; Galindo-Reyes, F.C.; Lopez-Arquillos, A. Work-related road safety: The impact of the low noise levels produced by electric vehicles according to experienced drivers. *Saf. Sci.* **2020**, *121*, 580–588. [CrossRef]
48. Kumar, M.S.; Revankar, S.T. Development scheme and key technology of an electric vehicle: An overview. *Renew. Sustain. Energy Rev.* **2017**, *70*, 1266–1285. [CrossRef]
49. Li, Y.; Chang, Y. Road transport electrification and energy security in the Association of Southeast Asian Nations: Quantitative analysis and policy implications. *Energy Policy* **2019**, *129*, 805–815. [CrossRef]
50. Department of Environmental Affairs. *National Climate Change Response White Paper*; Department of Environmental Affairs: Pretoria, South Africa, 2011; pp. 1–86.
51. Department of Roads and Transport. *Promoting Sustainable (Green) Transport in Gauten*; Department of Roads and Transport: Johannesburg, South Africa, 2014; p. 24.
52. Department of Trade Industry & Competition. *Auto Green Paper on the Advancement of New Energy Vehicles in South Africa (Public Consultation Version)*; Department of Trade Industry & Competition: Pretoria, South Africa, 2021; pp. 1–15.
53. SA Treasury. *Carbon Tax Bill*; National Treasury: Pretoria, South Africa, 2018; pp. 1–32.
54. Group1Nissan. South African Vehicles and Carbon Emissions. 2019. Available online: https://www.group1nissan.co.za/carbon-emissions/ (accessed on 2 December 2019).
55. National Treasury. Press Release Regarding CO2 Vehicle Emissions Tax. 2010. Available online: http://www.treasury.gov.za/comm_media/press/2010/2010082601.pdf (accessed on 13 December 2019).
56. Wang, N.; Tang, L.; Pan, H. A global comparison and assessment of incentive policy on electric vehicle promotion. *Sustain. Cities Soc.* **2019**, *44*, 597–603. [CrossRef]

Article

A Life Cycle Environmental Impact Comparison between Traditional, Hybrid, and Electric Vehicles in the European Context

Emiliano Pipitone *, Salvatore Caltabellotta and Leonardo Occhipinti**

Department of Engineering, University of Palermo, 90128 Palermo, Italy; salvatore.caltabellotta@unipa.it (S.C.); leo.occhipinti96@gmail.com (L.O.)
* Correspondence: emiliano.pipitone@unipa.it

Abstract: Global warming (GW) and urban pollution focused a great interest on hybrid electric vehicles (HEVs) and battery electric vehicles (BEVs) as cleaner alternatives to traditional internal combustion engine vehicles (ICEVs). The environmental impact related to the use of both ICEV and HEV mainly depends on the fossil fuel used by the thermal engines, while, in the case of the BEV, depends on the energy sources employed to produce electricity. Moreover, the production phase of each vehicle may also have a relevant environmental impact, due to the manufacturing processes and the materials employed. Starting from these considerations, the authors carried out a fair comparison of the environmental impact generated by three different vehicles characterized by different propulsion technology, i.e., an ICEV, an HEV, and a BEV, following the life cycle analysis methodology, i.e., taking into account five different environmental impact categories generated during all phases of the entire life of the vehicles, from raw material collection and parts production, to vehicle assembly and on-road use, finishing hence with the disposal phase. An extensive scenario analysis was also performed considering different electricity mixes and vehicle lifetime mileages. The results of this study confirmed the importance of the life cycle approach for the correct determination of the real impact related to the use of passenger cars and showed that the GW impact of a BEV during its entire life amounts to roughly 60% of an equivalent ICEV, while acidifying emissions and particulate matter were doubled. The HEV confirmed an excellent alternative to ICEV, showing good compromise between GW impact (85% with respect to the ICEV), terrestrial acidification, and particulate formation (similar to the ICEV). In regard to the mineral source deployment, a serious concern derives from the lithium-ion battery production for BEV. The results of the scenario analysis highlight how the environmental impact of a BEV may be altered by the lifetime mileage of the vehicle, and how the carbon footprint of the electricity used may nullify the ecological advantage of the BEV.

Keywords: life cycle analysis; passenger car; environmental impact; hybrid electric vehicle; battery electric vehicle

Citation: Pipitone, E.; Caltabellotta, S.; Occhipinti, L. A Life Cycle Environmental Impact Comparison between Traditional, Hybrid, and Electric Vehicles in the European Context. *Sustainability* **2021**, *13*, 10992. https://doi.org/10.3390/su131910992

Academic Editors: Tommi Inkinen, Tan Yigitcanlar and Mark Wilson

Received: 10 September 2021
Accepted: 24 September 2021
Published: 3 October 2021

1. Introduction

Worldwide vehicle production growth over the past decades has caused strong emissions increments which have affected both population and industrial sectors globally. EU-28's CO_2 emissions correspond to 10.8% of global CO_2 emissions [1]. In 2017, the transport sector contributed to 27.9% of the EU-28 CO_2 production, with a passenger cars participation of 43.5%, which hence represents about 12.1% of the total EU-28 CO_2 emissions [2]. To reduce air pollution, governments issued increasingly stringent regulations, pushing vehicle manufacturers towards innovative solutions. With a view to eco-sustainable mobility, battery electric vehicles (BEVs) and hybrid electric vehicles (HEVs) are nowadays proposed as clean or light-environmental impact technologies for road transport. In particular, BEVs are often promoted as zero-emitting vehicles since they are propelled by the use of electric energy, in contraposition to internal combustion engine vehicles (ICEVs) and HEVs, which instead make use of fossil fuels. If, on the one hand, it is true that a BEV does not produce tailpipe emissions while moving, on the other hand, it must be considered that the electric energy employed by the BEV may have been produced by the use of fossil fuels, thus contributing to air pollution and CO_2 emissions. A fundamental aspect, hence, in the evaluation of the environmental impact related to the use of a BEV is to know the source, or the mix of different sources, employed to produce the electric energy. It is rather obvious that if the electric energy is obtained only by renewable sources (e.g., hydraulic, solar, wind energy), no pollutant emissions are produced during the use of a BEV. Unfortunately, electric energy is not produced exclusively by means of renewable sources, but is still obtained by a mix of different sources which may have a carbon footprint (e.g., coal, natural gas, oil) or may produce different and hazardous waste, such as nuclear energy. Every single country is characterized by a particular mix of energy sources employed to generate electricity, which is usually referred to as the electricity mix (of the country). The pollutant emissions related to the use of a BEV hence depend on the particular mix employed for the electric energy production, and in turn, on the country in which the vehicle is employed. The considerations made up to this point regard only the energy transformation chain which is involved in the BEV propulsion phase. It is, however, widely recognized that a fair and complete evaluation of the pollutant emissions related to vehicle use should also take into consideration the vehicle production phases, since materials employed and manufacturing process may play an important role in determining the real environmental impact related to the use of a vehicle: this aspect is crucial when a comparison between the environmental footprint of different vehicle technology is performed. For this reason, several studies were carried out dealing with the evaluation of the environmental impact of vehicles through a life cycle assessment (LCA) methodology, that considers the entire life of the vehicle, from the production phase to the on-road use of the vehicle, and to the final disposal. In [3], the life cycle greenhouse gas emissions (GHG) of battery electric vehicles and conventional gasoline internal combustion engine vehicles are calculated and compared in different Chinese energy production mix scenarios (2010 and 2014). As a result, the ICEV revealed 34.9 tCO_2eq/vehicle with the 2010 electricity mix, and 29.7 tCO_2eq/vehicle with the 2014; as instead regards the BEV, 42.5 tCO_2eq/vehicle (i.e., +21.7%) were evaluated with the 2010 electricity mix, and 31.4 tCO_2eq/vehicle (i.e., +5.72%) with the 2014. In [4] a comparative LCA of European medium-sized passenger vehicles ("VW Passat class") was carried out, adopting Switzerland as a vehicle usage scenario, and taking into consideration different drive technologies and fuel supply chains, representing both the present and the modern future (2030) state of development. As a result, the CNG-fueled ICEV, the diesel hybrids, and the BEV charged employing the average European electricity mix, proved to generate the lowest life cycle greenhouse gas emissions, in the order of 210 g CO_2eq/km, while the highest levels (300 g CO_2eq/km) were calculated for the gasoline-fueled ICEV. In [5], the environmental impacts of a vehicle with an internal combustion engine (diesel, petrol, and CNG) is compared to a battery electric vehicle, considering the battery of the electric vehicle produced using the electricity from the Chinese, the European, and a 100% photovoltaic energy mix. In [6], a comparison based on real consumption data of two cars (Nissan Leaf BEV and Mercedes A-170 ICEV) on the New European Driving Cycle (UNECE 2005) is presented. In this study, great attention was paid to vehicle life cycle including both the high-voltage battery and the rest of the car components, based on well-detailed inventories and model parameters. All of these studies highlight how the vehicle production processes, traveled distance, and energy mix may substantially influence the real environmental impact of an electric vehicle during its entire life cycle [7]. According to these considerations, the authors of the present paper followed a life cycle approach to perform a fair comparative evaluation of the real environmental impact connected to the use of three different kinds of vehicle (i.e., ICEV, HEV, and BEV) derived from the 2019 market, considering different driving distances and different energy mixes, thus contributing to delimit the effective pollutant behavior of each kind of vehicle, even changing the conditions of use.

With the aim to faithfully represent the current conditions of production, use, and disposal, the analysis was performed employing the most up-to-date data present in the scientific literature (before the pandemic). Unlike other articles in the scientific literature, this work does not analyze only take into account greenhouse gas emissions, but also properly takes into consideration the terrestrial acidification, the particulate matter formation, and the deployment of mineral and fossil resources, thus highlighting, from several points of view, the real advantages, disadvantages, and limitations of modern car propulsion technologies.

2. Method and System Details

Life Cycle Assessment (LCA) is an analytical and systematic methodology that evaluates the environmental footprint of a product or service, along its entire life cycle ("from cradle to gate") [8]; hence, starting from the phases of extraction of the necessary raw materials, the analysis involves the production and distribution phases, the use and the final disposal. At the end of the calculation, the environmental impact of a product is quantified according to various environmental impact indicators. Worldwide, the LCA methodology is regulated by ISO standards 14040 [9,10] and is structured into the following phases:

- Goal and scope definition
- Life cycle inventory (LCI)
- Life cycle impact assessment (LCIA)
- Interpretation of results

3. Goal and Scope Definition

As already mentioned, the goal of this study is to compare, from a life cycle perspective, the ecological consequences related to the use of three vehicles characterized by different propulsion technologies, i.e., a gasoline ICEV, a gasoline HEV, and a BEV. Given the difficulty to analyze, in a single paper, all the different kinds of vehicles (from mini cars to SUV, nine different passenger cars categories may be identified, corresponding to the Euro car segments from A to J) the authors decided to focus on the most diffused kind of passenger cars in Europe, i.e., small-medium cars (corresponding to the B-C segments) which account for over half of total EU car sales [11]. Starting from raw materials supplying and processing, considering vehicle production and assembly, and arriving at the use and final disposal phase, the overall environmental impact of each vehicle was assessed employing GREET 2020.NET (Greenhouse gases, Regulated Emissions, and Energy use in Transportation), a widely adopted open-source suite for life cycle analysis developed by Argonne National Laboratory [12,13]; for the environmental impact assessment, the ReCiPe 2016 methodology was followed [14,15], taking into considerations five different environmental impact categories, each one represented by means of a proper characterization factor:

(1) Climate change: mainly due to the increase in greenhouse gases (GHG) in the atmosphere, it is accounted for by the global warming potential (GWP) which expresses the equivalent mass of CO_2 emitted to obtain a product or a service.

(2) Fine particulate matter formation: focusing on human population intake of $PM_{2.5}$, the change in ambient concentration of $PM_{2.5}$ after the emission of primary $PM_{2.5}$ or precursors (e.g., NH_3, NO_X, SO_2) is evaluated by means of the Particulate Matter Formation Potentials (PMFP), expressed in terms of the equivalent mass of $PM_{2.5}$ [14,15].

(3) Terrestrial acidification: soil acidity variation due to acid deposition is taken into account by means of the Terrestrial Acidification Potentials (TAP) which expresses the amount of acidifying emissions (e.g., NO_X, NH_3, and SO_2) introduced in the atmosphere in terms of the equivalent mass of SO_2 [14,15].

(4) Mineral resource scarcity: although minerals are available in almost infinite amounts in the world, the real availability of a mineral resource primarily depends on the grade, i.e., the concentration of the mineral within an ore. The primary extraction of a mineral resource leads to an overall decrease in the concentration of that resource in

ores worldwide. The lower is the grade of a mineral, the larger will be the amount of ore mined to extract the same amount of resource in the future, which obviously will imply larger land use, higher energy consumption, and waste production: this is the environmental impact related to the mineral resource depletion. The Surplus Ore Potential (SOP), expressed as the equivalent mass of copper, indicates the average extra amount of ore produced in the future caused by the extraction of a mineral resource considering all future production of that mineral resource [14–16].

(5) Fossil resource scarcity: with the same significance and approach followed for mineral resources, depletion of fossil energy resources is expressed in terms of the equivalent mass of oil using the Fossil Fuel Potential (FFP), which is calculated as the ratio between the higher heating value of the fossil resource and the higher heating value of crude oil [14,15].

Table 1 resumes the environmental impact categories considered, together with the characterization factors and the unit adopted [15], to quantify the contribution to each category.

Table 1. Impact categories and characterization factors employed for the LCIA.

Impact Category	Characterization Factor	Unit
Climate change	Global Warming Potential (*GWP*)	kg CO_2-eq
Terrestrial acidification	Terrestrial Acidification Potential (*TAP*)	kg SO_2-eq
Fine particulate matter formation	Particulate matter formation potential (*PMFP*)	kg $PM_{2.5}$-eq
Mineral resource scarcity	Surplus ore potential (*SOP*)	kg Cu-eq
Fossil resource scarcity	Fossil fuel potential (*FFP*)	kg Oil-eq

With the aim to give the obtained results a greater generality and applicability, the environmental impact comparison was carried out considering three fictitious vehicles (a gasoline ICEV, a gasoline HEV, and a BEV) representative of real market products, whose performance and characteristics, as shown further on, were determined on the basis of real vehicles available on the 2019 market. The functional unit [10] of this study is hence the distance traveled by each vehicle during its entire life. For the life cycle impact evaluation, the following assumptions were also made by the authors:

(6) The three vehicles compared were assumed to be produced in Germany in 2019: the reason for this assumption is that, as reported by ACEA [17], Germany is the European state with the largest production of passenger cars (4.66 million units in 2019, i.e., 30% of EU production) followed by Spain (2.18 million units in 2019, 14% of EU production).

(7) The lithium-ion batteries of both the hybrid and the electric vehicles were assumed to be manufactured in China, which is the largest producer in the world: considering, for example, lithium-ion bases batteries for electric vehicles, in 2017 about 70% of the world production (145 out of 206 GWh) came from China [18].

(8) A reference lifetime distance of 150,000 km traveled in Europe was assumed for each of the three vehicles considered: this was established considering that in the European Union, a passenger car travels an average distance of 12,529 km each year, and has an average useful life of 11.5 years [19]—it results in an average distance traveled by a passenger car during the lifetime of 144,085 km, rounded by the authors to 150,000 km.

According to the assumptions made, hence, the German electricity mix was considered for all the vehicle production phases, excluding the lithium-ion batteries, whose production was instead considered under the Chinese electricity mix. In contrast, in regards to the use phase, each vehicle was supposed to travel all over Europe for the whole lifetime distance. As a result of this assumption, the European gasoline production and distribution chains were assumed for the calculation of the impact related to the fuel employed in combustion engine vehicles, and the average European electricity mix was considered to account for the impact related to the electric energy used by the BEV.

As can be noted, in their evaluation, the authors followed the approach to represent the "most probable situation" making reference to the most diffused case; as a result, the assumptions made delimited the target of this study to the passenger cars belonging to the small-medium category, produced and assembled in Germany, endowed of Li-ion batteries produced in China, and fueled (or recharged) all over Europe.

Moreover, as will be shown further, with the aim to weigh the role of both electricity mix and traveled distance, a scenario analysis was also performed, adding to the comparative analysis two more lifetime mileages and taking into account the two extreme situations currently present in Europe, that is the Norwegian electricity mix, characterized by an almost null carbon footprint, and the Polish electricity mix, still dominated by the recourse to fossil sources. These cases have allowed extending the limits of the analysis, embracing all the European real possible scenarios.

4. Life Cycle Inventory

As already mentioned, the life cycle environmental impact comparison was carried out considering three fictitious vehicles representative of real market products; to this purpose, for each propulsion technology considered (i.e., ICEV, HEV, and BEV), the authors defined the characteristics of a plausible reference vehicle on the basis of the information available on five different vehicles belonging to the B-C segments and available in the 2019 European car market. With regards, for example, to the traditional ICEV, Table 2 reports the technical data provided by the manufacturer of the five commercial vehicles (*ICEV1* to *ICEV5*)—as shown, the last column, reports the technical specification of the representative ICEV used in the comparative environmental impact analysis. The assumptions made and the calculation performed to obtain the data of the reference ICEV are described in detail in Appendix A. Following a similar approach, the reference HEV and BEV were defined on the basis of the technical data available on five commercial products for each kind of propulsion; as a result, Tables 3 and 4 report the specification of the real vehicles considered and of the reference vehicles adopted for the comparison.

Table 2. Technical data of the gasoline internal combustion engine vehicles.

	Gasoline Internal Combustion Engine Vehicle (ICEV)					
Vehicle ref. code	ICEV1	ICEV2	ICEV3	ICEV4	ICEV5	Reference ICEV
Make and model	Volkswagen Polo 1.0 TSI 115 CV	Peugeot 208 PureTech 130 Stop and Start	Opel Corsa 1.2 130 CV	Renault Clio TCe 130 CV	Citroen C3 PureTech 83 Stop and Start Van Live	
Tank capacity [L]	45	44	44	42	45	44
Vehicle mass [kg]	1190	1233	1233	1323	1165	1228.8
Displacement [cm^3]	999	1199	1199	1333	1199	1176.6
Max power [kW]	85	96	96	96	81	90.6
Standard emission	Euro 6d temp	Euro 6d temp	Euro 6d temp	Euro 6d temp	Euro 6d temp	Euro 6d temp
WLTP consumption [km/L]	17.2	17.4	16.7	15.9	17.2	16.93

Table 3. Technical data of the gasoline hybrid electric vehicles.

Gasoline Hybrid Electric Vehicle (HEV)						
Vehicle ref. code	HEV1	HEV2	HEV3	HEV4	HEV5	Reference HEV
Make and model	Renault Clio Hybrid E-Tech 140 CV	Toyota Corolla 1.8 Hybrid Touring Sport	Hyundai Ioniq FL Hybrid 1.6	Toyota Prius 1.8 AWD	Kia Niro 1.6 GDI	
Tank capacity [L]	39	43	45	43	45	43.0
Vehicle mass [kg]	1398	1430	1436	1440	1490	1439
Displacement [cm^3]	1618	1798	1580	1798	1580	1676
Max power [kW]	103	90	104	90	104	98.2
Standard emission	Euro 6d	Euro 6d	Euro 6d	Euro 6d	Euro 6d	Euro 6d
WLTP consumption [km/L]	19.6	23.0	20.8	22.7	20.8	21.4
Battery capacity [kWh]	1.2	0.75	1.56	1.3	1.56	1.27
Battery technology	Li-ion	Li-ion	Li-ion polymer	NiMH	Li-ion polymer	NMC
Battery mass [kg]	38	//	//	//	33	26.9

Table 4. Technical data of the battery electric vehicles.

Battery Electric Vehicle (BEV)						
Vehicle ref. code	BEV1	BEV2	BEV3	BEV4	BEV5	Reference BEV
Make and model	Peugeot e-208	Renault Zoe R110 2019	Volkswagen e-Golf 2019	Nissan Leaf S 2019	Hyundai Ioniq EV 2019	
Vehicle mass [kg]	1500	1500	1615	1558	1575	1550
Max power [kW]	100	80	100	110	100	97.9
Battery capacity [kWh]	50	45.61	35.8	40	38.3	42.1
Battery warranty [km]	160,000 (70%)	160,000 (66%)	160,000	160,000	200,000	160,000
Battery technology	//	NMC 712	NMC	NMC	NMC 622	NMC 622
Battery mass [kg]	//	305	318	303	340	374
WLTP Driving range [km]	340	300	232	270	311	291
WLTP Consumption [kWh/km]	0.164	0.178	0.176	0.171	0.138	0.166

The evaluation performed to determine the main characteristics of the reference HEV and BEV are also reported in Appendix A. The average composition of each electricity mix considered in this study [20] is instead reported in Table 5, according to the assumption made in the Goal and Scope section.

Table 5. Composition of the electricity mixes considered in the LCA [20].

	China (2018)	EU-28 (2019)	Germany (2019)	Norway (2019)	Poland (2019)
Coal	66.4%	15.4%	30.0%	0.121%	73.72%
Oil	0.153%	1.64%	0.822%	0.013%	1.09%
Natural gas	3.28%	21.9%	15.3%	1.732%	9.18%
Nuclear	4.09%	25.3%	12.1%	0%	0%
Hydro	17.1%	10.9%	4.24%	93.4%	1.63%
Wind	5.07%	13.3%	20.4%	4.1%	9.20%
Solar PV	2.45%	4.07%	7.69%	0.010%	0.44%
Biofuels	1.26%	5.27%	7.22%	0.03%	4.30%
Waste	0.187%	1.60%	2.03%	0.31%	0.38%
Geothermal	0.002%	0.206%	0.0317%	0%	0%
Solar thermal	0.004%	0.178%		0%	0%
Tide	0.0002%	0.0152%		0%	0%
Other sources		0.141%	0.168%	0.244%	0.05%

5. Life Cycle Impact Assessment

5.1. Production Stage

As already mentioned in the Goal and Scope section, the environmental impact related to the production phase of each standard vehicle was estimated using the *GREET 2020.Net* suite. Being this software developed in the United States of America, the processes related to materials production (e.g., steel production) and to mining (e.g., iron extraction) are referred to as the default geographical context of the USA. With the aim to adapt the GREET model evaluation to the assumption made (i.e., the vehicles were supposed to be produced in Germany), the authors replaced the built-in electricity mix of the USA with the electricity mix of Germany (Table 5). It was also assumed an equal demand of thermal energy and materials during the vehicle production phases, which in actual fact corresponds to assume the same industrial technological level between Germany and the USA. It is worth mentioning, however, that the assumption on the geographical localization of the vehicles production phases in Germany did not regard rare materials produced only in few parts of the planet. In these cases, in effect, the source is considered the same for all the production companies all over the world.

According to the GREET model, the mass of each vehicle is divided into three categories [21]:

- Components
- Fluids
- Battery

For the "Vehicle components" category, in turn, the GREET software estimates the percentage mass distributions reported in Table 6 for each kind of vehicle considered; as will be shown, these mass distributions will be necessary to evaluate the environmental impact associated with the production of the components of each vehicle considered. Moreover, the "Vehicle Assembly" function of the GREET model takes into account the assembly, welding, and painting processes necessary for each vehicle. In this phase, the energy necessary for the end-of-life disposal of the vehicle (battery excluded) is also considered and evaluated. The emissions related to the production of each individual component will be hence summed to the emissions caused by the assembly of the vehicle, thus allowing estimating the overall emissions connected to the production of the entire vehicle (battery excluded).

Table 6. Percentage mass distribution for each kind of passenger car [21].

Component	ICEV	HEV	BEV
Body	44.1%	45.3%	53.5%
Powertrain	25.7%	17.0%	1.70%
Transmission	6.30%	7.20%	3.30%
Chassis	23.9%	24.5%	28.9%
Traction motor	0	2.10%	6.70%
Generator	0	2.10%	0
Controller/Inverter	0	1.80%	5.90%

5.1.1. Battery Production

This paragraph refers to the production of the batteries employed in fully electric and hybrid vehicles, while the production of lead-acid batteries usually adopted in ICEV is already accounted for in the vehicle production phase. BEVs and HEVs are mainly equipped with lithium-ion batteries [22] of various types, depending on the different compositions of cathode materials; at present, the most diffused cathode chemistries for lithium-ion batteries are [22,23]:

- LCO—Lithium Cobalt Oxide (LiCoO2)
- LMO—Lithium Manganese Oxide ($LiMn_2O_4$)
- NMC—Lithium Nickel Manganese Cobalt Oxide (LiNiMnCoO2)
- LFP—Lithium Iron Phosphate (LiFePO4)
- NCA—Lithium Nickel Cobalt Aluminum Oxide ($LiNiCoAlO_2$)

In connection with the reported cathode materials, graphite is the most commonly used anode material [22,24,25]. Lithium-ion polymer batteries (often referred to as LIPO) employ a polymer-based composite material as electrolyte and are produced with all the different cathode chemistries already listed. With the aim to represent the most probable situation, the authors focused on NMC batteries, which currently constitutes the most diffused technology among the electric vehicles registered in the United States, Europe, and Japan [26] due to its high energy density and long cycle life, as is confirmed by the data reported in Table 4; NMC batteries are increasingly used compared to LFP technology, and continued growth is expected in years to come. In the present paper, hence, the GREET model was used to estimate the environmental impact associated with the production of the batteries of each BEV and HEV considered, with reference to the production technologies and to updated data collected by manufacturers of lithium-ion batteries for the automotive sector in China [27,28]; as already explained in the Goal and scope section, the China electricity mix was introduced in the GREET model for the evaluations of the impact related to Li-ion battery production, which is substantially carried out in two separate steps. In the first, the bill of materials (BOM) necessary for the battery realization is compiled based on the battery type and weight, and the environmental impact associated to the amount of each element of the bill is computed; in the second step, instead, all the manufacturing processes necessary for the battery assembly are considered and their environmental impact evaluated on the basis of the battery type and capacity.

5.1.2. Vehicle Production Results

This paragraph describes the procedures and calculations carried out to estimate the environmental impact due to the production of each vehicle. The masses of the three reference vehicles compared in this work are indicated in Tables 2–4. The values reported, however, refer to the complete vehicle kerb masses [29], including hence the necessary fluids for vehicle operation (e.g., lubricants, coolant, washer fluid, fuel, etc.) and the batteries. The empty mass of each vehicle (i.e., related to vehicle components only) was hence determined subtracting the masses of fluids and batteries:

$$m_{\text{empty}} = m_{\text{kerb}} - \left(\sum m_{\text{fluids}} + m_{\text{battery}}\right) \quad (1)$$

The masses of the different fluids [21] are reported in Table 7 for each reference vehicle considered; in regards to the batteries, a mass of 16.3 kg was adopted for the ICEV traditional 12V lead-acid model [21], while for the reference HEV and BEV, as already reported in Tables 3 and 4, the evaluation carried out in Appendix A led to 26.9 and 374 kg, respectively.

Table 7. Fluids masses for each kind of vehicle.

	ICEV	HEV	BEV
Engine lubricant [kg]	4.1	4.1	0.0
Power steering fluid [kg]	0.0	0.0	0.0
Brake fluid [kg]	0.91	0.91	0.91
Transmission fluid [kg]	10.9	0.91	0.91
Powertrain coolant [kg]	10.4	10.4	7.2
Windshield wiper fluid [kg]	2.7	2.7	2.7
Adhesives [kg]	13.6	13.6	13.6

Applying the percentage mass distributions reported in Table 6 to each vehicle's empty mass allowed for determining the mass of each vehicle component, i.e., body, powertrain, transmission, chassis, motor/generator, and controller/inverter. The mass of each vehicle component was then introduced as input to the GREET model, which returned the related polluting emissions and resources used. The influence of each vehicle component on each environmental impact category was hence evaluated applying the ReCiPe characterization factors of Table 1. As a final result, Tables 8–10 report the environmental impact related to the production of each component (including fluids and batteries) of the reference ICEV, HEV, and BEV respectively. As can also be observed, the last column reports the energy consumption (expressed in MJ) related to the component—this information has been added with the aim to highlight the energy impact of every single component, which, therefore, affects some of the impact categories considered.

Table 8. Environmental impact and resource deployment related to the reference ICEV components production.

	GWP [kg CO_2 eq]	TAP [kg SO_2 eq]	PMFP [kg $PM_{2.5}$ eq]	SOP [kg Cu eq]	FFP [kg Oil eq]	Energy Consumption [MJ]
Body	1303.0	7.1	2.2	17.7	334.2	20,890.3
Powertrain	659.1	3.9	1.3	24.7	191.2	12,435.9
Transmission	163.3	0.6	0.2	5.4	46.1	3081.5
Chassis	666.1	3.3	1.1	10.1	133.5	9339.3
Assembly	1127.5	1.8	0.6	0.0	300.4	20,413.5
Oil	13.4	0.1	0.0	0.0	4.0	219.8
Brake fluid	3.0	0.0	0.0	0.0	0.9	48.8
Transmission fluid	35.8	0.2	0.1	0.0	10.7	586.1
Coolant	18.0	0.1	0.0	0.0	4.1	205.9
Adhesives	48.8	0.3	0.1	0.0	23.3	1190.8
Windshield wiper fluid	4.9	0.0	0.0	0.0	0.9	41.7
Battery	11.7	0.2	0.1	5.1	4.1	251.6

Table 9. Environmental impact and resource deployment related to the reference HEV components production.

	GWP [kg CO_2 eq]	TAP [kg SO_2 eq]	PMFP [kg $PM_{2.5}$ eq]	SOP [kg Cu eq]	FFP [kg Oil eq]	Energy Consumption [MJ]
Body	1574.7	8.5	2.70	18.8	403.9	25,246.1
Powertrain	523.0	3.0	0.97	16.6	142.5	9373.7
Transmission	317.3	3.9	1.20	20.1	61.7	5155.5
Chassis	803.3	3.9	1.32	10.5	161.0	11,263.5
Traction Motor	74.1	1.4	0.41	9.8	16.3	1188.9
Generator	74.1	1.4	0.41	9.8	16.3	1188.9
Controller/Inverter	63.4	0.5	0.16	3.9	20.5	1263.2
Assembly	1320.2	2.1	0.70	0.0	351.8	23,902.1
Oil	13.4	0.1	0.03	0.0	4.0	219.8
Brake fluid	3.0	0.0	0.01	0.0	0.9	48.8
Transmission. Fluid	3.0	0.0	0.01	0.0	0.9	48.8
Coolant	18.0	0.1	0.02	0.0	4.1	205.9
Adhesives	48.8	0.3	0.07	0.0	23.3	1190.8
Windshield wiper fluid	4.9	0.0	0.00	0.0	0.9	41.7
Battery BOM	214.0	1.6	0.50	6.4	45.0	2949.6
Battery assembly	20.0	0.0	0.01	0.0	5.7	289.2

Table 10. Environmental impact and resource deployment related to the reference BEV components production.

	GWP [kg CO_2 eq]	TAP [kg SO_2 eq]	PMFP [kg $PM_{2.5}$ eq]	SOP [kg Cu eq]	FFP [kg Oil eq]	Energy Consumption [MJ]
Body	1627.6	8.8	2.80	22.1	417.4	26,094.7
Powertrain	57.9	0.8	0.25	4.4	17.3	1035.7
Transmission	127.8	1.6	0.48	8.2	24.8	2075.8
Chassis	832.4	4.1	1.36	12.7	166.8	11,671.7
Traction Motor	213.9	4.0	1.18	28.6	47.1	3431.6
Controller/Inverter	182.5	1.5	0.46	11.1	59.1	3637.3
Assembly	1421.8	2.2	0.75	0.0	378.9	25,742.7
Brake fluid	3.0	0.0	0.01	0.0	0.9	48.8
Transmission fluid	3.0	0.0	0.01	0.0	0.9	48.8
Coolant	12.5	0.1	0.01	0.0	2.8	143.3
Adhesives	48.8	0.3	0.07	0.0	23.3	1190.8
Windshield wiper fluid	4.9	0.0	0.00	0.0	0.9	41.7
Battery BOM	3008.0	50.3	14.92	180.7	683.3	43,210.8
Battery assembly	658.5	0.9	0.23	0.0	187.6	9519.7

5.2. Use Stage

The environmental impact related to the use phase of the vehicles is, as a general rule, composed of two different contributions. The first is related to the energy source employed by the vehicle, whose production is characterized by a certain environmental impact; the second contribution, instead, accounts for the exhaust emissions produced by the vehicle during on-road operation.

5.2.1. Energy Consumption

Regarding the first contribution, both reference ICEV and HEV use as the energy source the fuel introduced in the internal combustion engine (i.e., gasoline in the cases here considered), while the BEV employs electric energy previously produced by means of different primary energy sources, as for example natural gas, coal, hydro, wind, solar energy, etc. As declared in the Goal and scope section, in this comparative LCA analysis, a lifetime traveling distance of 150,000 km was assumed for each reference vehicle, entirely run in Europe; for this reason, when evaluating the impact related to the fuel consumed by both ICEVs and HEVs, the European fuel production and distribution chain was properly considered through the use of *Ecoinvent V3*, a widely employed life cycle inventory database [30,31], in place of the GREET model. The reason for this change is that, as already explained, the processes related to materials production comprised in the GREET model refer to the geographical context of the USA. Moreover, even if in accordance with the European Parliament regulation 2009/30/EC, European gasoline can contain up to 10% of ethanol, on average it results in 5% of ethanol present in the gasoline distributed in the European Union [32]. For this reason, the environmental impact related to gasoline production (including manufacturing and transportation) was computed assuming gasoline with 5% of ethanol from biomass, referred to as BE5: Table 11 shows the pollutant emission and the resources use associated with the production of a single kg of BE5. For each of the two reference ICEV and HEV, the total mass of fuel consumed by the vehicle during its entire life m_{fuel} was deduced on the basis of the WLTP fuel consumption F (reported in Tables 2 and 3), of the total driving distance D_{tot} (150,000 km), and of the fuel density ρ_{fuel} (0.752 kg/L for the BE5 [33]):

$$m_{fuel} = F_{[L/km]} \cdot D_{tot\,[km]} \cdot \rho_{fuel\ [kg/L]} \tag{2}$$

to which correspond the energy required for vehicle traction:

$$E_{trac} = m_{fuel} \cdot LHV \tag{3}$$

being LHV the fuel Lower Heating Value (41.7 MJ/kg for the BE5 [33]). Indicating with x the generic impact category, the characterization factor $I_{x,source}$ connected to the production of the total mass of fuel employed by the ICEV or by the HEV was obtained as:

$$I_{x,source} = \phi_x \cdot m_{fuel} \tag{4}$$

being ϕ_x the specific impact factor referred to the production of 1 kg of gasoline, reported in Table 11. As can be noted, in the last row, the energy required for the production of 1 kg of BE5 is reported: this is the source of production energy and is responsible for the related GWP. In addition, the electric energy used by the BEV was supposed to be entirely produced in Europe, and hence, when computing the environmental impact related to the use of the reference electric vehicle, the mean European electricity mix reported in the third column of Table 5 was assumed.

Table 11. Specific impact factors related to the production of 1 kg of gasoline with 5% ethanol from biomass (BE5) [30]).

GWP [kg CO_2 eq/kg]	0.596896
TAP [kg SO_2 eq/kg]	0.00529478
PMFP [kg $PM_{2.5}$ eq/kg]	0.00169847
SOP [kg Cu eq/kg]	0.00157793
FFP [kg Oil eq/kg]	1.14230
Energy consumption [MJ/kg]	7.0463

The total traction energy required by the BEV during its entire life $E_{trac,BEV}$ was deduced on the basis of WLTP energy consumption F, reported in the last row of Table 4, and of the total driving distance D_{tot}:

$$E_{trac,BEV[kWh]} = F_{[kWh/km]} \cdot D_{tot[km]} \quad (5)$$

Adopting the same symbol x to denote the generic impact category, hence, the characterization factor $I_{x,source}$ connected to the production of the total amount of electric energy $E_{trac,BEV}$ consumed by the BEV was obtained as:

$$I_{x,source} = \varphi_x \cdot E_{trac,BEV} \quad (6)$$

where the specific impact factor φ_x referred to each impact category x and associated to the production of 1 kWh of electric energy is reported in Table 12 and was evaluated by means of the GREET model employing the mean EU-28 electricity mix (already presented in Table 5).

Table 12. Specific impact characterization factors related to the production of 1 kWh of electric energy (EU-28 average electricity—Table 5).

GWP [kg CO_2 eq/kWh]	0.2994
TAP [kg SO_2 eq/kWh]	0.0006227
PMFP [kg $PM_{2.5}$ eq/kWh]	0.0001971
SOP [kg Cu eq/kWh]	0.0000
FFP [kg Oil eq/kWh]	0.07252
Energy consumption [MJ/kWh]	6.8368

As shown, the last row of Table 12 reports the source production energy, i.e., the energy input required for the production of 1 kWh of electric energy. The resulting environmental impact and resource deployment associated with the energy consumed in the use phase of each reference vehicle, during its entire life, is reported in Table 13.

Table 13. Environmental impact and resource deployment related to the energy consumed in the use phase of each reference vehicle (whole vehicle life = 150,000 km).

	ICEV	HEV	BEV
Total fuel consumed [kg]	6662.4	5274.9	0
Traction energy [kWh]	77,173	61,101	24,841
Source production energy [kWh]	13,040	10,325	47,176
GWP [kg CO_2eq/kWh]	3976.7	3148.6	7436.5
TAP [kg SO_2eq/kWh]	35.276	27.930	15.469
PMFP [kg $PM_{2.5}$eq/kWh]	11.316	8.9594	4.8965
SOP [kg Cu eq/kWh]	10.513	8.3235	0
FFP [kg Oil eq/kWh]	7610.4	6025.6	1801.4

5.2.2. On-Road Emissions

Being the BEV free from on-road emissions, its road environmental impact was coherently considered null. As instead concerns the other two reference vehicles (ICEV and HEV), the authors evaluated the impact related to the emissions of CO_2 produced during the use phase, as well as the impact associated with the other relevant emissions produced (e.g., CO, PM, NOx, etc.). For the evaluation of the carbon dioxide emissions due to the fuel combustion, coherently with the BE5 gasoline assumed (i.e., 95% petrol and 5% bio-ethanol), the authors considered only the carbon participation of petrol, being null the carbon cycle of the bio-ethanol—it resulted in a carbon mass fraction of 82% (instead of the

86.5% usually adopted for petrol) [33], according to which the amount of CO_2 emitted by the combustion of each liter of BE5 is:

$$f_{CO_2} = \rho_{FUEL} \cdot 0.82 \cdot 1000 \cdot \frac{44}{12} = 2261 \left[\frac{g_{CO_2}}{L_{FUEL}} \right] \tag{7}$$

being 12 and 44 the molecular masses of carbon and carbon dioxide, respectively. As a result, the CO_2 emission factor (g/km) related to both reference ICEV and HEV was evaluated on the basis of the WLTP fuel consumptions F (already shown in Tables 2 and 3):

$$e_{CO_2[g/km]} = f_{CO_2[g/L]} \cdot F_{[L/km]} \tag{8}$$

and is reported in the 4th row of Table 14. As instead regards the impact related to other relevant exhaust emissions produced by both the ICEV and the HEV, the authors referred to the emission inventory data of the European Environment Agency [34] which contains average values of the emission factor (i.e., grams of pollutant per kilometer of distance traveled) related to different kinds of vehicles (passenger cars, light commercial trucks, heavy-duty vehicles including buses, mopeds, and motorcycles), belonging to different categories (mini, small medium, large, executive, etc.), using different fuels (diesel, petrol, LPG, and CNG), and recorded on several different standard tests (starting from pre-ECE, up to Euro 6 d-temp); for the reference ICEV, the authors considered the emission factor reported for a small petrol passenger car, Euro 6 d-temp, while the reference HEV was considered as a small petrol Hybrid passenger car, Euro 6, thus obtaining the values shown in Table 14.

Table 14. Emission factors related to the reference vehicles ICEV and HEV (European Environment Agency [34]).

	ICEV	HEV
Type of car	Petrol Small	Hybrid Petrol Small
Technology	Euro 6 d-temp	Euro 6
CO_2 [g/km]	133.5	105.7
CO [g/km]	0.69	0.042
NMHC [g/km]	0.048	0.001
NO_X [g/km]	0.056	0.013
N_2O [g/km]	0.0013	0.0002
NH_3 [g/km]	0.0123	0.0328
Pb [g/km]	1.82×10^{-5}	1.82×10^{-5}
CO_2 lube [g/km]	0.398	0.398

On the basis of the emission factors and of the total driving distance, the authors could establish the total mass of each pollutant emitted during the life cycle of both the reference ICEV and HEV, which, according to the ReCiPe 2016 methodology described in [14,15], allowed evaluation of the characterization factor for each impact category considered, as reported in Table 15. Summing the contribution of the energy source production to the contribution derived from the exhaust emissions, the total environmental impact and resource deployment related to the use phase of each reference vehicle is reported in Table 16 for the entire vehicle's life.

Table 15. Environmental impact and resource deployment related to the exhaust emissions of both reference ICEV and HEV produced during the entire life (i.e., 150,000 km).

	ICEV	HEV
GWP [kg CO_2 eq]	20,032	15,860
TAP [kg SO_2 eq]	3.024	0.7020
PMFP [kg $PM_{2.5}$ eq]	0	0
SOP [kg Cu eq]	0.2400	0.03000
FFP [kg Oil eq]	0	0

Table 16. Environmental impact and resource deployment related to the use phase of the three reference vehicles on their entire life (i.e., 150,000 km).

	ICEV	HEV	BEV
GWP [kg CO_2 eq]	24,008	19,009	7437
TAP [kg SO_2 eq]	38.300	28.632	15.47
PMFP [kg $PM_{2.5}$ eq]	11.556	8.9894	4.896
SOP [kg Cu eq]	10.513	8.3235	0
FFP [kg Oil eq]	7610.4	6025.6	1801

5.3. End-of-Life

The disposal and recycling phases of all the components of the ICEV are already taken into consideration in the "vehicle assembly phase" of the GREET model. Moreover, the disposal and recycling of most of the components of both HEV and BEV are included in the GREET "vehicle assembly phase", remaining out of this evaluation only the batteries, due to the existence of different processes for the disposal of the several different kinds of batteries available. On account of this, the disposal phase of the batteries of both HEV and BEV was expressly carried out by the authors.

Electric Vehicle Battery Disposal

The battery disposal and recycling process was modeled through the environmental impact indicators provided for lithium-ion batteries in the scientific literature [7]. To account for the different battery capacities (i.e., different amounts of materials and hence different environmental impact), a mass-based proportionality was assumed thus adapting the literature available data to the batteries considered in this study. Each battery cell was assumed to be recycled through a pyrometallurgical process, which is commonly used in Europe for vehicle battery recycling [24]. The pyrometallurgical process, however, does not allow the recovery of materials such as graphite, plastic materials, aluminum, lithium, and manganese; in particular, the last three elements are retained in the slag produced during the process [25]. The metal alloy and slag obtained from the pyrometallurgical process, which represent about 55% of the initial battery mass, are hence further refined through the hydrometallurgical process, to recover the metal sulphates, which can be used again to produce the cathode of lithium-ion batteries [35]. The resulting impact indicators related to the disposal of the batteries of both reference HEV and BEV are shown in Table 17.

Table 17. Impact characterization factors related to the End of Life (EoL) of the lithium-ion batteries of both reference HEV and BEV.

	HEV	BEV
GWP [kg CO_2 eq]	28.46	396.0
TAP [kg SO_2 eq]	0.07371	1.026
PMFP [kg $PM_{2.5}$ eq]	0.02063	0.2870
SOP [kg Cu eq]	0	0
FFP [kg Oil eq]	6.610	91.97
Energy consumption [MJ]	488.2	6794

Finally, summing the three contributions (production phase, use phase, and disposal phase), the life cycle environmental impact, and resource deployment for each reference vehicle considered was obtained—these results will be discussed in the following section.

6. Results of the LCA Analysis and Discussion

This section deals with the results obtained from the life cycle environmental impact assessment carried out for each of the three vehicles considered.

6.1. Global Warming Effect

Being the most significant indicator related to climate changes and to its causes, GWP is one of the most diffused environmental impact indicators among the various life cycle impact assessments. shows the GWP generated by the three reference vehicles during each phase of their life.

As can be observed, the total amount of equivalent CO_2 emitted during a vehicle's life is reported on the top of each bar, while the percentage with respect to the ICEV case is also reported on the top of HEV and BEV bars. The values reported inside the bar refer instead to the single phase (production or use phases, negligible values are not reported). The results of the GWP analysis show that, at the end of its life, traditional vehicles with gasoline-fueled internal combustion engines are responsible for an average specific CO_2 equivalent emission of 187 g/km, which is about 40% higher than the "on-road" CO_2 emission related to the fuel consumption (i.e., 133.5 g/km in Table 15), due to vehicle production impact. The HEV is characterized by a lower impact (−14.1%) with respect to the ICEV, with an overall average specific CO_2eq emission of 160 g/km, which is +52% higher than the on-road emission of 105.7 g/km (Table 15) mainly due to the impact of the production phase. Finally, the BEV revealed a CO_2 equivalent emission of 109.6 g/km, which represents about 58.6% of the traditional ICEV, and confirms the relevant role played by the vehicle production phase, as well as by the energy source production processes, in determining the real overall environmental impact of a passenger car [4–6]. The graph in Figure 1 also shows that the production of the traditional vehicle is the less impacting among the three production phases, thanks to processes and manufacturing technologies optimized and refined over a long time; BEV and HEV, instead, share the burden of the lithium-ion battery production, which, in the case of the electric vehicle, implies 4215 kg of CO_2eq (indicated as BP in Figure 1) causing +112% higher CO_2 emissions with respect to the ICEV production phase. The diagram in Figure 1 also shows that the use phase represents the major source of greenhouse gas emissions for both the reference ICEV (85.5% of total CO_2eq) and the reference HEV (78.8% of total CO_2eq), while for the BEV the production phase involves the major part of the total CO_2 emissions (52.3% of total CO_2eq), coming from the battery production process the most relevant contribution.

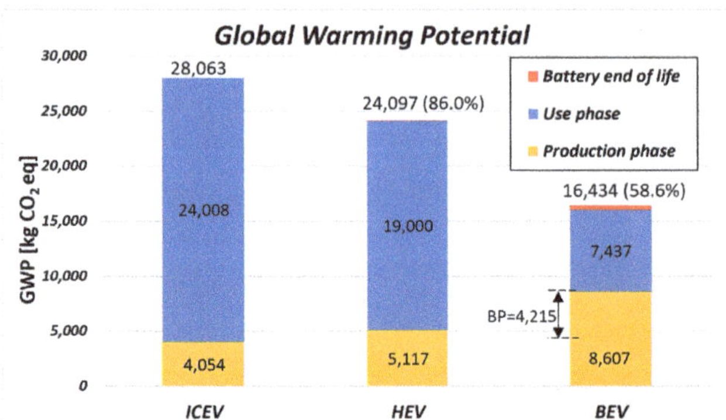

Figure 1. Global warming potential related to the entire lifecycle of the three reference vehicles (BP = Battery Production).

The comparison carried out until now is referred to a lifetime distance traveled of 150,000 km and confirms the BEV to be the less impactful solution among the three alternatives considered with a −41.4% cut with respect to the ICEV and −31.8% with respect to HEV. However, to understand the effect of the vehicle lifetime mileage (or usage time) on overall greenhouse gas emissions, the authors evaluated the GWP impact factor as a function of the distance traveled, as shown in Figure 2. Besides confirming the higher starting impact (i.e., for a null distance traveled) of both HEV and BEV, the diagram in Figure 2 also shows that the reference ICEV reveals to be the less greenhouse gas emitting vehicle up to a mileage of 32,500 km (i.e., roughly the first 2.6 years of vehicle usage, if the already mentioned average European lifetime distance traveled of 12,529 km is considered), and remains cleaner than the BEV up to 41,250 km (i.e., up to 3.3 years of vehicle usage), while the advantage of the HEV on the BEV extends to 46,250 km (i.e., a vehicle usage period of 3.7 years).

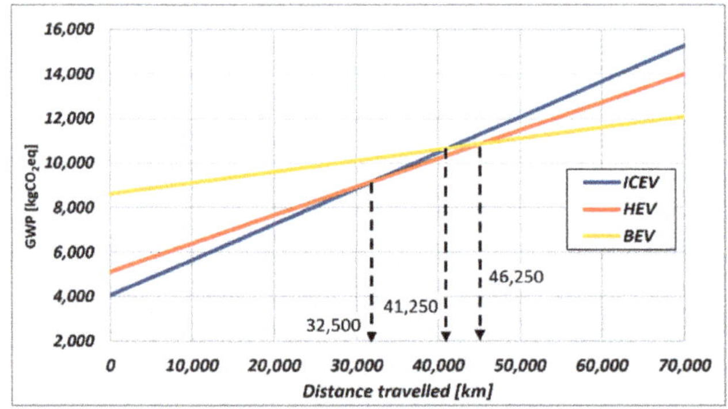

Figure 2. Global warming potential comparison as a function of the lifetime distance traveled.

6.2. Terrestrial Acidification

Anthropogenic terrestrial acidification is primarily caused by atmospheric deposition of acidity, mainly through acid rain originated by the emissions into the atmosphere of substances such as nitrogen oxides (NO_X), ammonia (NH_3), and sulphur dioxide (SO_2). The evaluation carried out in terms of Terrestrial Acidification Potentials (TAP) for the whole life cycle of the three reference vehicles is shown in Figure 3. The first notable result is that the production phase of the BEV causes a very high level of terrestrial acidification giving an overall final result of 661 mg/km which exceeds +78% of the impact related to the reference ICEV (372 mg/km). This is principally referring to the production of the

Lithium-ion battery (60.3 kg of SO_2eq, as indicated in Figure 3, i.e., 73% of the total impact generated in the production phase), which causes huge emissions of sulphur and nitrogen oxides (SO_X and NO_X) for the extraction and refining of nickel, copper, and aluminum, for cell production and synthetic graphite processes [26]. Moreover, a further contribution to the high acidification impact related to the battery production processes is provided by the Chinese electricity mix, which is dominated by coal-fired plants (66% of total electric energy produced, as resumed in Table 5), and hence characterized by high levels of SO_X emissions. For the same reason, the overall impact of the HEV (386 mg/km) also results in slightly higher than ICEV (+4%), due to the production of the small Lithium-ion battery. In regards to the ICEV, most of the terrestrial acidification is caused by the exhaust emissions produced during the use phase (68.3% of total), while for the HEV and for the BEV the main impact is related to the production phase, which accounts for the 50.8% and for the 83.3% of the total, respectively.

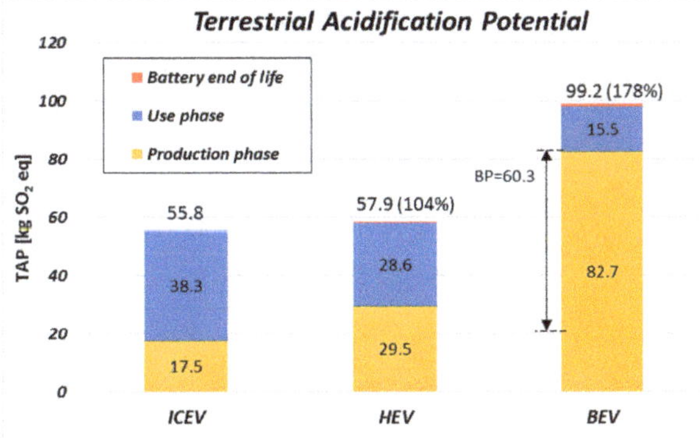

Figure 3. Terrestrial acidification caused during the entire lifecycle of the three reference vehicles (BP = Battery Production).

6.3. Particulate Matter Formation

The results obtained for this environmental impact category are shown in the diagram of Figure 4. Apart from the unit of measurement, a close resemblance between Figures 3 and 4 can also be noted, i.e., almost identical relations exist between the bars of each diagram. This is easily explained since the main substances involved in the formation of secondary $PM_{2.5}$ (i.e., SO_2, NH_3, and NO_X) are also responsible for the terrestrial acidification—the battery production process, hence, also has a high impact in terms of particulate matter formation, due to both the extraction and refining of the materials used for NMC powders and to the phenomenon caused by the outdoor storage of copper-cobalt minerals in Congo [36]. Considering as reference the impact generated by the reference ICEV, the impact of BEV was substantially stronger (+75%), while HEV remains were slightly higher (+6%). Moreover, the distribution within the results of the different phases was similar to the terrestrial acidification, being 67.4% of the portion of the particulate impact caused by the use phase of the ICEV, while the production phase of both HEV and BEV still represents the most impacting phase with 51.0% and 82.7% of the total emissions produced, respectively.

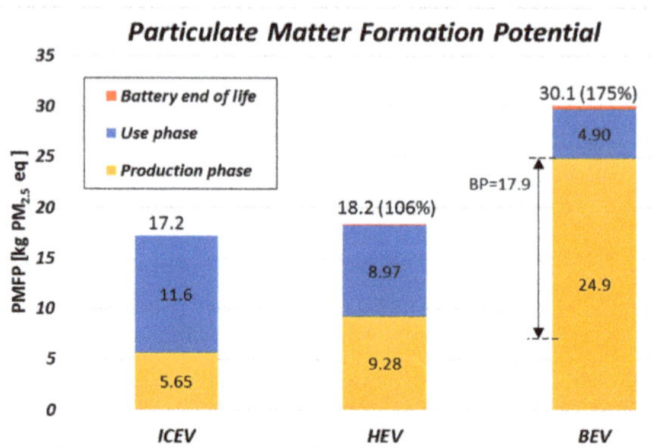

Figure 4. Particulate matter formation related to the entire lifecycle of the three reference vehicles (BP = Battery Production).

6.4. Mineral Resource Deployment

Due to the extensive use of rare materials such as lithium, nickel, cobalt, and copper [14,24] required for the production of lithium-ion batteries, the mineral resource deployment (reported in Figure 5) related to the BEV results abundantly higher than the impact caused by the ICEV (about four times), while HEV revealed "only" a + 54% increment. As expected, apart from the reference vehicle considered, almost the entire impact is generated during the production phase. The contribution due to electric vehicle battery production is indicated as BP in the graph of Figure 5 and amounts to 72% of the total surplus ore potential related to the battery electric vehicle.

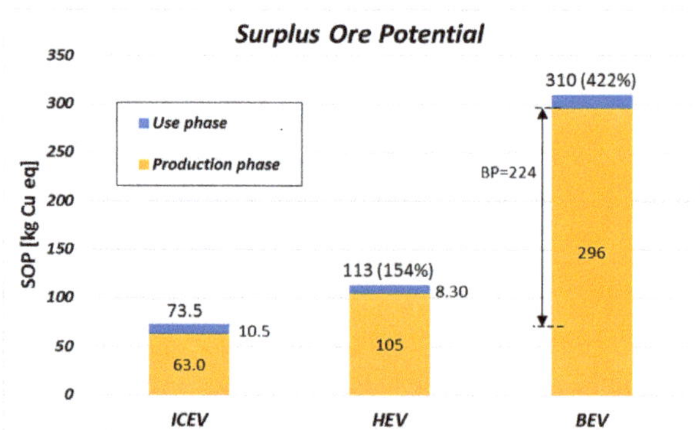

Figure 5. Mineral resource depletion related to the entire lifecycle of the three reference vehicles (BP = Battery Production).

6.5. Fossil Resource Deployment

Regarding the deployment of fossil fuel sources, the results obtained in this evaluation are reported in Figure 6. Several observations can be made and are worthy of note: firstly, as expected, the highest overall fossil source consumption is generated by the ICEV (i.e., 57.8 g/km), followed by the HEV (48.4 g/km, i.e., 84.1% with respect to the ICEV) and by the BEV, which, even if "fully electric" vehicle, implies a fossil fuel consumption of 26.6 g/km (i.e., 46% with respect to the ICEV); moreover, both ICEV and HEV cause most of their fossil fuel consumption during the use phase of the vehicle (87.8% and 82.7% of the total respectively), being their main energy source for traction a fossil-derived fuel, while, when the BEV is concerned, the most of fossil resource consumption takes place in the production phase (52.6%), mainly due to the high consumption of electricity energy required for battery production, and to the fossil source domination in the Chinese electricity mix. It can also be observed that the fossil resource consumption generated in the use phase by the BEV (1801 kg Oil eq.) constitutes 23.6% of the consumption caused by the ICEV in the same phase (7610 kg Oil eq). This result, however, depends on the particular electricity mix considered for the supply of the BEV, and, as shown further on, may substantially change from one country to another.

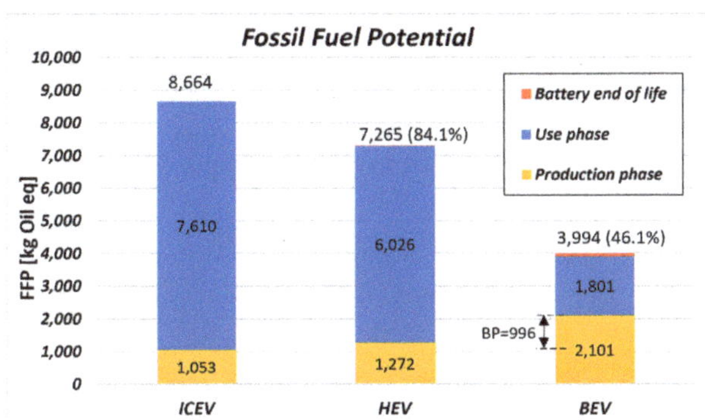

Figure 6. Fossil fuel depletion related to the entire lifecycle of the three reference vehicles (BP = Battery Production).

7. Scenario Analysis

The results of the analysis carried out to this point are obviously related to the reference scenario adopted which is characterized by several assumptions. Some of these assumptions may have a general validity (i.e., vehicles produced in Germany may be distributed and used all over Europe, the impact related to petrol production is almost the same in Europe and the USA, most of the lithium-ion batteries used worldwide come from China) while other may easily change and produce considerable variations in the results obtained, such as the lifetime mileage and the country where the vehicles are supposed to be employed. The total distance traveled during a vehicle's lifetime obviously influences the amount of both energy resource consumed and pollutant emitted, while the country where the vehicle is used has a strict correlation with the electricity mix (i.e., the composition of primary energy resources employed to generate electric energy) and hence with the impact related to the electric energy consumed. With the aim to highlight the importance of these two variables (the most susceptible of variations) on the lifecycle environmental impact of the three reference vehicles considered, a scenario analysis was performed. In regards to the lifetime mileage of each vehicle, two different cases were added to the evaluation, considering a ±30% deviation from the average European lifetime distance traveled, i.e., 105,000 and 195,000 km. Concerning the second variable, two particular countries were considered for the vehicles traveling, characterized by substantially different electricity mixes: Norway, where 97.5% of electric energy is produced by means of renewable sources (mainly hydroelectric, as shown in Table 5), and hence with a near-zero carbon footprint, and, on the other hand, Poland, where 84% of electric energy is produced by fossil sources (above all coal, as shown Table 5). The introduction of these two countries has the meaning to observe how different electricity mixes may influence the environmental impact related to the use of the battery electric vehicles, compared to the gasoline-fueled ICEV and HEV. It is worth repeating that, regarding petrol production, no substantial differences could be traced on the technologies and processes adopted among the different European countries, as, therefore, confirmed by both the database consulted (i.e., GREET and Ecoinvent v.3). For this reason, the impact and resources deployment related to gasoline production in Norway and in Poland was considered equal to the average European assumed in the previous section. In addition, the other production steps were considered unchanged, i.e., the vehicles were assumed to be produced in Germany and the lithium-ion batteries employed in the HEV and in the BEV were supposed to be produced in China. Moreover, since lithium-ion batteries have a limited duration, their replacement was also taken into consideration. Due to a lack of literature references dealing with the longevity of electric vehicle batteries in real conditions of use, all life cycle analyses usually consider the duration of a battery equal to the manufacturer's warranty [7]. Since the batteries of most of the electric vehicles considered for the characterization of the reference BEV (as reported in Table 4) are guaranteed for 160,000 km, therefore, when exceeding this traveling distance (i.e., only in the scenarios with a lifetime distance traveled of 195,000 km), a battery replacement was introduced in the calculation. Moreover, with the purpose to suppress any difference related to the three lifetime mileages considered and make the results of all scenarios comparable, for each impact category, the authors evaluated the specific factor dividing each characterization factor by the lifetime distance traveled, obtaining hence the impact per km of traveled distance.

8. Results of the Scenario Analysis and Discussion

The results obtained from this scenario analysis, in regards to the specific global warming potential (expressed as gCO_2eq/km), are summarized in the graph of Figure 7. Specific global warming potential (gCO_2eq/km) related to the three reference vehicles in the scenario analysis. As can be observed, three series of histograms are reported, one for each lifetime mileage considered; each series of histograms, in turn, represents the lifecycle impact evaluated for the ICEV, the HEV, the BEV employed in Norway (BEV-NOR), the BEV employed in Poland (BEV-POL), and the BEV employed using the average European electricity mix (BEV-EU28). As already shown in the previous graphs of this paper, the impacts related to the production and to the use phases are reported inside the colored bar (negligible values are not reported), while the percentage ratio with respect to the ICEV case is reported on the top of both HEV and BEV bars. Starting from the reference ICEV, Figure 7 shows that, the increase in the lifetime distance traveled causes a light specific impact reduction due to the reduction in the specific impact related to the production phase of the vehicle; quite a similar situation occurs for the HEV, which is characterized by a slightly higher impact in the production phase (on account of the lithium-ion battery production) and lower greenhouse gas emissions in the use phase, thanks to the higher vehicle efficiency. It can also be observed that, independent from the lifetime mileage, its global warming impact remains between 85.5% and 88.5% of the impact caused by the ICEV. The specific global warming caused by the BEV, as expected, is instead strongly dependent on the electricity mix adopted by the country where the vehicle is employed, and on the lifetime mileage; more specifically, the specific emissions of CO_2eq of the BEV range between 33% and 44% with respect to the ICEV when the vehicle is operated in Norway. In this case, the environmental impact is almost entirely due to the production phase of the vehicle, being negligible the carbon footprint of the Norwegian electricity mix. It is also worth pointing out that in the third scenario (i.e., 195,000 km) the battery replacement required at 160,000 km causes a sharp increment of the specific impact related to the production phase. The advantage of the BEV on the ICEV, in terms of greenhouse gas emissions, however, reduces if the average European electricity mix is considered (with percentage ratio from 58.6% to 68.1%) and reveals null if the electric vehicle is operated in Poland. In this case, the high carbon footprint of the electricity mix causes the greenhouse gas emissions of the BEV to become even higher than the emissions of the ICEV, with a percentage ratio between 108% to 117%, depending on the lifecycle mileage of the vehicle.

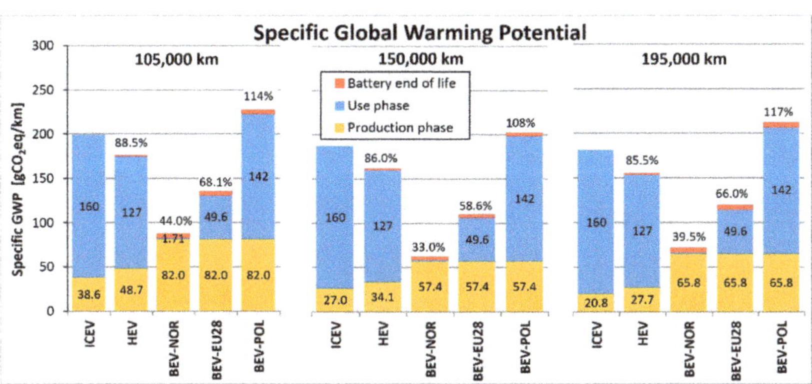

Figure 7. Specific global warming potential (gCO_2eq/km) related to the three reference vehicles in the scenario analysis.

Concerning the second impact category, i.e., the specific terrestrial acidification (expressed as $mgSO_2eq/km$), the results obtained from the scenario analysis, reported in Figure 8, confirm the dominant role of the lithium-ion battery production, which causes the HEV and the BEV to have higher environmental impacts than ICEV, apart from the vehicle mileage and the country of utilization; more specifically, the HEV slightly exceeds the ICEV, with percentage ratios between 104% and 112%, while the BEV, whose battery pack has a substantially higher capacity, reveals a percentage ratio in the range of 150–216% in the case of the Norwegian electricity mix, moving up to a range of 246–319% in the case of the Polish electricity mix.

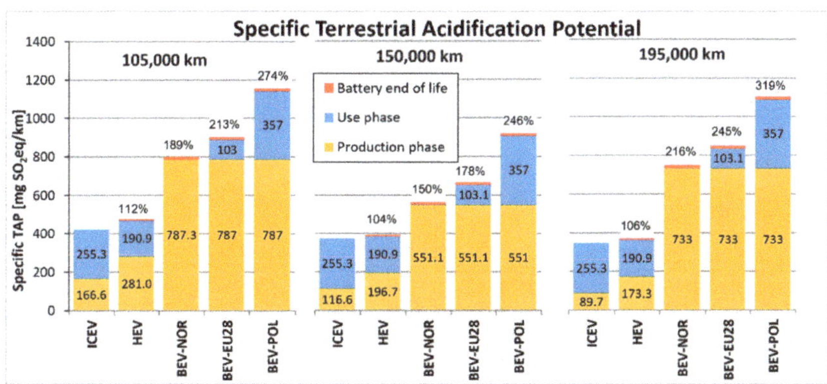

Figure 8. Specific terrestrial acidification potential ($mgSO_2eq/km$) related to the three reference vehicles in the scenario analysis.

It is worth pointing out that, due to the battery replacement at 160,000 km, the BEV gives the worst results in the longer mileage scenario.

As already observed in the previous section, the scenario analysis confirms that the specific environmental impact due to primary and secondary particulate matter has a trend quite similar to the terrestrial acidification (as shown in Figure 9), being involved the same chemical species. The BEV is confirmed to be the most impacting vehicle, principally due to the battery production processes and resources, with percentage ratios between 147% and 210% (with respect to ICEV) if used in Norway, and between 240% and 311% if the vehicle is instead operated in Poland. The HEV confirms a slightly higher impact with respect to the ICEV, with a percentage ratio between 106% and 114%.

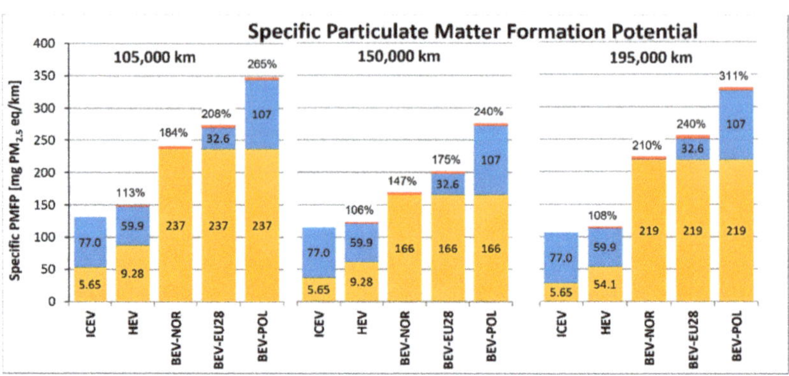

Figure 9. Specific particulate matter formation potential (mgPM$_{2.5}$eq/km) related to the three reference vehicles in the six different scenarios.

Moving on to the impact on resources, the results obtained by the scenario analysis with regard to the mineral resource deployment are reported in Figure 10 in terms of mg Cu-eq/km. As already highlighted in the previous section, the production of lithium-ion batteries involves a wide use of uncommon metals such as copper, nickel, and cobalt, which, therefore, explains the very high impact of BEV with respect to both ICEV and HEV, apart from the scenario adopted. The percentage ratio ranges from 408% to 422% in the best cases (i.e., for a lifetime distance traveled of 150,000 km) and moves to values higher than 600% for the longer traveled distance due to the battery replacement. The lower capacity of the hybrid electric vehicle battery involves a lower deployment of mineral resources, which results in a lower impact, with a percentage ratio in the range 154–171%.

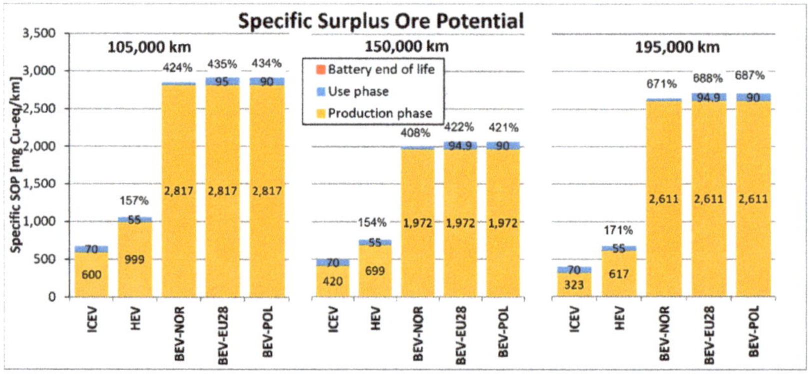

Figure 10. Specific surplus ore potential (mgCu-eq/km) related to the three reference vehicles in the six different scenarios.

Finally, the results regarding the deployment of fossil fuel resources are reported in Figure 11 in terms of consumption of oil per km of distance traveled (gOil-eq/km). As evidenced, the highest impacts are caused by the ICEV, followed by the HEV, whose percentage ratio remains around 85%, apart from the scenario considered. The recourse to fossil sources of the BEV instead has a strict correlation to the country that utilizes the vehicle, as shown in the graph; with respect to the ICEV, the impact of the BEV remains around 31% in the fossil-free Norway, rising to an average percentage ratio of 51% when the EU-28 electricity mix is considered, and arriving at an average percentage ratio of 71% if the vehicle is employed in Poland. This scenario analysis, hence, reveals quantitatively (also with the help of percentage ratio referred to the traditional internal combustion engine vehicle) the non-explicit recourse to fossil sources of BEV, and how its real impact strictly depends on the fossil source exploitation in the country that utilizes the vehicle. Finally, the specific impact factors obtained by the scenario analysis are reported, for each scenario considered and for each phase evaluated, in Table 18.

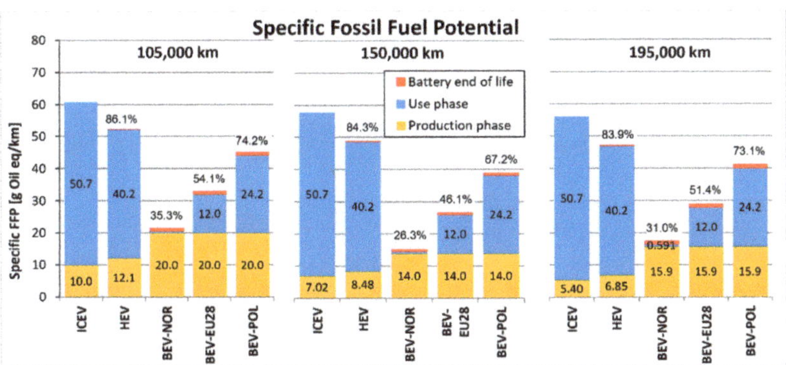

Figure 11. Specific fossil fuel potential (gOil-eq/km) related to the three reference vehicles in the six different scenarios.

According to the results obtained by this scenario analysis, a consideration can be made: if, at the end of its life, an electric vehicle has caused a global warming impact comparable to a traditional petrol or hybrid vehicle (as in the average European and in the Polish case), from an environmental point of view, this could be considered a failure, given that, with respect to a traditional petrol vehicle, an electric vehicle gives rise to roughly two times the acidifying and particulate emissions, and requires the extraction of five times the amount of minerals. This means that, given the current manufacturing and production processes and technologies, promoting the use of lithium-ion-based electric vehicles in geographical areas where the energy mix used to produce electricity still relies on fossil fuels may be counterproductive. In this case, in effect, the substitution of a large part of the existing traditional vehicles with current technology electric vehicles would lead to a change in the worrying environmental impact category, moving from the global warming problem to the other relevant environmental issues such as acidifying and particulate emissions. Since an overall reduction in all the impact categories is instead desirable, it is necessary to promote research towards the development of more efficient and less polluting battery production processes, reducing significantly the recourse to rare minerals and to fossil energy sources.

The scenario analysis revealed great variations in the impacting behavior of the reference BEV, with considerable deviation also from the results obtained in the first analysis based on the EU-28 average electricity mix. On account of this observation, the authors repeated the breakeven analysis already performed in the previous section (see Figure 2), with the aim to evaluate the traveled distance which makes one vehicle more attractive than another from the perspective of global warming potential. The new breakeven analysis was carried out considering both the Norwegian and the Polish electricity mixes, and the results are shown in the two graphs of Figure 12. As can be seen, if the vehicles are operated in Norway, the BEV exhibits an almost constant GWP with varying the lifetime traveled distance, while both ICEV and HEV are characterized by a lower initial impact (production phase), which hence linearly increases with the traveled distance. The result is that both ICEV and HEV reveal a lower global warming impact up to approximately 29,000 km (which means roughly 2.3 years of vehicle usage), and the ICEV remains less impacting than HEV up to 32,500 km (i.e., 2.6 years of vehicle usage). This means that, even in the best possible electricity mix scenario (the Norwegian case is in effect rare to the point of being unique), the global warming caused during BEV production makes both ICEV and HEV more respective of the environment for at least two years of usage.

Table 18. Specific impact indicator results of the scenario analysis.

		spec. GWP [gCO_2eq/km]			spec. TAP [$mgSO_2eq/km$]			spec. PMFP [$mgPM_{2.5}eq/km$]			spec. SOP [mgCu-eq/km]			FFP [gOil-eq/km]		
		Prod.	Use	Batt. EoL	Prod.	Use	Batt. EoL	Prod.	Use	Batt. EoL	Prod.	Use	Batt. EoL	Prod.	Use	Batt. EoL
EUROPE 105,000 km	**ICEV**	38.6	160	0.00	167	255	0.00	53.8	77.0	0.00	600	70.1	0.00	1053	50.7	0.00
	HEV	48.7	127	0.271	281	191	0.702	88.4	59.9	0.196	999	55.5	0.00	1272	40.2	0.0629
	BEV	82.0	49.6	3.77	787	103	9.77	237	32.6	2.73	2817	94.9	0.00	2101	12.0	0.876
EUROPE 150,000 km	**ICEV**	27.0	160	0.00	117	255	0.00	37.7	77.0	0.00	420	70.1	0.00	1053	50.7	0.00
	HEV	34.1	127	0.190	197	191	0.491	61.9	59.9	0.138	699	55.5	0.00	1272	40.2	0.0441
	BEV	57.4	49.6	2.64	551	103	6.84	166	32.6	1.91	1972	94.9	0.00	2101	12.0	0.613
EUROPE 195,000 km	**ICEV**	20.8	160	0.00	89.7	255	0.00	29.0	77.0	0.00	323	70.1	0.00	1053	50.7	0.00
	HEV	27.7	127	0.292	173	191	0.756	54.1	59.9	0.212	617	55.5	0.00	1335	40.2	0.0678
	BEV	65.8	49.6	4.06	733	103	10.5	219	32.6	2.94	2611	94.9	0.00	3097	12.0	0.943
NORWAY 105,000 km	**ICEV**	38.6	160	0.00	167	255	0.00	53.8	77.0	0.00	600	70.1	0.00	1053	50.7	0.00
	HEV	48.7	127	0.271	281	191	0.702	88.4	59.9	0.196	999	55.5	0.00	1272	40.2	0.0629
	BEV	82.0	1.71	3.77	787	1.48	9.77	237	0.354	2.73	2817	27.2	0.00	2101	0.591	0.876
NORWAY 150,000 km	**ICEV**	27.0	160	0.00	117	255	0.00	37.7	77.0	0.00	420	70.1	0.00	1053	50.7	0.00
	HEV	34.1	127	0.190	197	191	0.491	61.9	59.9	0.138	699	55.5	0.00	1272	40.2	0.0441
	BEV	57.4	1.71	2.64	551	1.48	6.84	166	0.354	1.91	1972	27.2	0.00	2101	0.591	0.613
NORWAY 195,000 km	**ICEV**	20.8	160	0.00	89.7	255	0.00	29.0	77.0	0.00	323	70.1	0.00	1053	50.7	0.00
	HEV	27.7	127	0.292	173	191	0.756	54.1	59.9	0.212	617	55.5	0.00	1335	40.2	0.0678
	BEV	65.8	1.71	4.06	733	1.48	10.5	219	0.354	2.94	2611	27.2	0.00	3097	0.591	0.943
POLAND 105,000 km	**ICEV**	38.6	160	0.00	167	255	0.00	53.8	77.0	0.00	600	70.1	0.00	1053	50.7	0.00
	HEV	48.7	127	0.271	281	191	0.702	88.4	59.9	0.196	999	55.5	0.00	1272	40.2	0.0629
	BEV	82.0	142	3.77	787	357	9.77	237	107	2.73	2817	89.8	0.00	2101	24.2	0.876
POLAND 150,000 km	**ICEV**	27.0	160	0.00	117	255	0.00	37.7	77.0	0.00	420	70.1	0.00	1053	50.7	0.00
	HEV	34.1	127	0.190	197	191	0.491	61.9	59.9	0.138	699	55.5	0.00	1272	40.2	0.0441
	BEV	57.4	142	2.64	551	357	6.84	166	107	1.91	1972	89.8	0.00	2101	24.2	0.613
POLAND 195,000 km	**ICEV**	20.8	160	0.00	89.7	255	0.00	29.0	77.0	0.00	323	70.1	0.00	1053	50.7	0.00
	HEV	27.7	127	0.292	173	191	0.756	54.1	59.9	0.212	617	55.5	0.00	1335	40.2	0.0678
	BEV	65.8	142	4.06	733	357	10.5	219	107	2.94	2611	89.8	0.00	3097	24.2	0.943

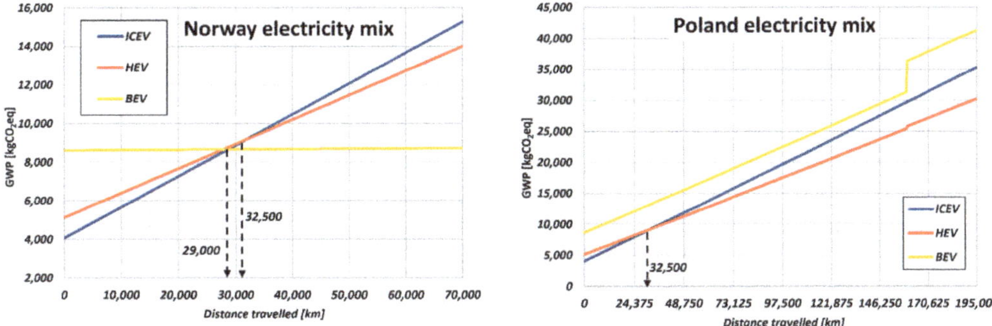

Figure 12. Global warming impacts comparison as function of the lifetime traveled distance, with reference to the Norwegian (**left**) and to the Polish (**right**) electricity mix.

The results strongly change if the vehicles are supposed to travel in Poland; in this case, as shown on the right graph of Figure 12, the high carbon footprint of the Polish electricity mix causes the BEV to have such a GWP gradient that even the ICEV remains less polluting up to distance in the order of 180,000 km. Moreover, as can also be noted in Figure 12, the intersection between the global warming curves of the ICEV and the BEV is prevented by the sharp increase caused by the battery replacement on the BEV curve. As a result, in this scenario, the lifecycle global warming impacts of both ICEV and HEV result are always lower than the lifecycle impact caused by the BEV.

9. Conclusions

In this paper, the authors performed a comparative evaluation of the life cycle impact of three different vehicles of different technologies, namely an internal combustion engine vehicle (ICEV), a hybrid electric vehicle (HEV), and a battery electric vehicle (BEV). The study was carried out considering three fictitious vehicles representative of existing products, whose performance and characteristics were determined on the basis of real vehicles available on the 2019 market for the B-C segments. The life cycle impact was evaluated by the use of the GREET model following the ReCiPe 2016 methodology and taking into consideration five different impact categories represented by their characterization factors: Global Warming Potential (GWP), Terrestrial Acidification Potential (TAP), Particulate matter formation potential (PMFP), Surplus ore potential (SOP), and Fossil fuel potential (FFP). The impact evaluation was properly performed taking into account all the phases of the vehicle life, from its production to its use, and to the final disposal. To this purpose, some assumptions were made, justified, and corroborated by proper references, which defined the reference scenario adopted for the comparison—the vehicles were supposed to be produced in Germany and used all-over Europe, while the lithium-ion batteries of both the BEV and the HEV were assumed to be produced in China. The assessment of the environmental impact associated with the production of the fuel for the ICEV and for the HEV was carried out by means of the Ecoinvent v3 database. The procedure adopted by the authors is hence a general and "blind" procedure, which, starting from objective data and through some assumptions (clearly stated in the "Goal and scope definition" section), allows evaluating the lifetime environmental impact of a selected kind of vehicle in a properly defined scenario; the same procedure could be hence repeated considering different vehicles, or different energy scenarios, or making different assumptions, for the evaluation of the lifetime environmental impact caused by vehicles according to LCA methodology. With the aim to extend the limit of the first analysis, the authors also performed a scenario analysis by changing the lifetime traveled distance and the country of utilization of the vehicles.

Several conclusions and observations can be drawn on the basis of the results of this study. First of all, the results obtained clearly pointed out the fundamental role played by the production and the disposal phases of vehicles on the evaluation of the real lifetime impact generated by their use. The production phase of the electric vehicle, and specifically the production of its lithium-ion battery, revealed a very critical phase, with a strong impact in terms of terrestrial acidification, particulate matter formation, and mineral resource deployment. As a result, the environmental impact generated in these categories by the BEV resulted abundantly higher than in the ICEV (from +50% to +500%). The main reasons for this high environmental impact can be found in the high energy required for lithium-ion battery production, in the relevant emissions of primary and secondary particulate which characterize the lithium-ion battery production process, in the large recourse to metals such as cobalt, copper, and nickel, and in the coal-dominated electricity mix of the largest lithium-ion battery producer of the world, i.e., China. In contrast, in regard to the global warming effect and the fossil sources deployment, the BEV is confirmed to be the least impacting vehicle, if the electricity used for vehicle propulsion has been generated by an adequate recourse to renewable sources. For example, in the case of the average European electricity mix (34% from renewable sources, 25% from Nuclear), the GWP impact caused by the BEV in its entire life revealed (for a lifetime distance traveled of 150,000 km) 58.6% of the impact produced by the ICEV; assuming instead to employ the vehicles in Norway (where 97% of electric energy is obtained from renewable sources), the GWP impact generated by the BEV reduced to 33% with respect to the ICEV, while when considering the vehicles used in Poland (where 73% of electric energy is obtained from coal-fired power plants), the BEV reveals to be the most impacting vehicle, regardless of the lifetime distance traveled. The results obtained by the analysis carried out highlight how carefully the real overall impact generated by a vehicle during its entire life must be evaluated, and how this impact may be affected by the fossil source exploitation of the country that utilizes the vehicle.

A further observation must be made on the basis of the very different entities of the impact generated by the BEV in the five impact categories considered—the introduction of electric vehicles on the market should be carefully monitored with life cycle analysis tools, avoiding focus on a single environmental impact category that is currently particularly problematic and known (the global warming) at the risk of causing huge impact increments on less considered but equally harmful categories.

The results of the environmental impact comparison also confirmed the hybrid vehicle as an excellent alternative to ICEV, being capable to achieve a good compromise between all the categories of environmental impact—on all the scenarios considered in this study, the HEV revealed GWP and FFP impacts in the order of 85% with respect to an equivalent ICEV, while maintaining acidifying and particulate emissions well below the high levels of the BEV.

In the study presented in this paper, the effect of the variation of the lifetime distance travelled by the three vehicles on their global warming impact was analyzed. The results show that, apart from the country that utilizes the vehicles, the BEV is always the more impacting vehicle in the lower mileage range, due to the high global warming emissions generated during the production phase of the lithium-ion battery; in the same distance range, the HEV, endowed of a smaller battery, shows slightly higher impact than the ICEV. Being the rate of increase in the GWP impact related to both ICEV and HEV was exclusively dependent on their fuel consumption, it was found that the HEV becomes less impacting than the ICEV after 32,500 km, which, according to the average European annual distance traveled by a passenger car, means after about 3 years of utilization of the vehicles. In the case of the BEV, instead, the rate of increase in the GWP impact with the traveled distance depends also on the particular electricity mix of the country where the vehicle is employed; on account of this, it was found that, according to the average European electricity mix, the BEV reveals less impacting than the ICEV after 41,250 km and less impacting than the HEV after 46,250 km (i.e., roughly after 3.5–4 years of utilization of the vehicles); obviously, the low carbon footprint of the Norwegian electricity mix reduces these distances, which become both equal to 29,000 km. On the contrary, in a country with a high carbon footprint such as Poland, the GWP impact of both ICEV and HEV remains always lower than BEV whichever is the lifetime traveled distance (also due to the lithium-ion battery replacement after 160,000 km, which, in turn, causes a sharp and considerable increment of the environmental impact caused by the BEV).

Author Contributions: E.P. contributed to the conceptualization and supervision of the work, to editing and writing of the paper; S.C. and L.O. contributed to the acquisition and analysis of data, to the interpretation of results, and drafted the paper. All authors have read and agreed to the published version of the manuscript.

Funding: The research was funded by Università degli Studi di Palermo.

Institutional Review Board Statement: Not applicable.

Informed Consent Statement: Not applicable.

Data Availability Statement: All data generated or analyzed during this study are included in this published article or are publicly available.

Conflicts of Interest: The authors declare no conflict of interest.

Abbreviations and Symbols

B_{BEV}	Electric vehicle battery capacity
BE5	Gasoline with 5% ethanol from biomass
BEV	Battery electric vehicle
B_{HEV}	Hybrid electric vehicle battery capacity
BOM	Bill of material
C_{HEV}	Hybrid electric vehicle fuel tank capacity
C_{ICEV}	Internal combustion engine vehicle fuel tank capacity
CNG	Compressed natural gas
D_{tot}	Electric vehicle total driving distance on the WLTP cycle
e_{CO2}	CO_2 emitted per kilometer on the WLTP cycle [g/km]
EoL	End-of-life
E_{trac}	Total traction energy required by ICEV or HEV during its entire life
$E_{trac,BEV}$	Total traction energy required by the BEV during its entire life
$F_{[kWh/km]}$	Electric vehicle energy consumption per kilometer on the WLTP cycle
$F_{[l/km]}$	Fuel consumption per kilometer on the WLTP cycle
f_{CO2}	CO_2 emitted by the combustion of a liter of BE5 [g/km]
FFP	Fossil fuel potential [kg Oil-eq]
GHG	Greenhouse gases
GWP	Global warming potential [kg CO2-eq]
HEV	Hybrid electric vehicle
ICEV	Internal combustion engine vehicle
$I_{x,source}$	Characterization factor connected to the production of the total mass of fuel employed by the ICEV or by the HEV
LCA	Life cycle assessment
LCI	Life cycle inventory
LCIA	Life cycle impact assessment
LCO	Lithium Cobalt Oxide (LiCoO2)
LFP	Lithium Iron Phosphate (LiFePO4)
LHV	Fuel Lower Heating Value
LMO	Lithium Manganese Oxide ($LiMn_2O_4$)
LPG	Liquefied petroleum gas
$m_{battery}$	Mass of vehicle battery
m_{BEV}	Kerb mass of the battery electric vehicle

m_{empty}	Empty mass of each vehicle (i.e., related to vehicle components only)
m_{fluids}	Mass of vehicle fluids
m_{fuel}	Total mass of fuel consumed by the vehicle during its entire life
m_{HEV}	Kerb mass of the hybrid electric vehicle
m_{ICEV}	Kerb mass of the internal combustion engine vehicle
m_{kerb}	Kerb mass of the generic vehicle
NCA	Lithium Nickel Cobalt Aluminum Oxide ($LiNiCoAlO_2$)
NMC	Lithium Nickel Manganese Cobalt Oxide (LiNiMnCoO2)
P	Maximum power of the generic vehicle
P_{BEV}	Battery electric vehicle maximum power
P_{HEV}	Hybrid electric vehicle maximum power
P_{ICEV}	Internal combustion engine vehicle maximum power
$PMFP$	Particulate Matter Formation Potentials [kg PM2.5-eq]
R_{BEV}	WLTP driving range of the battery electric vehicle
SOP	Surplus ore potential [kg Cu-eq]
TAP	Terrestrial acidification potential [kg SO2-eq]
$UNECE$	New European Driving Cycle
V_{HEV}	Hybrid electric vehicle engine displacement
V_{ICEV}	Internal combustion engine vehicle engine displacement
W	Mass of the Li-ion polymer battery
$WLTP$	Worldwide harmonized light vehicles test procedure
ϕ_x	Specific impact factor referred to each impact category x and associated to the production of 1 kg of gasoline
φ_x	Specific impact factor referred to each impact category x and associated to the production of 1 kWh of electric energy
$\beta_{B,k}$	Ratio between battery capacity and WLTP driving range
$\beta_{F,k}$	Ratio between WLTP consumption and vehicle mass
$\beta_{P,k}$	Ratio between vehicle maximum power and vehicle mass
β_W	Capacity-to-mass ratio of the BEV battery
$\theta_{C,i}$	Ratio between fuel tank capacity and vehicle mass
$\theta_{F,i}$	Ratio between vehicle consumption and vehicle mass
$\theta_{P,i}$	Ratio between engine maximum power and vehicle mass
$\theta_{V,i}$	Ratio between engine maximum power and engine displacement
ρ_{fuel}	Fuel density
$\psi_{B,j}$	Ratio between battery capacity and fuel tank capacity
$\psi_{C,j}$	Ratio between fuel tank capacity and vehicle mass
$\psi_{F,j}$	Ratio between vehicle consumption and vehicle mass
$\psi_{P,j}$	Ratio between maximum vehicle power and vehicle mass
$\psi_{V,j}$	Ratio between engine displacement and vehicle mass
$\psi_{W,j}$	Ratio between battery capacity and mass battery

Appendix A

Reference Vehicles Specifications and Characteristics

The main characteristics of the three reference vehicles considered in the environmental impact comparison were determined on the basis of coefficients derived from vehicle fundamental parameters. Starting with the ICEV, the authors evaluated, for each real vehicle considered, the following coefficients:

(1) $\theta_{C,i} = \frac{C_i}{m_i}\left[\frac{L}{kg}\right]$ = ratio between fuel tank capacity and vehicle mass

(2) $\theta_{P,i} = \frac{P_i}{m_i}\left[\frac{kW}{kg}\right]$ = the ratio between engine maximum power and vehicle mass

(3) $\theta_{V,i} = \frac{P_i}{V_i}\left[\frac{kW}{L}\right]$ = the ratio between engine maximum power and engine displacement

(4) $\theta_{F,i} = \frac{F_i}{m_i}\left[\frac{km/L}{kg}\right]$ = the ratio between vehicle consumption and vehicle mass

where the subscript i refers to the generic real internal combustion engine vehicle (i ranges from 1 to 5), m represents the vehicle mass, C the fuel tank capacity of the vehicle, P the engine maximum output power, and F the vehicle fuel consumption on the WLTP cycle (i.e., km/L). The sense of the selected coefficients can be explained by simple considerations: for the first coefficient, the authors considered that the higher is the vehicle mass, the larger will be the necessary fuel tank to allow the vehicle a certain operating range; for the second coefficient, it is easy to consider that the higher is the vehicle mass, the higher will be the necessary power output to produce a certain vehicle acceleration or speed; the third coefficient was deduced considering that engine of similar technological development will exhibit similar specific power; the fourth coefficient is based on the simple consideration that higher vehicle mass will cause higher fuel consumption for the same driving cycle; the last coefficient is based on the proportionality between the amount of CO_2 emitted and the amount of fuel burned. It is worth mentioning that, to ascertain the significance of the selected coefficients, their dispersion was evaluated in terms of the range of variation: a range of ±11% was found in the worst case, which means that the selected coefficients have a limited variation from one vehicle to another. Focusing hence on the standard representative ICEV, its mass m_{ICEV} was simply determined as the average value of the masses of the five vehicles considered, while the other characteristics were determined employing the above coefficients:

$$\text{Vehicle mass} = m_{ICEV} = \frac{1}{5}\sum_{i=1}^{5} m_i$$

$$\text{Fuel tank capacity} = C_{ICEV} = m_{ICEV} \cdot \frac{1}{5}\sum_{i=1}^{5} \theta_{C,i}$$

$$\text{Maximum output power} = P_{ICEV} = m_{ICEV} \cdot \frac{1}{5}\sum_{i=1}^{5} \theta_{P,i}$$

$$\text{Consumption} = F_{ICEV} = m_{ICEV} \cdot \frac{1}{5}\sum_{i=1}^{5} \theta_{F,i}$$

$$\text{Engine displacement} = V_{ICEV} = P_{ICEV} / \frac{1}{5}\sum_{i=1}^{5} \theta_{V,i}$$

A similar approach was followed for the determination of the standard representative HEV. In this case, the coefficients taken into consideration were:

(1) $\psi_{C,j} = \frac{C_j}{m_j}\left[\frac{L}{kg}\right]$ = ratio between fuel tank capacity and vehicle mass

(2) $\psi_{P,j} = \frac{P_j}{m_j}\left[\frac{kW}{kg}\right]$ = ratio between maximum vehicle power and vehicle mass

(3) $\psi_{V,j} = \frac{V_j}{m_j}\left[\frac{L}{kg}\right]$ = ratio between engine displacement and vehicle mass

(4) $\psi_{B,j} = \frac{B_j}{C_j}\left[\frac{kWh}{L}\right]$ = ratio between battery capacity and fuel tank capacity

(5) $\psi_{F,j} = \frac{F_j}{m_j}\left[\frac{km/L}{kg}\right]$ = ratio between vehicle consumption and vehicle mass

(6) $\psi_{W,j} = \frac{B_j}{W_j}\left[\frac{kWh}{kg}\right]$ = ratio between battery capacity and battery mass

where the subscript j refers to the generic real hybrid electric vehicle (j ranges from 1 to 5), the parameters m, C, P, V, and F have the same meaning adopted for the ICEV, B represents the battery capacity, which, as represented in the fourth coefficient, was considered proportional to the capacity of the fuel tank, and W the battery mass. According to the selected coefficients, the main characteristics of the standard representative HEV were determined as follows:

$$\text{Vehicle mass} = m_{HEV} = \frac{1}{5}\sum_{j=1}^{5} m_j$$

$$\text{Fuel tank capacity} = C_{HEV} = m_{HEV} \cdot \frac{1}{5}\sum_{j=1}^{5} \psi_{C,j}$$

$$\text{Maximum output power} = P_{HEV} = m_{HEV} \cdot \frac{1}{5}\sum_{j=1}^{5} \psi_{P,j}$$

$$\text{Consumption} = F_{HEV} = m_{HEV} \cdot \frac{1}{5}\sum_{j=1}^{5} \psi_{F,j}$$

$$\text{Engine displacement} = V_{HEV} = m_{HEV} \cdot \tfrac{1}{5}\sum_{j=1}^{5} \psi_{V,j}$$

$$\text{Battery capacity} = B_{HEV} = C_{HEV} \cdot \tfrac{1}{5}\sum_{j=1}^{5} \psi_{B,j}$$

As regards the battery of the std. HEV, it must be pointed out that an Li-ion polymer model was adopted. Since no information was available on the battery of the vehicle HEV3, the only available capacity-to-mass ratio of the battery of vehicle HEV5 was adopted for the coefficient ψ_W, hence:

$$\psi_W \equiv \psi_{W,5} = \frac{B_5}{W_5} = \frac{1.56}{33} = 0.473\left[\frac{kWh}{kg}\right]$$

According to this coefficient, the mass of the Li-ion polymer battery of the standard HEV was evaluated as:

$$W_{HEV} = B_{HEV}/\psi_W$$

Concerning the standard representative BEV, the coefficients taken into consideration were:

(1) $\beta_{P,k} = \frac{P_k}{m_k}\left[\frac{kW}{kg}\right]$ = ratio between vehicle maximum power and vehicle mass =

(2) $\beta_{F,k} = \frac{F_k}{m_k}\left[\frac{kWh}{km\cdot kg}\right]$ = the ratio between WLTP consumption and vehicle mass

(3) $\beta_{B,k} = \frac{B_k}{R_k}\left[\frac{kWh}{km}\right]$ = the ratio between battery capacity and WLTP driving range

(4) $\beta_{W,k} = \frac{B_k}{W_k}\left[\frac{kWh}{kg}\right]$ = ratio between battery capacity and battery mass

where the subscript k refers to the generic real battery electric vehicle (k ranges from 1 to 5), the parameters m, B, P, F, and W have the same meaning of previous vehicles, while R represents the WLTP driving range of the vehicle, which, as reported by the third coefficient, was assumed to be related to the capacity of the battery. As in the previous cases, the mass of the standard representative BEV was evaluated as the average value among the 5 commercial vehicles:

$$\text{Vehicle mass} = m_{BEV} = \frac{1}{5}\sum_{k=1}^{5} m_k$$

Moreover, considering that the std. BEV should also have an average operating range with respect to the five commercial vehicles, its WLTP driving range was evaluated as average value:

$$\text{Driving range (WLTP)} = R_{BEV} = \frac{1}{5}\sum_{k=1}^{5} R_k$$

According to the selected coefficients, the other main characteristics of the standard representative BEV were determined as follows:

$$\text{Maximum output power} = P_{BEV} = m_{BEV} \cdot \tfrac{1}{5}\sum_{k=1}^{5} \beta_{P,k}$$

$$\text{Battery capacity} = B_{BEV} = R_{BEV} \cdot \tfrac{1}{5}\sum_{k=1}^{5} \beta_{B,k}$$

$$\text{Consumption} = F_{BEV} = m_{BEV} \cdot \tfrac{1}{5}\sum_{k=1}^{5} \beta_{F,k}$$

As reported in Table 4, an Li-ion NMC622 battery was adopted for the std. BEV, hence its mass was deduced on the basis of the only available capacity-to-mass ratio of the vehicle BEV5 battery:

$$\beta_W \equiv \beta_{W,5} = \frac{B_5}{W_5} = \frac{38.3}{340} = 0.112\left[\frac{kWh}{kg}\right] \Rightarrow \text{Battery mass} = W_{BEV} = B_{BEV}/\beta_W$$

References

1. European Commission. Statistical Pocketbook 2019: EU Energy. 2019. Available online: https://op.europa.eu/it/publication-detail/-/publication/e0544b72-db53-11e9-9c4e-01aa75ed71a1 (accessed on 23 September 2021).
2. European Commission. Statistical Pocketbook 2019: EU Transport. 2019. Available online: https://ec.europa.eu/transport/sites/default/files/pocketbook-2019.pdf (accessed on 23 September 2021).
3. Wu, Z.; Wang, M.; Zheng, J.; Sun, X.; Zhao, M.; Wang, X. Life cycle greenhouse gas emission reduction potential of battery electric vehicle. *J. Clean. Prod.* **2018**, *190*, 462–470. [CrossRef]
4. Bauer, C.; Hofer, J.; Althaus, H.-J.; Del Duce, A.; Simons, A. The environmental performance of current and future passenger vehicles: Life cycle assessment based on a novel scenario analysis framework. *Appl. Energy* **2015**, *157*, 871–883. [CrossRef]
5. Helmers, E.; Dietz, J.; Weiss, M. Sensitivity Analysis in the Life-Cycle Assessment of Electric vs. Combustion Engine Cars under Approximate Real-World Conditions. *Sustainability* **2020**, *12*, 1241. [CrossRef]
6. Hawkins, T.R.; Singh, B.; Majeau-Bettez, G.; Strømman, A.H. Comparative Environmental Life Cycle Assessment of Conventional and Electric Vehicles. *J. Ind. Ecol.* **2012**, *17*, 53–64. [CrossRef]
7. Cusenza, M.A.; Bobba, S.; Ardente, F.; Cellura, M.; Di Persio, F. Energy and environmental assessment of a traction lithium-ion battery pack for plug-in hybrid electric vehicles. *J. Clean. Prod.* **2019**, *215*, 634–649. [CrossRef] [PubMed]
8. Hauschild, M.; Goedkoop, M.; Guinee, J.; Heijungs, R.; Huijbregts, M.; Joillet, O.; Margni, M.; De Schryver, A.; Pennington, D.; Pant, R.; et al. *Recommendations for Life Cycle Impact Assessment in the European Context, International Reference Life Cycle Data System—ILCD Handbook, EUR 24571 EN.*; Publications Office of the European Union: Luxembourg, 2011; JRC61049.
9. ISO 14040. *International Standard, Environmental Management—Life Cycle Assessment—Principles and Framework*, 2nd ed.; 389 Chiswick High Road: London, UK, 2006; EN ISO 14040.
10. Guinee, J.B. Handbook on life cycle assessment operational guide to the ISO standards. *Int. J. Life Cycle Assess.* **2002**, *7*, 311–313. [CrossRef]
11. ACEA. The Automobile Industry Pocket Guide 2019/2020. 2020. Available online: https://www.acea.auto/publication/automobile-industry-pocket-guide-2019-2020 (accessed on 23 September 2021).
12. GREET Model—The Greenhouse Gases, Regulated Emissions, and Energy Use in Technologies Model. Available online: https://greet.es.anl.gov/index.php (accessed on 23 September 2021).
13. Elgowainy, M.; Wang, A. Overview of Life Cycle Analysis (LCA) with the GREET Model, Argonne National Laboratory. 2019. Available online: https://greet.es.anl.gov/files/workshop_2019_overview (accessed on 23 September 2021).
14. Huijbregts, M.A.J.; Steinmann, Z.J.N.; Elshout, P.M.F.; Stam, G.; Verones, F.; Vieira, M.; Zijp, M.; Hollander, A.; van Zelm, R. *ReCiPe2016: A Harmonized Life Cycle Impact Assessment Method at Midpoint and Endpoint Level*; RIVM Report 2016–2014; Ministry of Health: Bilthoven, The Netherlands, 2016; Available online: www.rivm.nl/en (accessed on 23 September 2021).
15. Huijbregts, M.A.J.; Steinmann, Z.; Elshout, P.M.F.; Stam, G.; Verones, F.; Vieira, M.; Zijp, M.; Hollander, A.; van Zelm, R. ReCiPe2016: A harmonised life cycle impact assessment method at midpoint and endpoint level. *Int. J. Life Cycle Assess.* **2016**, *22*, 138–147. [CrossRef]
16. Vieira, M.D.M.; Ponsioen, T.C.; Goedkoop, M.J.; Huijbregts, M. Surplus Ore Potential as a Scarcity Indicator for Resource Extraction. *J. Ind. Ecol.* **2016**, *21*, 381–390. [CrossRef]
17. ACEA. Economic and Market Report—Full Year 2019. 2020. Available online: www.acea.be (accessed on 23 September 2021).
18. IEA. Commissioned EV and Energy Storage Lithium-Ion Battery Cell Production Capacity by Region, and Associated Annual Investment, 2010–2022. Available online: https://www.iea.org/data-and-statistics/charts/commissioned-ev-and-energy-storage-lithium-ion-battery-cell-production-capacity-by-region-and-associated-annual-investment-2010-2022 (accessed on 23 September 2021).
19. ACEA. Vehicles in Use Report. 2021. Available online: www.acea.be/statistics/tag/category/vehicles-in-use (accessed on 23 September 2021).
20. IEA. Data and Statistics. Available online: www.iea.org/data-and-statistics (accessed on 23 September 2021).
21. Burnham, A. *Updated Vehicle Specifications in the GREET Vehicle-Cycle Model*; Center for Transportation Research; Argonne National Laboratory: Lemont, IL, USA, 2012.
22. Ellingsen, L.A.-W.; Majeau-Bettez, G.; Singh, B.; Srivastava, A.K.; Valøen, L.O.; Strømman, A.H. Life Cycle Assessment of a Lithium-Ion Battery Vehicle Pack. *J. Ind. Ecol.* **2013**, *18*, 113–124. [CrossRef]
23. Blomgren, G.E. The Development and Future of Lithium Ion Batteries. *J. Electrochem. Soc.* **2016**, *164*, A5019–A5025. [CrossRef]
24. Mathieux, F.; Ardente, F.; Bobba, S.; Nuss, P.; Blengini, G.; Alves Dias, P.; Blagoeva, D.; Torres De Matos, C.; Wittmer, D.; Pavel, C.; et al. *Critical Raw Materials and the Circular Economy—Background Report, JRC Science-for-Policy Report, EUR 28832 EN.*; Publications Office of the European Union: Luxembourg, 2017; ISBN 978-92-79-74282-8. [CrossRef]
25. Dunn, J.B.; Gaines, L.; Barnes, M.; Wang, M.; Sullivan, J.S. *Material and Energy Flows in the Materials Production, Assembly, and End-of-Life Stages of the Automotive Lithium-Ion Battery Life Cycle. ANL/ESD/12-3*; Argonne National Lab: Argonne, IL, USA, 2012.
26. Dai, Q.; Kelly, J.C.; Gaines, L.; Wang, M. Life Cycle Analysis of Lithium-Ion Batteries for Automotive Applications. *Batteries* **2019**, *5*, 48. [CrossRef]
27. Dai, J.Q.; Dunn, J.; Kelly, J.; Elgowainy, A. Update of Life Cycle Analysis of Lithium-Ion Batteries in the GREET Model. 2017. Available online: https://greet.es.anl.gov/publication-Li_battery_update_2017 (accessed on 23 September 2021).

28. Dai, Q.; Kelly, J.C.; Dunn, J.; Benavides, P.T. Update of Bill-of-Materials and Cathode Mate-Rials Production for Lithium-Ion Batteries in the GREET Model. 2018. Available online: https://greet.es.anl.gov/publication-update_bom_cm (accessed on 23 September 2021).
29. ISO 1176:1990(en), Road Vehicles-Masses-Vocabulary and Codes. Available online: https://www.iso.org/obp/ui/#iso:std:iso:24444:ed-1:v1:en (accessed on 23 September 2021).
30. Wernet, G.; Bauer, C.; Steubing, B.; Reinhard, J.; Moreno-Ruiz, E.; Weidema, B. The ecoinvent database version 3 (part I): Overview and methodology. *Int. J. Life Cycle Assess.* **2016**, *21*, 1218–1230. [CrossRef]
31. Althaus, H.-J.; Bauer, C.; Doka, G.; Dones, R.; Hischier, R.; Hellweg, S.; Humbert, S.; Kollner, T.; Loerinick, Y.; Margni, M.; et al. *Implementation of Life Cycle Impact Assessment Methods. Ecoinvent Report No. 3, v2.2*; Swiss Centre for Life Cycle Inventories: Dübendorf, Switzerland, 2010.
32. European Environment Agency. *Fuel Quality in the EU in 2016*; Publications Office of the European Union: Luxembourg, 2018. [CrossRef]
33. Gnansounou, E.; Dauriat, A.; Panichelli, L.; Villegas, J. Energy and greenhouse gas balances of biofuels: Biases induced by LCA modelling choices. *J. Sci. Ind. Res.* **2008**, *67*, 885–897. Available online: http://nopr.niscair.res.in/handle/123456789/2418 (accessed on 23 September 2021).
34. Leonidas, N.; Zissis, S. EMEP/EEA Air Pollutant Emission Inventory Guidebook 2019—Update Oct. 2020. European Environmental Agency. 2020. Available online: http://eea.europa.eu/emep-eea-guidebook (accessed on 23 September 2021).
35. Recharge. The Advanced Rechargeable & Lithium Batteries Association PEFCR—Product Environmental Footprint Category Rules for High Specific Energy Rechargeable Batteries for Mobile Applications Version: H Time of Validity: 31 December 2020. 2018. Available online: https://ec.europa.eu/environment/eussd/smgp/pdf/PEFCR_Batteries.pdf (accessed on 23 September 2021).
36. Golder Associates. Environmental Impact Assessment—Tenke Fungurume Project-Volume A: ESIA Introduction and Project Description. 2007. Available online: https://www3.opic.gov/environment/eia/tenke/Volume%20A%20ESIA%20Introduction%20and%20Project%20Description.pdf (accessed on 23 September 2021).

MDPI

Article

Possibility of a Solution of the Sustainability of Transport and Mobility with the Application of Discrete Computer Simulation—A Case Study

Nikoleta Mikušová [1,*], Gabriel Fedorko [1], Vieroslav Molnár [2], Martina Hlatká [3], Rudolf Kampf [3] and Veronika Sirková [1]

1 Faculty of Mining, Ecology, Process Control and Geotechnologies, Technical University of Košice, Letná 9, 042 00 Košice, Slovakia; gabriel.fedorko@tuke.sk (G.F.); sirkova1982@azet.sk (V.S.)
2 Faculty of Manufacturing Technologies, Technical University of Košice, Bayerova 1, 080 01 Prešov, Slovakia; vieroslav.molnar@tuke.sk
3 Faculty of Technology, The Institute of Technology and Business in České Budějovice, 370 01 České Budějovice, Czech Republic; hlatka@mail.vstecb.cz (M.H.); kampf@mail.vstecb.cz (R.K.)
* Correspondence: nikoleta.mikusova@tuke.sk

Abstract: The paper is focused on an example of a solution for the sustainability of transport and mobility with the application of discrete computer simulation. The obtained results from the realized simulation were complemented with the selected multi-criteria decision-making method, namely the analytic hierarchy process (AHP) method. The paper describes the use of the simulation model for obtaining characteristics of alternative solutions that were designed for the needs of transport sustainability. The aim is to address the problem of traffic congestion in urban agglomerations. The simulation model serves as a means to provide information for the needs of their analysis by multi-criteria evaluation by the AHP. The methodology is based on a combination of computer simulation and multi-criteria decision-making and presents a useful tool that can be used in the field of transport sustainability. The paper notes methods to implement analysis of alternative solutions in transport. However, this procedure can also be used to solve other problems in the field of logistics systems. The paper compares five possible solutions for the organization of transport at intersections. Multi-criteria decision-making was realized based on 12 criteria. The result was the solution that reduced the length of congestion in almost all directions, with a maximum shortening of 69 m and a shortening of the average delay by 26 s compared to the current state.

Keywords: transport; sustainability; mobility; simulation

Citation: Mikušová, N.; Fedorko, G.; Molnár, V.; Hlatká, M.; Kampf, R.; Sirková, V. Possibility of a Solution of the Sustainability of Transport and Mobility with the Application of Discrete Computer Simulation—A Case Study. *Sustainability* **2021**, *13*, 9816. https://doi.org/10.3390/su13179816

Academic Editor: Elżbieta Macioszek

Received: 3 August 2021
Accepted: 28 August 2021
Published: 1 September 2021

Publisher's Note: MDPI stays neutral with regard to jurisdictional claims in published maps and institutional affiliations.

1. Introduction

Sustainable transport within urban agglomerations creates conditions that ensure reliable satisfaction of transport needs and the functioning of individual transport systems. The aim is to ensure the smooth travel of the population, promote public passenger transport, improve the environment, and increase safety and the flow of traffic [1]. To achieve sustainable cities, it is essential to create and sustain changes in people's social behavior through new approaches to mobility, from inefficient, uneconomical, and motorized means, to cleaner, greener, healthier, and more economical means. One of the solutions implemented in connection with sustainable mobility is the support of public passenger transport.

This relates to increasing its attractiveness, with the aim to encourage passengers to switch from individual motoring to public transport. To achieve this aim, it is necessary to create such preconditions within the transport infrastructure that public passenger transport runs continuously, without undue delay, and allows a continuous form of transport [2].

Sustainable mobility within urban agglomerations is closely linked to the effect of transport on the environment. In this regard, it is possible to talk about the importance of green city transport [3]. One of the key tasks is to realize measures to reduce emissions of toxic substances into the environment, the interaction of the car with the environment [4], and if it is possible to reduce the traffic volume in cities [5].

In addition, environmentally friendly vehicles are a key element of green transportation in modern economies [6]. However, it must be emphasized that the use of these types of vehicles can depend on the financial and economic situation of cities. As such, their use faces challenges in the countries of Eastern Europe.

This problem can be solved using traffic planning, whose importance is due to the growth of car traffic in cities. For this reason, cities are beginning to implement town planning measures aimed at improving the traffic situation, reducing congestion with a focus on public transport, and improving the environment [6]. One of the possible tools can be a Transportation Management Information Systems (TMIS) with the overall social and economic development, such as improvement of regional conditions and optimization of the environment, promoting communication and accelerating development [7]. It is very important for the cities to guarantee the efficiency and also the accuracy of the transportation system. One of the ways how to achieve this aim is to analyze the real-time status of transportation network, which determines the urban distribution, travel activities, and development of the urban systems [8]. In relation to the discussion above, real-time traffic prediction based on highly accurate spatio-temporal datasets of traffic sensors is a major challenge for intelligent transportation systems and sustainability. However, this is challenging due to complex topological dependencies and high dynamics associated with changes in road conditions [9]. An important topic of research in the field of modern intelligent traffic systems (ITSs) is path planning. Complex and changeable factors, for example, traffic congestion and traffic accident, should be considered by planning paths, and path point planning schemes can improve the reliability of path planning and also ensure several services needed for transportation process [10].

Town planning measures can be based on decision-making processes using a comparison of variants and partial solutions. The aim is to find a realistic transport scenario that can meet all the basic requirements associated with congestion.

One approach to traffic planning is the well-known microscopic traffic simulator, Simulation of Urban Mobility (SUMO), which is used to design traffic scenarios and present their parameters, in addition to the evaluation and validation of traffic requirements and mobility patterns [11]. The interdependencies among multimodal modes of transport significantly contribute to effective urban transport planning [11].

Due to the complexity of traffic problems, new approaches based on computer simulation and traffic modeling are increasingly being introduced.

It is possible to effectively use simulation models with different levels of detail [11]. For example, it is possible to use macroscopic traffic simulations, which focus on traffic streams but do not take into account the vehicles of the traffic stream. By contrast, microsimulation is a form of traffic simulation capable of accurately modeling the behavior of vehicles in a defined environment [11]. The importance of real data for microsimulation purposes of urban mobility monitoring can be noted by examining the mobility of vehicles at two daily peak times at a roundabout [12]. For the need of micro and macrosimulations, it is possible to use Internet of Things (IoT). IoT-based solution also presents an interesting tool for traffic problems solution. Internet of Things can be used in the daunting task of quickly identifying vulnerable network sections [13]. IoT can provide data transmission and their storage in the form of Big Data [14]. The obtained data are possible to use for real-time planning. Another interesting technology related with the data storage is a blockchain technology, which is a combination of distributed data storage, timestamp technology, and peer-to-peer network. This technology also can provide a solution for the secure distributed cloud data storage system. [15]

Several software packages are used for traffic simulation. One interesting example is the PTV Vissim software that realistically simulates complex vehicles interactions at the microscopic level [16]. This software can be applied to the simulation of future traffic and transit conditions [17], and the obtained results can present the potential to reduce vehicle travel times and delays [17]. This software can be used for testing several scenarios relating to the effectiveness of increased safety of toll plazas [18]. This software and the resultant simulation models allow the optimal efficiency of the road network to be determined—for example, in a viaduct—hence allowing traffic management proposals to be made to reduce delays [19]. PTV Vissim can be used to simulate the effects of congestion and delays on a motorway network due to an accident, and then to apply a quantified regression formula to predict the time of traffic recovery [20]. This software and simulation models can evaluate road improvements or new traffic management strategies in different weather conditions [21]. With the help of the standard microscopic simulation platform of PTV Vissim, it is possible to compare the efficiency characteristics of algorithms for autonomous intersection control [22].

In addition to computer simulation tools, it is suitable and effective to implement other methods to support the complex and difficult problems of the sustainability of transport, particularly in the field of research [23]. One possible approach is the use of multi-criteria decision-making (MCDM). It must be emphasized that the MCDM regarding public transportation presents a complicated task that involves environmental, economic, and socio-political issues [24]. However, several studies have applied this approach to solve different transport problems, for example, the use of MCDM for decision-making relating to alternative fuel public transport buses [25], selection of sustainable urban transportation alternatives using fuzzy multi-criteria decision-making (FMCDM) [26], use of analytical hierarchical process (AHP) for the selection of suitable vehicles [27], use of AHP to determine the best solutions by traffic planning [28], and use of AHP for design and evaluate highway routes [29].

This literature review provides interesting combinations of MCDM and the simulation approach. This combination can be also used for the solution of transport sustainability and transport problems, for example, modeling and testing of different intersections using Vissim software, followed by generation of the AHP model using the PTV Vissim results or using AHP to solve traffic issues that arise due to ad hoc urban planning by changing road geometries and signaling model alternative solutions via PTV Vissim software [30–32]. This idea is also presented in this paper.

The goal of this paper is to note the possibilities of using the information provided by experts, using the simulation software PTV Vissim, and implementation of the simulation results in the proposed AHP model. By experts, we mean specialists for realization and evaluation of traffic survey. These are specialists who carried out a traffic survey at the examined transport hub, and subsequently, their data were used for the creation of a simulation model.

A novelty of the paper is the application of AHP for the selection of a suitable alternative based on the results of simulation experiments. In comparison to a recently published paper [30], which focuses on evaluating the quality of public passenger transport using AHP, the present paper focuses on the solution associated with changing the transport organization, which will not only have an impact on the quality of public transport but will also benefit the environment. The paper also demonstrates its application in the field of transport problems. A similar issue is also presented in [31], but this research study uses multi-criteria decision-making for evaluation and diagnostics of urban streets through an integrated multi-criteria model of a sustainable nature.

2. Materials and Methods

Addressing the issue of transport sustainability is dependent on the relevance of the criteria used, which is one of the most critical points of the many techniques available to derive decision-making solutions [32]. The selection of criteria and the sequence of steps in the analysis of transport sustainability is a challenging process [33]. The approach to the solution of transport sustainability includes several steps, of which the first is the identification of the sustainability assessment criteria. Computer simulation is often used in this process.

The use of computer simulation to tackle transport problems is described by the algorithm in Figure 1. This suggests that computer simulation can be used twice. The first use is within the analytical phase when the simulation model is amassing data and materials to suggest a solution. The second use of the simulation model is implemented in the designing phase when it is necessary to conduct a large number of experiments and evaluations with the intent to assess individual alternative solutions. This simulation method can also be used to compare the original state with the suggested solution. In the future, the resulting model may be applied to the traffic of the transport process and to answer "what if?" questions.

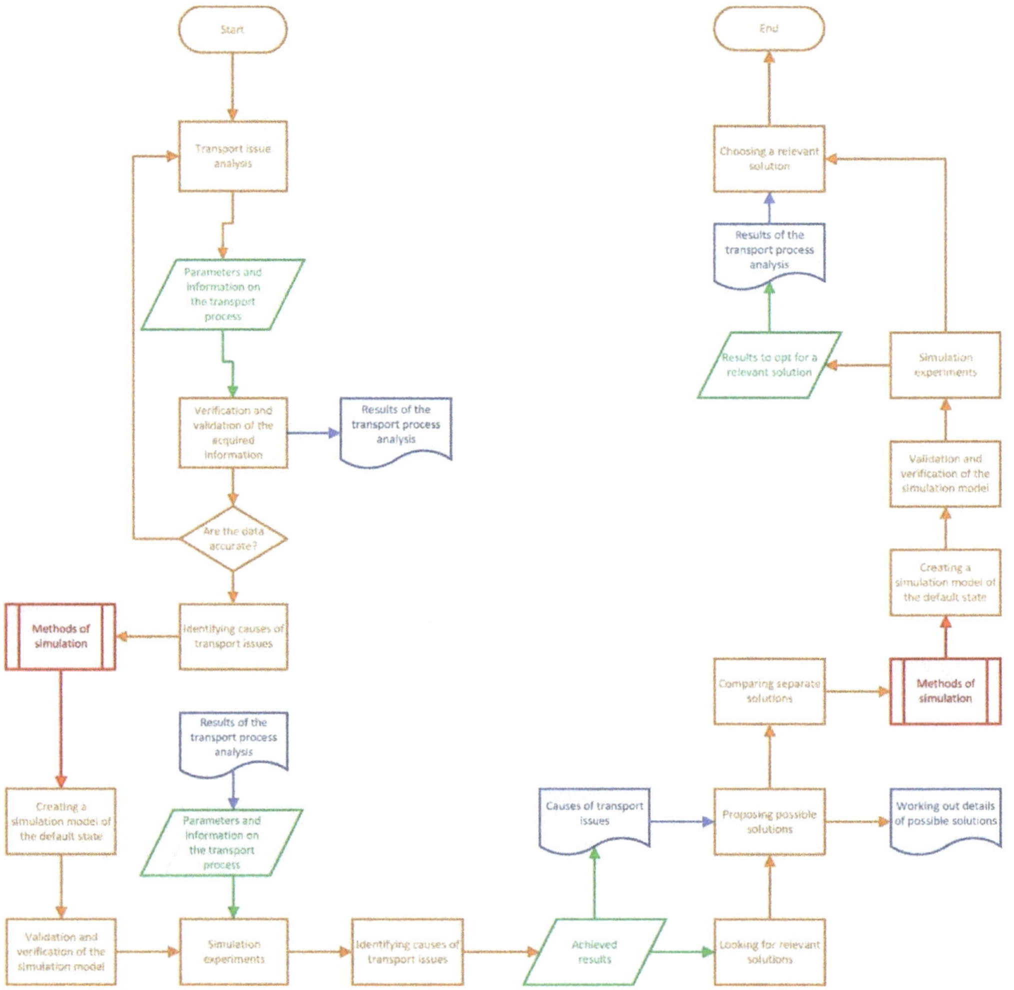

Figure 1. The algorithm solving the transport problem using computer simulation on a micro-level.

The use of the simulation model in the area of transport based on a case study referring to an actual traffic hub in the city of Uherské Hradiště in the Czech Republic is presented in the next section of the article.

Described Solution of the Traffic Hub

The traffic hub was constructed as a level intersection with light signaling to manage the traffic. This research was conducted in the territory of Uherské Hradiště, Kunovice, and Staré Město, using the transport plan of the city from 2015. The data from research carried out in different directions show a cartogram of a transport network loaded by individual car transport within 8 h of the research's duration in the above-mentioned territory. Figure 2 shows a section of the surroundings of the designated traffic hub.

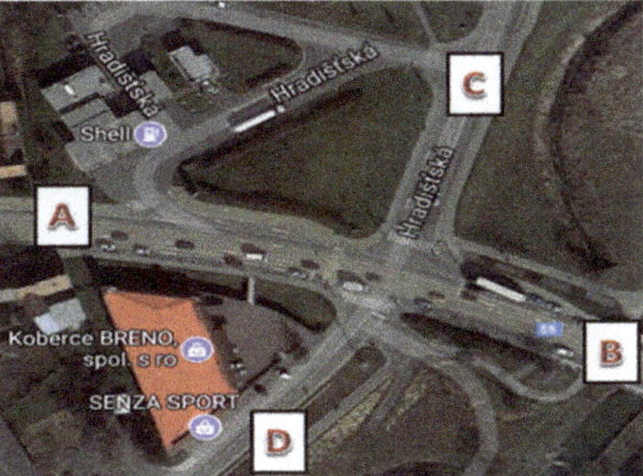

Figure 2. The surroundings of the analyzed traffic hub.

The research duration was 8 h; from 6:00 to 10:00 a.m. and from 2:00 to 6:00 p.m. Passing vehicles were recorded within a time interval of 30 min. The time and scope of the research were chosen in terms of when the researched transport hub was characterized by the creation of traffic congestions.

The traffic hub is situated on the traffic artery connecting Kunovice, Uherské Hradiště, and Staré Město (Figure 2). The traffic volume is greatest on the east–west axis in both directions. The traffic situation is complicated in this hub by severe traffic congestions during the rush hour.

Vehicles approach the hub from the eastern side (B) over bridges leading across the arm of the river. In this direction, the intersection consists of three lanes:

- straight direction with a possibility to turn left,
- straight direction,
- short turning lane to the right.

The other side (A) has two lanes:

- straight direction with a possibility to turn left,
- straight direction with a possibility to turn right.

The southern part of the intersection (D) has only two opposite lanes. The northern part (C) is an industrial and commercial quarter. There are a shopping center and different business networks—food, fashion shops, electronics, or hobby shops. There is also a company whose business relates to spare parts for different types of cars. In the immediate surroundings of the intersection, there is an industrial zone with a large number of passing lorries even though a new bypass was built especially so as not to overload the traffic hub.

The total number of passing vehicles in a specific direction is recorded as $\sum$ vehicles in Table 1. The number of lorries and buses from the total number of passing vehicles is expressed in L + B. Figures 3 and 4 represent passing vehicles moving in the prescribed directions. The figures were drawn according to the general city transport plan.

Table 1. The traffic intensity of the Hradišťská–Východní intersection.

	∑Vehicles	L + B	∑Vehicles	L + B	∑Vehicles	L + B	Together	
Entry A	A–B		A–C		A–D		∑vehicles	L + B
	4815	212	358	5	32	0	5205	217
Entry B	B–A		B–C		B–D			
	3499	247	2530	125	121	0	6150	372
Entry C	C–A		C–B		C–D			
	404	7	2732	108	38	0	3174	115
Entry D	D–A		D–B		D–C			
	18	0	83	1	19	1	120	2

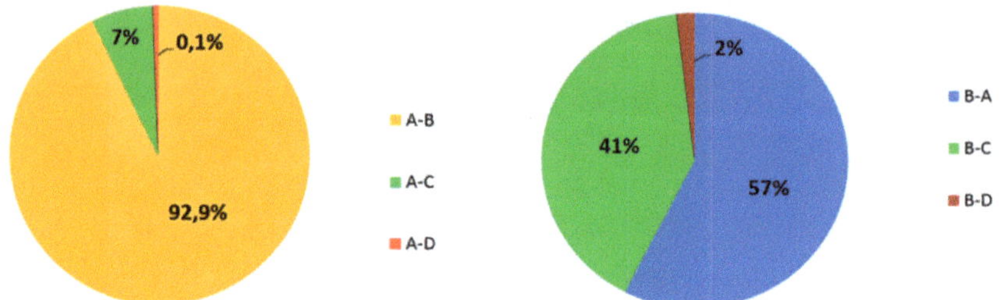

Figure 3. Vehicles' dispersal from direction A and direction B.

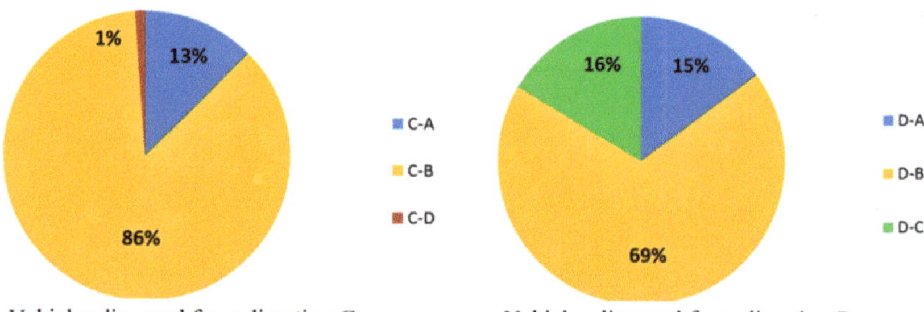

Figure 4. Vehicles' dispersal from direction C and direction D.

Each direction is colored as follows:

- A—Staré Město—blue;
- B—Uherské Hradiště—yellow;
- C—Commercial and industrial zone—green;
- D—Quay—red.

Most vehicles coming from Staré Město continue on the main road to Uherské Hradiště. Figure 3 shows that only 7% of the vehicles turn to the shopping centers, and it does not cause significant delay at the traffic light.

The B direction (Figure 3) experiences a more significant dispersal than that of A. Most vehicles from Uherské Hradiště go on to Staré Město; however, 41% of the vehicles turn to the shopping centers. The vehicles turning at the light signal slow the traffic because there is a short turning lane, and when the light is green, the vehicles have to let pedestrians cross the road, as suggested in Figure 4.

Upon leaving the shopping centers, most vehicles continue to Uherské Hradiště (Figure 4). Traffic congestions are caused by a large number of vehicles heading to the shopping center and the short duration of the green signal (18 s). The number of vehicles from direction D (Figure 4) is negligible (120 vehicles) considering the total number of vehicles at the intersection. No traffic congestion negatively influences the situation at the specific traffic hub.

Figures 3 and 4 suggest that the most problematic directions are as follows (considering traffic congestion):

- Uherské Hradiště to shopping centers direction;
- shopping centers to Uherské Hradiště direction;
- the main route: Uherské Hradiště—Staré Město, in both directions.

The analysis results suggested four possible solutions to the existing situation at the traffic hub. These solutions were compared with a computer simulation method. The following solutions were proposed:

- change in vehicle composition, excluding lorries from the intersection;
- change in the traffic route from the Staré Město direction;
- using a roundabout;
- change in the cycle of light signalization.

3. Results

To create a simulation model, the PTV Vissim program was applied. The program is based on a multipurpose microscopic simulation of traffic. The program PTV Vissim allows to realize a multipurpose microscopic traffic simulation based on the behavior of participants and also allows to examine and optimize traffic flows. This program contains a wide range of applications for modeling urban and motorway traffic and the integration of public and passenger transport. Visualization of operating conditions is at a high level [34].

The road network is represented by nodes located at intersections and connectors that are on-road segments. Within the model, the road has defined the following properties:

- start and end coordinates,
- number of traffic lanes for sections,
- width of the traffic lane, and
- type of vehicles passing on the road.

Roads and connectors are the basic building blocks for adding more infrastructure objects. System elements are divided into different classes and spatial resolutions. Modeling of the transport system in this program is dependent on the specification of vehicles that will be used in the simulation model. Vehicles have the option of choosing the route. The vehicles are divided into categories in the model. Each category has a specific model of vehicle with mandatory technical characteristics, namely length, width, maximum speed, and deceleration and acceleration of the vehicle. The vehicles are generated randomly at the beginning by the function "vehicle inputs."

The program PTV Vissim analyzes and optimizes traffic flows. This program comprises a large scale of applications for modeling the city and highway traffic and integrating public and passenger transport. The visualization of traffic ratios is undertaken at a high standard. The workstation was equipped with the processor Intel Core i9-8950HK and the graphic card nVidia GeForce GTX 1080 8GB DDR5 [34].

The road network in the simulation model is represented by traffic hubs placed at intersections and by connectors at road segments. In this case, four routes are integrated with the use of connectors.

The model is defined by these characteristics:

- beginning and end coordinates;
- number of traffic lanes for individual segments;
- lane width;
- type of vehicles passing through a specific route.

Routes and connectors decide whether and which infrastructure objects should be added. Elements of the system are classified into various classes and 3D definitions. Placing an object is related to a specific traffic lane, which means that objects relevant to a specific traffic hub must be implemented in all segments. A local object does not have a physical length allocated, which means the object must be allocated in one specific point of a traffic lane. These point objects were used for the model:

- traffic signs "Give way" and "Stop";
- traffic sign "Main road";
- light signalization.

Vehicles randomly stop 0.5–1.5 m before the specific signal. Three-dimensional objects with defined lengths emerge at a specific position of the lane. The objects used to define the infrastructure are:

- detectors;
- speed areas.

When modeling the system, it is necessary to specify vehicles located in a particular infrastructure. Passenger vehicles may choose a route. The model is classified into:

- passenger transport and vans;
- haulage.

Individual categories include a specific vehicle model with mandatory technical characteristics, which are length, width, maximal speed, deceleration, and acceleration of the vehicle. Vehicles are randomly generated at the beginning of routes using the Vehicle Inputs function.

The numbers of vehicles included in the model and dispersal of the traffic flow through the Relative Flows function are suggested in Table 2.

Table 2. Number of vehicles and dispersal of the traffic flow in the simulation model.

Source	Destination	Dispersal of the Traffic Flow	Number of Vehicles	Number of Vehicles on the Specific Route	Composition of the Traffic Flow
A	B D C	0.925 0.006 0.069	5205 veh/h	4988 217	car lorry
B	A C D	0.569 0.411 0.020	6150 veh/h	5778 372	car lorry
C	A D B	0.127 0.012 0.861	3174 veh/h	3059 115	car lorry
D	C A B	0.158 0.150 0.692	120 veh/h	118 2	car lorry

"Volume" represents the number of vehicles generated on the specific route. "Vehcomp" refers to the composition of the traffic flow, in which 3 means passenger transport and 2 is haulage. For A and D directions, the total should equal 1; this result shows that all included vehicles were dispersed into the designated directions.

Vissim includes the light signalization at intersections in the infrastructure. The cycle length of the light signalization is influenced by dispositions of the traffic hub, traffic load, number of cycle phases, form of turning, length of the pedestrian crossing, and construction work at the intersection. The length I determined by fixed split times.

The cycle length at the Hradišťská–Východní–Zrezavice intersection is modified according to rush hours in the morning and the afternoon and on weekends and holidays. The morning cycle lasts 65 s, the afternoon cycle 80 s, and the cycle during holidays is 100 s.

Figure 5 suggests a signal program for the intersection in Uherské Hradiště. The cycle length is 65 s in the basic model. The duration of the color green for individual directions is as follows:

- direction from Uherské Hradiště—33 s;
- direction from Staré Město—33 s;
- direction from the industrial and commercial zone—18 s;
- direction from the quay—5 s.

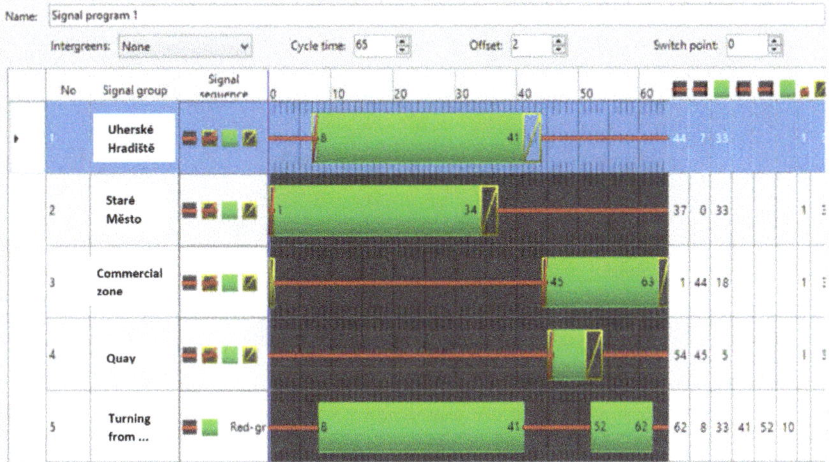

Figure 5. A signal program for a modeled traffic hub in Uherské Hradiště.

When the green light is on, it is possible to turn right from the industrial area in the direction of Uherské Hradiště. The specific layout of the intersection (Figure 6) overlaps the green wave from the Uherské Hradiště direction with the green wave from Staré Město, which results in downtime when turning from Staré Město toward the commercial and industrial zone.

Simulation experiments were realized using the simulation models, the results of which present the basic characteristics of variants (Tables 3–5).

(**a**) turning right from the industrial zone in the direction of Uherské Hradiště (when the green line is on)

(**b**) downtime in turning from the direction Staré Město toward the commercial and industrial zone;

Figure 6. *Cont.*

(c) overlaps the green wave from the Uherské Hradiště direction with the green wave from Staré Město

Figure 6. Model of the Hradišťská–Východní–Zrezavice intersection. (**a**) turning right from the industrial zone in the direction of Uherské Hradiště (when the green line is on); (**b**) downtime in turning from the direction Staré Město toward the commercial and industrial zone; (**c**) overlaps the green wave from the Uherské Hradiště direction with the green wave from Staré Město.

Table 3. Numbers of vehicles and vehicle travel time—current state.

Direction	A–B		B–C		C–B	
Simulation Time (s)	Number of Vehicles	Vehicle Travel Time (s)	Number of Vehicles	Vehicle Travel Time (s)	Number of Vehicles	Vehicle Travel Time (s)
0–900	383	34.73	214	24.74	169	68.57
900–1800	409	33.22	227	26.08	176	67.34
1800–2700	314	40.70	213	26.27	176	71.92
2700–3600	332	41.27	215	25.18	157	67.32
Average	359.5	37.50	217.25	25.57	169.5	68.80
Minimum	314	33.22	213	24.74	157	67.32
Maximum	409	41.27	227	26.27	176	71.92

Table 4. Numbers of vehicles and vehicle travel time—changing route from Staré Město.

Direction	A–B		B–C		C–B	
Simulation Time (s)	Number of Vehicles	Vehicle Travel Time (s)	Number of Vehicles	Vehicle Travel Time (s)	Number of Vehicles	Vehicle Travel Time (s)
0–900	543	25.19	219	22.22	187	58.10
900–1800	553	25.18	252	22.48	203	61.70
1800–2700	587	25.41	222	24.51	168	60.54
2700–3600	562	25.91	244	22.66	198	58.47
Average	561.3	25.40	234.3	23.00	189	59.70
Minimum	553	25.18	219	22.22	187	58.10
Maximum	562	25.91	222	24.51	203	61.70

Table 5. Numbers of vehicles and vehicle travel time—modified traffic lights.

Direction	A–B		B–C		C–B	
Simulation Time (s)	Number of Vehicles	Vehicle Travel Time (s)	Number of Vehicles	Vehicle Travel Time (s)	Number of Vehicles	Vehicle Travel Time (s)
0–900	366	35.80	215	24.52	170	66.33
900–1800	393	34.49	233	24.47	194	64.12
1800–2700	330	39.92	211	25.29	179	64.24
2700–3600	359	41.66	228	23.85	201	63.27
Average	362	38.00	221.8	24.50	186	64.50
Minimum	393	34.49	228	23.85	201	63.27
Maximum	359	41.66	211	25.29	170	66.33

This structure of results allows variants to be compared, taking into account the lengths of congestion in each direction (Figure 7) and the vehicle delay (Figure 8).

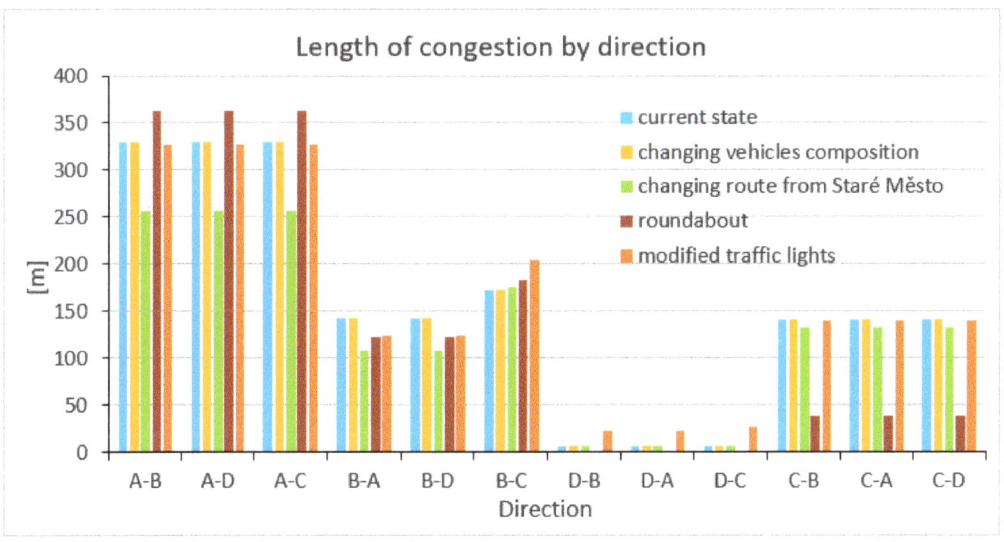

Figure 7. Length of congestion by direction.

The results of simulation experiments for alternative solutions were realized based on data sets that most concisely characterized the undesirable condition at the transport node—traffic congestion. First of all, the length of columns in all directions was monitored. This indicator presents the capacity of traffic directions. The next indicator was the time of the vehicles' passage through the transport hub. It was another set of data describing whether the vehicles would not pass slowly within the individual variants, which would have bad environmental impacts. The final monitored indicator was the delay of vehicles at the intersection. It was the time that included, in addition to driving time, the waiting of vehicles at a transport hub due to traffic congestion.

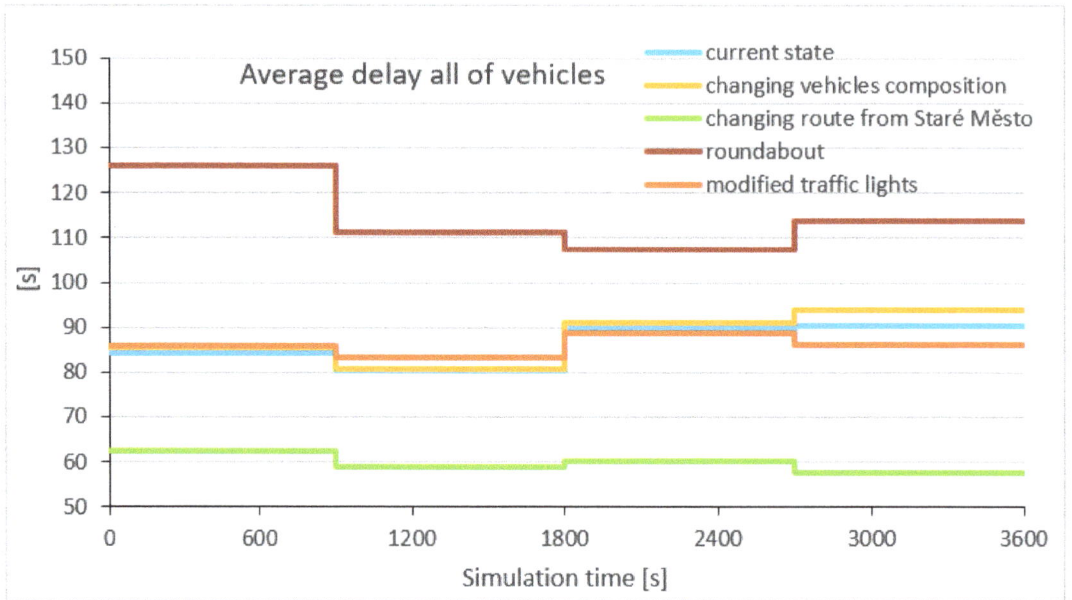

Figure 8. Average delay of all vehicles.

4. Discussion

To provide a comprehensive solution and comparison of individual alternatives, four independent microscopic simulation models were created. Subsequently, careful simulation experiments were implemented and the results were then compared. The key objective was to derive a solution that represented a substantial improvement compared to the existing, uncomfortable situation of the transport process.

4.1. The Experiment of Changing the Composition of Vehicles

This experiment consists of excluding lorries from the intersection. Lorries can use a bypass without slowing down the traffic in this prominent hub. The model preserved random numbers of buses because the exact numbers were not the subject of the research. During the simulation, 4638 vehicles passed through the traffic hub, which was 126 vehicles greater than in the existing state. The transit time of vehicles when excluding lorries did not exceed 25 s for 218 vehicles. Compared to the existing state, there was a slight difference. The main difference in the transit time and the number of vehicles coming from the commercial zone to Uherské Hradiště compared to the current state is that an additional 15 vehicles can pass through the specific segment within the approximate same period. The transit from Staré Město to Uherské Hradiště upon excluding lorries is not significantly influenced, which means that there is no major difference compared to the current state. Tailbacks in individual directions are usually longer than in the default state of the intersection. The average delay of cars at the traffic hub approximately equals that of the basic model, with the difference not exceeding 7 s of delay within one hour's simulation.

4.2. Changing the Route from Staré Město Direction

This experiment focuses on vehicles turning toward commercial centers in the direction of Staré Město. At this traffic hub, it means that this direction will no longer provide the possibility of turning toward the commercial zone. The only possible ways are to go straight or turn toward the quay. Vehicles traveling to the commercial zone must turn onto Huštěnovská Street at the previous junction and then take the direction to Luční District. As a result, the overall traffic capacity increased by 958 vehicles compared to the default state. In comparison to the current state, the amount of emissions produced was reduced by 23% after re-calculating the ratio of passing vehicles and the quantity of emissions. Because this experiment excluded turning left from the Staré Město direction, the light signalization was modified, which resulted in a shorter transit time in the given directions and an increase in several vehicles passing through this segment. As contrasted with the basic model simulation, in the model with changed composition, approximately 20 more vehicles were able to pass in the C–B direction in a shorter transit time. The most significant change occurred in vehicles transiting in the A–B direction, where the number of passing vehicles increased from the original 360 to 561 and the transit time was reduced to 12 s. This represents an improvement of 62.5% after the total period was calculated. The tailback from the Staré Město direction was considerably shorter than in the original situation. Other directions did not see such a large difference. Overall, we may argue that the situation at the traffic hub improved because the overall delay at the intersection dropped from 344.5 to 239.45 s within one hour of the simulation. After re-calculating, the interval fell from 86 s to approximately 60 s.

4.3. Implementing a Roundabout

In this experiment, the construction and type of the intersection were changed to a two-lane roundabout (Figure 9). During the design, a minimal roundabout inner diameter of 5.3 m was preserved; the outer diameter was 25 m, which enables a smooth transit of longer semi-trailers.

The number of vehicles that passed through the given intersection within one hour's simulation equaled 545. The total amount of emissions that were released into the atmosphere during the simulation process decreased. Nonetheless, after re-calculating the emissions concerning the total amount of vehicles that had passed through the given segment, this ratio increased by 19.36% compared to the current state.

The transit time of vehicles from Uherské Hradiště to the commercial zone suggested significantly increased compared to the basic model. By contrast, upon implementing the roundabout, the transit time from the commercial zone decreased and the traffic capacity from the respective zone expanded. The vehicles heading to Uherské Hradiště passed through within 46 s on average. However, vehicles coming from Staré Město experienced an increased transit time in this experiment. Tailbacks in separate directions after implementing the roundabout and in the current state were approximately equal; however, the tailback from the commercial–industrial zone was reduced by 72%.

4.4. Changing the Cycle and Type of Light Signalization

Another possible solution focused on modifying the light signalization at the traffic hub. The cycle of 65 s was preserved, and the alternatives for turning from the quay and commercial zone were added. The green interval from the C direction was changed from 18 s to 16 s. The green for the quay and commercial zone did not start at the same time—the green for the commercial zone was delayed by 2 s.

After modifying the light signalization, 4649 vehicles passed through the specific intersection; i.e., 137 vehicles more than in the current state. The transit of vehicles in the B–C direction improved by 1 s on average and the traffic capacity expanded by four vehicles. The transit time from the commercial zone was reduced, which resulted in a larger number of vehicles passing through the given traffic hub compared to the default setting of the light signalization. The same applied to the transit time of vehicles from Staré Město. The length of congestion was approximately equal; the only significant difference was from the quay direction, with a tailback 27 m long compared to the current 6 m. The waiting time at the traffic hub was roughly the same as in the default state even in Experiment 3. Vehicles wait a maximum of 89 s; i.e., 1.5 s more than before.

(**a**) view on the two-lane roundabout from the direction of industrial zone

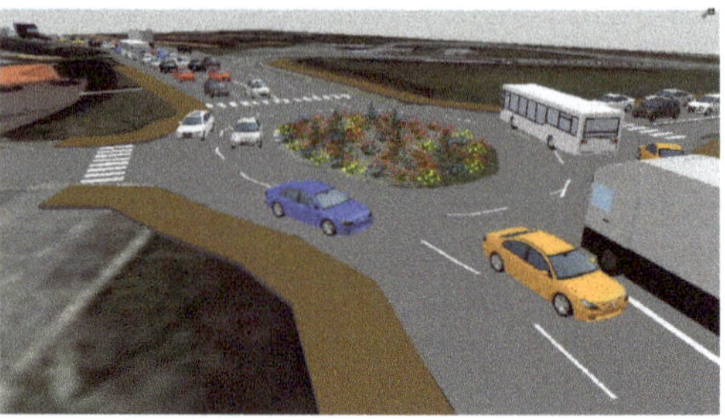

(**b**) view on the two-lane roundabout to the direction of the Staré Město

(**c**) view on the two-lane roundabout to the direction of Uherské Hradiště

Figure 9. Model of the roundabout. (**a**) view on the two-lane roundabout from the direction of industrial zone; (**b**) view on the two-lane roundabout to the direction of the Staré Město; (**c**) view on the two-lane roundabout to the direction of Uherské Hradiště.

4.5. Evaluation of Experiments

The results of simulation experiments provide a wide range of information that characterizes in detail the individual variants designed to ensure the sustainability of transport in a given urban area. Figure 10 presents an example of part of the obtained results.

From the results, it is necessary to choose the optimal solution based on the obtained parameters. Because the comparison is based on several criteria, it is appropriate in this case to apply the method of multi-criteria decision-making.

There are currently a large number of multi-criteria decision-making methods. One of these methods is the analytical hierarchy process (AHP). AHP is a structured technique used to solve complex decisions. It is based on mathematical procedures and human psychology [35]. AHP provides a complex and logical concept for structuring a problem, quantifying its elements that are related to the overall objectives and evaluating alternative solutions. From the factors that make the AHP perhaps the most popular decision-making method, it can be emphasized that it adapts to fixed data, such as the speed of delivery, price, and personal experiences [36]. It allows mathematical derivation of the weight of criteria, instead of the subjective selection of criteria weights, as used by other decision-making methods [37].

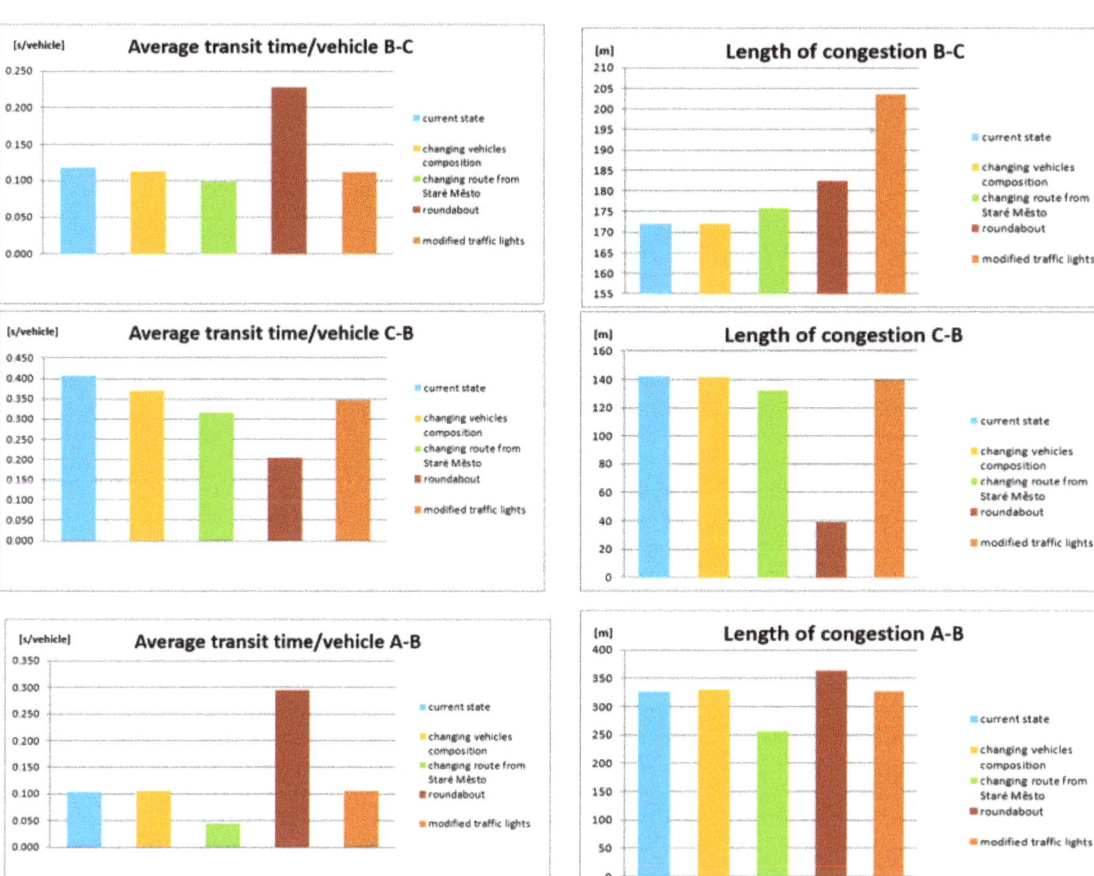

Figure 10. Transit time and congestion length.

Several studies have used AHP for the solution of transport problems in the direction of sustainability and environmental impact; for example, using AHP for the development of a decision support framework to assess quantitative risk in multimodal green logistics [38]; the multiple criteria decision-making approach for route selection in the multimodal supply chain, which is based on the combination of AHP, data envelopment analysis (DEA), and the techniques for the order of preference by similarity to ideal solution (TOPSIS) [39]; and the evaluation and diagnosis of urban streets using an integrated multi-criteria model of a sustainable nature [31].

Based on published research, this method has good preconditions for its use in combination with the computer simulation method.

The model of AHP was created for comparison of variants of transport sustainability on the basis of the results of simulation experiments. The basic structure is presented in Figure 11.

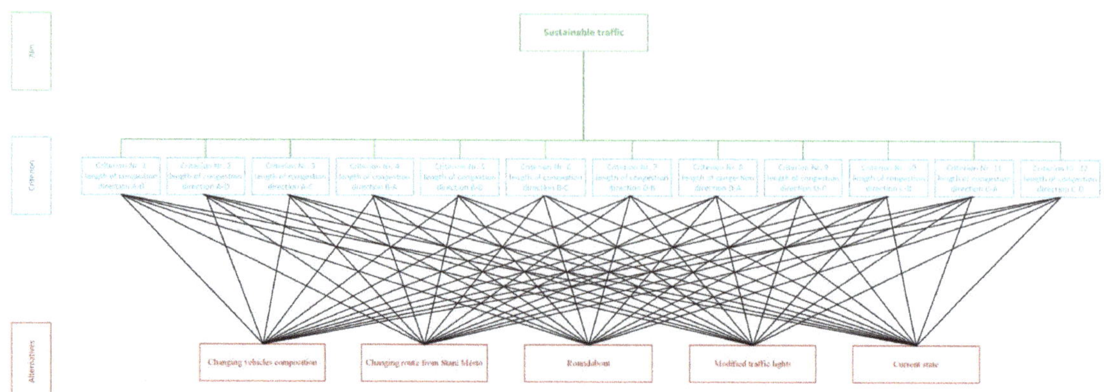

Figure 11. The structure of the analytical hierarchy process (AHP) model.

The structure of the proposed AHP model consists of three hierarchical levels. At the highest level is the goal of choosing a suitable solution for the needs of transport sustainability. Its achievement is realized based on 12 criteria, which were based on traffic directions connected with the analyzed transport hub. The aforementioned 12 criteria form the second level of the model. The third level of the model is represented by individual alternatives, which were part of the realized simulation experiments.

The evaluation criteria and the determination of their weights was realized by a group of experts. The role of this group was to help determine the weights of the criteria, to determine the weights of the objectives, to organize the objectives, and to determine the weights of the decision criteria. It must be said that the selection of the evaluation criteria and their weights was based on published knowledge [40,41]. Experts were selected so that their evaluation was considered objective. We tried to achieve this by concentrating several specialists from areas that are related to the problem of the transport hub and its operation.

The criterion of the competence of the experts was their professional knowledge, knowledge of the transport hub in terms of its operation and its impact on the environment. The experts have more than 10 years of experience in the researched issue. The structure of the experts is presented in Table 6.

Table 6. The criterion of the competence of the experts.

	Expert	Experiences Practical	Experiences Theoretical	Male/Female
1.	The expert from the field of transport	2–5 y	more than 10 y	was not considered
2.	Residents living near the transport hub	2–5 y	-	50/50
3.	Public transport drivers passing through a hub	0.5–1 y	more than 10 y	was not considered
4.	Drivers of vehicles supplying operations in the area	0.5–1 y	more than 10 y	was not considered
5.	People working in the area	0.5–1 y	more than 10 y	50/50
6.	Persons passing through a transport hub	0.25–0.5 y	-	50/50

y means years.

The individual directions at the transport hub were chosen as criteria. These criteria made it possible to evaluate the individual variants. Criteria that represent the directions of traffic that directly affect the traffic situation at the main nodes have higher weight than the directions that provide the diversion of traffic to areas that are adjacent to the traffic hub and are located on commercial premises and residential blocks. The definition of weights (Table 7) was based on the recommendation published in [35].

Table 7. The definition of weights.

Intensity of Importance	Definition	Explanation
1	Equal importance	Two variants are also involved in the intervention of the goals.
3	Less importance of one variant compared to another	Experience and opinions gently prefer one attribute over another.
5	Substantial or strong importance	Experience and opinions strongly prefer one attribute over another.
7	Demonstrable importance	One attribute is highly preferred, and its dominance is demonstrated in practice.
9	Absolute importance	The obvious favoring of one attribute over another is at the highest possible level of expression.
2, 4, 6, 8	Mean values between two adjacent variants	A compromise is needed due to the ambiguity of the assignment in relation to the above definitions of importance.

We used three options for the selection of the method to evaluate the criteria. It was specifically about brainwriting, brainstorming, and the Delphic method. This selection was realized on the basis of the selected literature review, for example [42] and [43]. Based on the above facts, the brainwriting method was chosen. The main reason was the fact that it was possible to harmonize the selected experts to negate their fear of expressing their views in public, and this method made it possible to address that [44]. Brainwriting is a method that is suitable to use in the case that we assume that some participants may be shy in expressing their ideas [45]. It is a method that, in contrast to brainstorming, can effectively involve all its participants and, as a result, often offers more options [45]. The comparison of individual methods and their possibilities of use within AHP is offered by [42,43].

The evaluation of criteria in terms of significance is shown in Table 8. In their evaluation, the impact of individual roads on the creation of traffic congestion was taken into account.

Table 8. Evaluation of criteria in term of significance.

Criteria Preferences	K1 Route A–B	K2 Route A–D	K3 Route A–C	K4 Route B–A	K5 Route B–D	K6 Route B–C	K7 Route D–B	K8 Route D–A	K9 Route D–C	K10 Route C–B	K11 Route C–A	K12 Route C–D
K1 Route A–B	1	5	3	8	6	7	6	8	9	7	6	9
K2 Route A–D	1/5	1	2	1/6	3	1/2	1/5	1/4	8	1/5	4	3
K3 Route A–C	1/3	1/2	1	1/7	5	6	1/7	1/6	6	3	4	5
K4 Route B–A	1/8	6	7	1	7	5	4	8	8	1/6	5	8
K5 Route B–D	1/6	1/3	1/5	1/7	1	1/3	1/6	1/4	5	1/4	1/7	6
K6 Route B–C	1/7	2	1/6	1/5	3	1	1/6	1/7	3	1/5	1/7	5
K7 Route D–B	1/6	5	7	1/4	6	6	1	5	5	3	5	5
K8 Route D–A	1/8	4	6	1/8	4	7	1/5	1	1/6	1/7	1/3	5
K9 Route D–C	1/9	1/8	1/6	1/8	1/5	1/3	1/5	6	1	1/6	1/6	1/2
K10 Route C–B	1/7	5	1/3	6	4	5	1/3	7	6	1	7	7
K11 Route C–A	1/6	1/4	1/4	1/5	7	7	1/5	3	6	1/7	1	8
K12 Route C–D	1/9	1/3	1/5	1/8	1/6	1/5	1/5	1/5	2	1/7	1/8	1

In the next step, a pairwise comparison of the alternative solutions considered in the context of individual criteria was realized (Table 9). In this comparison, the lengths of traffic congestion were taken into account, which were identified in the implementation of simulation experiments. This means that based on routes (criteria), variants were evaluated and compared with each other. The lower the length of the traffic congestion, the higher the rating assigned to the alternative. At the same time, the principle of universal axiom was observed. This created a reciprocal matrix in which all elements on the main diagonal were equal to 1.

Table 9. Pairwise comparison of alternative solutions.

K1 Route A–B	Changing Vehicle Composition	Changing Route from Staré Město	Roundabout	Modified Traffic Lights	Current State
Changing vehicle composition	1	1/8	6	1/2	1/4
Changing route from Staré Město	8	1	9	7	5
Roundabout	1/6	1/9	1	1/3	1/4
Modified traffic lights	2	1/7	3	1	1/2
Current state	4	1/5	4	2	1
CI: 0.1151		CR: 0.1037		λ: 5.4606	
K2 Route A–D	**Changing Vehicle Composition**	**Changing Route from Staré Město**	**Roundabout**	**Modified Traffic Lights**	**Current State**
Changing vehicle composition	1	1/8	6	1/2	1/4
Changing route from Staré Město	8	1	9	7	5
Roundabout	1/6	1/9	1	1/3	1/4
Modified traffic lights	2	1/7	3	1	1/2
Current state	4	1/5	4	2	1
CI: 0.1151		CR: 0.1037		λ: 5.4606	
K3 Route A–C	**Changing Vehicle Composition**	**Changing Route from Staré Město**	**Roundabout**	**Modified Traffic Lights**	**Current State**
Changing vehicle composition	1	1/8	6	1/2	1/4
Changing route from Staré Město	8	1	9	7	5
Roundabout	1/6	1/9	1	1/3	1/4
Modified traffic lights	2	1/7	3	1	1/2
Current state	4	1/5	4	2	1
CI: 0.1151		CR: 0.1037		λ: 5.4606	
K4 Route B–A	**Changing Vehicle Composition**	**Changing Route from Staré Město**	**Roundabout**	**Modified Traffic Lights**	**Current State**
Changing vehicle composition	1	4	3	3	3
Changing route from Staré Město	1/4	1	3	3	3
Roundabout	1/3	1/3	1	2	3
Modified traffic lights	1/3	1/3	1/2	1	2
Current state	1/3	1/4	1/3	1/2	1
CI: 0.1078		CR: 0.0971		λ: 5.4310	
K5 Route B–D	**Changing Vehicle Composition**	**Changing Route from Staré Město**	**Roundabout**	**Modified Traffic Lights**	**Current State**
Changing vehicle composition	1	4	3	3	3
Changing route from Staré Město	1/4	1	3	3	3
Roundabout	1/3	1/3	1	2	3
Modified traffic lights	1/3	1/3	1/2	1	2
Current state	1/3	1/4	1/3	1/2	1
CI: 0.1078		CR: 0.0971		λ: 5.4310	

Table 9. *Cont.*

K6 Route B–C	Changing Vehicle Composition	Changing Route from Staré Město	Roundabout	Modified Traffic Lights	Current State
Changing vehicle composition	1	2	3	5	2
Changing route from Staré Město	1/2	1	2	6	1/2
Roundabout	1/3	1/2	1	4	1/3
Modified traffic lights	1/5	1/6	1/4	1	1/6
Current state	1/2	1/5	4	2	1
CI: 0.0453		CR: 0.0408		λ: 5.1810	
K7 Route D–B	**Changing Vehicle Composition**	**Changing Route from Staré Město**	**Roundabout**	**Modified Traffic Lights**	**Current State**
Changing vehicle composition	1	1/2	1/4	7	1/2
Changing route from Staré Město	2	1	1/3	8	1/3
Roundabout	4	3	1	9	6
Modified traffic lights	1/7	1/8	1/9	1	1/4
Current state	2	3	1/6	4	1
CI: 0.1484		CR: 0.1337		λ: 5.5934	
K8 Route D–A	**Changing Vehicle Composition**	**Changing Route from Staré Město**	**Roundabout**	**Modified Traffic Lights**	**Current State**
Changing vehicle composition	1	1/2	1/4	7	1/2
Changing route from Staré Město	2	1	1/3	8	1/3
Roundabout	4	3	1	9	6
Modified traffic lights	1/7	1/8	1/9	1	1/4
Current state	2	3	1/6	4	1
CI: 0.1484		CR: 0.1337		λ: 5.5934	
K9 Route D–C	**Changing Vehicle Composition**	**Changing Route from Staré Město**	**Roundabout**	**Modified Traffic Lights**	**Current State**
Changing vehicle composition	1	1/2	1/4	7	1/2
Changing route from Staré Město	2	1	1/3	8	1/3
Roundabout	4	3	1	9	6
Modified traffic lights	1/7	1/8	1/9	1	1/4
Current state	2	3	1/6	4	1
CI: 0.1484		CR: 0.1337		λ: 5.5934	
K10 Route C–B	**Changing Vehicle Composition**	**Changing Route from Staré Město**	**Roundabout**	**Modified Traffic Lights**	**Current State**
Changing vehicle composition	1	1/3	1/8	1/2	2
Changing route from Staré Město	3	1	1/8	3	4
Roundabout	8	8	1	8	9
Modified traffic lights	2	1/3	1/8	1	3
Current state	1/2	1/4	1/9	1/3	1
CI: 0.0742		CR: 0.0669		λ: 5.2970	

Table 9. *Cont.*

K11 Route C-A	Changing Vehicle Composition	Changing Route from Staré Město	Roundabout	Modified Traffic Lights	Current State
Changing vehicle composition	1	1/3	1/8	1/2	2
Changing route from Staré Město	3	1	1/8	3	4
Roundabout	8	8	1	1/8	1/9
Modified traffic lights	2	1/3	8	1	3
Current state	1/2	1/4	9	1/3	1
CI: 1.4521		CR: 1.3082		λ: 10.8085	
K12 Route C–D	**Changing Vehicle Composition**	**Changing Route from Staré Město**	**Roundabout**	**Modified Traffic Lights**	**Current State**
Changing vehicle composition	1	1/3	1/8	1/2	2
Changing route from Staré Město	3	1	1/8	3	4
Roundabout	8	8	1	8	9
Modified traffic lights	2	1/3	1/8	1	1/3
Current state	1/2	1/4	1/9	3	1
CI: 0.1614		CR: 0.1454		λ: 5.6464	

The basic scale of pairwise comparison was used in the comparison. In total, 12 pairwise comparisons were created according to the criteria. In addition to alternative proposals, the existing current situation at a researched transport hub was also included in the pairwise comparison. The aim was to avoid a situation in which none of the compared solutions was worse than the current situation.

The study used values that were the result of simulation experiments. Specifically, it was the length of traffic congestion on individual traffic directions. These characterized the number of means of transport on a given track profile. As another value, the transit time of the given section was used, which was used to monitor how favorable the given transport solution was. The application of these values in the evaluation was able to compare the individual sections and remove any advantage because the section is less frequent in terms of its importance. The value of consistency ratio (CR): 0.0933. The main limit in the analysis is the capacity of the transport infrastructure. We speculated, excepting the roundabout, about the use of the existing transport infrastructure, without increasing its capacity. This solution was chosen because the extension of the road infrastructure would be quite demanding in terms of securing the available space and high investment costs.

Based on the presented data, a synthesis of partial evaluations of alternative solutions was realized. Its result is presented in Figure 12.

Figure 12. The result of alternatives compared with the help of the AHP method.

5. Conclusions

The sustainability of transport in urban agglomerations is currently a key challenge to which constant attention is paid in the field of logistics. Approaches based on the use of computer simulation are widely used in the solution of partial needs related to transport sustainability.

Computer simulation presents a highly useful tool to resolve transport issues and should be applied as much as possible. Experts in transport and infrastructure should consider simulation models as invaluable working tools.

However, computer simulation is not a universal solution that can automatically solve all transport sustainability issues. Computer simulation in the field of transport is often used in the form of a support tool for obtaining broad-spectrum information, whose level of detail can be specified according to the requirements.

The ability of experts to effectively use and work with simulation models, their modification with the view to efficiently resolving transport issues should be a steady trend in the area of transport. In particular, microscopic simulation models, which may be applied to a large scale of transport problems and issues, should be employed at the outset.

In using computer simulation as a tool for obtaining a set of information covering different variants of the solutions to traffic problems, it is necessary to realize a final decision and select a suitable variant of the solutions. In these cases, multi-criteria decision-making appears to be an effective application, in which it is suitable to use the AHP method. The combination of computer simulation and multi-criteria decision-making is an effective analytical tool for solving traffic tasks related to the field of transport suitability.

This paper verifies, based on a practical example of a real traffic problem, possible solutions to the traffic problem using microscopic computer simulation and the AHP method. The investigated transport problem is associated with a change in the organization of the traffic hub and the selection of a suitable final variant.

The traffic hub was simulated using the PTV Vissim program, which is based on a multipurpose microscopic simulation of road traffic. After entering the required data (i.e., the number of vehicles, the traffic flow distribution, the permitted speed, and the light signalization interval) into the program, the output data were recorded. The applied results included information about the passing time of the vehicles, the length of the congestion, and the average delay of the vehicles at the intersection per simulated hour. Four simulation experiments were performed using the basic model, and the results obtained from these experiments were compared using multi-criteria decision-making via the AHP method.

The first experiment was oriented to changing the composition of the vehicles, thus excluding trucks from the intersection. The conditions of traffic were not improved significantly; rather, the conditions worsened.

Experiment no. 2 was focused on a change in traffic management in the direction from Staré Město, i.e., the vehicles were not allowed to turn from this direction toward the commercial zone. Comparing the results with the original situation, the traffic-carrying capacity improved by almost 1000 vehicles. The transit time was also more efficient and the length of the line of vehicles in the direction from Staré Město shortened.

Another experiment was focused on the application of a roundabout equipped with two traffic lanes. Passage of the vehicles through this intersection significantly worsened. The only positive result was the shortening of the line of vehicles in the direction from the commerce–industrial zone by 72%.

The last experiment consisted of a change in the traffic light signalization. Signals were added for the turn options and the length of the green light was adjusted. The result of this last experiment was that more vehicles passed through the traffic hub. The passage of vehicles and the length of the column of cars changed slightly.

The evaluation of the experiments was based on a comparison of the length of congestion. The AHP method was used. Twelve criteria were used in multi-criteria decision-making, which was used to compare individual variants based on the results of simulation experiments and to select the optimal solution.

According to the performed experiments, it is evident that the best solution for improvement of the traffic situation for the given traffic hub was derived from the experiment with the change in traffic management in the direction from Staré Město.

The presented procedure is a productive and reliable methodology that can be applied to solve a wide range of traffic problems. This enables us to make decisions, perform different types of analyses, and investigate the functioning of different types of transport processes.

The application of the model makes it possible to create guidelines for the priority measures in the field of public investment in urban infrastructure. The appropriateness of AHP in the field of transport is also presented by the research study [46], in which AHP is used for evaluation of the public passenger transport quality. In conclusion, it can be stated that the main novelty of the current paper is the combination of the simulation approach and the AHP model to select a suitable transport alternative. This is a solution that, according to the available information, has not yet been applied in this way to the area of urban transport. The paper was written to show a practical example of how it is possible to use a combination of discrete simulation and multi-criteria decision-making for the selection of solutions for the needs of transport sustainability. In terms of the combination of both methods for the field of transport sustainability, such a solution has not been obtained. The results presented follow several studies that have already been published. Creation of the discrete simulation model applied knowledge about the calibration and validation of the microscopic simulation model in the program PTV Vissim [47]. Application of the set of the optimal solution by the AHP method is related to the study focused on the creation of a traffic model of an intersection [48]. The paper also points out that multi-criteria decision-making can be also used for modeling a mixed traffic flow that passes a traffic junction and causes traffic congestions. New knowledge about this research problem is presented in [49]. The paper further indicates which data need to be focused on if there are several possible solutions in terms of transport sustainability within urban agglomerations. The paper thus extends the results that [50] presents marginally. The results presented extend the knowledge base presented in [51]. In contrast to the use of the AHP method for the process of transport organization in road tunnels, the present paper extends this issue to the area of intersections in built-up areas.

Author Contributions: Conceptualization, G.F. and V.M.; methodology, M.H. and N.M.; validation, G.F. and V.M.; formal analysis, N.M.; resources, R.K.; data curation, M.H.; writing—original draft preparation, G.F.; writing—review and editing, V.M. and V.S.; visualization, N.M. and V.S.; supervision, V.M.; project administration, R.K.; funding acquisition, G.F. All authors have read and agreed to the published version of the manuscript.

Funding: This work was supported by the projects of the Scientific Grant Agency of the Ministry of Education, Science, Research and Sport of the Slovak Republic and the Slovak Academy of Sciences (project No. VEGA 1/0403/18, VEGA 1/0638/19, and VEGA 1/0600/20) and by the projects of the Cultural and Educational Grant Agency of the Ministry of Education, Science, Research and Sport of the Slovak Republic and the Slovak Academy of Sciences (project No. KEGA 012TUKE-4/2019, KEGA 013TUKE-4/2019, KEGA 049TUKE-4/2020, and APVV SK-SRB-18-0053).

Institutional Review Board Statement: Not applicable.

Informed Consent Statement: Not applicable.

Data Availability Statement: The presented data can be obtained by email from corresponding author.

Conflicts of Interest: The authors declare no conflict of interest.

Nomenclature

AHP	analytic hierarchy process
SUMO	Simulation of Urban Mobility
MCDM	multi-criteria decision-making
FMCDM	fuzzy multi-criteria decision-making
veh	vehicles
h	hour
DEA	data envelopment analysis
CR	consistency ratio
CI	confidence interval
IoT	Internet of Things
λ	maximal eigenvalue
TMIS	Transportation Management Information Systems

References

1. Balasubramaniam, A.; Paul, A.; Hong, W.H.; Seo, H.C.; Kim, J.H. Comparative analysis of intelligent transportation systems for sustainable environment in Smart Cities. *Sustainability* **2017**, *9*, 1120. [CrossRef]
2. Vujadinović, R.; Jovanović, J.Š.; Plevnik, A.; Mladenovič, L.; Rye, T. Key challenges in the status analysis for the sustainable urban mobility plan in podgorica, montenegro. *Sustainability* **2021**, *13*, 1037. [CrossRef]
3. Barkenbus, J.N. Eco-driving: An overlooked climate change initiative. *Energy Policy* **2010**, *38*, 762–769. [CrossRef]
4. Zhao, L.; Jia, Y. Intelligent transportation system for sustainable environment in smart cities. *Int. J. Electr. Eng. Educ.* **2021**. [CrossRef]
5. Hu, J.; Zhang, Z.; Xiong, L.; Wang, H.; Wu, G. Cut through traffic to catch green light: Eco approach with overtaking capability. *Transp. Res. Part C Emerg. Technol.* **2021**, *123*, 102927. [CrossRef]
6. Hamurcu, M.; Eren, T. Electric bus selection with multicriteria decision analysis for green transportation. *Sustainability* **2020**, *12*, 2777. [CrossRef]
7. Yan, B. Improvement of the economic management system based on the publicity of railway transportation products. *Intell. Autom. Soft Comput.* **2020**, *26*, 539–547. [CrossRef]
8. Liu, J.; Kang, X.; Dong, C.; Zhang, F. Simulation of real-time path planning for large-scale transportation network using parallel computation. *Intell. Autom. Soft Comput.* **2019**, *25*, 65–77. [CrossRef]
9. Deng, D.; Shahabi, C.; Demiryurek, U.; Zhu, L.; Yu, R.; Liu, Y. Latent space model for road networks to predict time-varying traffic. In Proceedings of the 22nd ACM SIGKDD International Conference on Knowledge Discovery and Data Mining, San Francisco, CA, USA, 13–17 August 2016; pp. 1525–1534.
10. Gao, H.; Huang, W.; Yang, X. Applying probabilistic model checking to path planning in an intelligent transportation system using mobility trajectories and their statistical data. *Intell. Autom. Soft Comput.* **2019**, *25*, 547–559. [CrossRef]
11. Codeca, L.; Frank, R.; Faye, S.; Engel, T. Luxembourg SUMO Traffic (LuST) Scenario: Traffic Demand Evaluation. *IEEE Intell. Transp. Syst. Mag.* **2017**, *9*, 52–63. [CrossRef]
12. Lèbre, M.A.; Le Mouël, F.; Ménard, E. On the importance of real data for microscopic urban vehicular mobility trace. In Proceedings of the 2015 14th International Conference on ITS Telecommunications (ITST 2015), Copenhagen, Denmark, 2–4 December 2016; pp. 22–26.
13. Liu, W.; Tang, Y.; Yang, F.; Zhang, C.; Cao, D. Internet of things based solutions for transport network vulnerability assessment in intelligent transportation systems. *Comput. Mater. Contin.* **2020**, *65*, 2511–2527. [CrossRef]
14. Wang, J.; Yang, Y.; Wang, T.; Sherratt, R.; Zhang, J. Big Data Service Architecture: A Survey. *J. Internet Technol.* **2020**, *21*, 393–405. [CrossRef]
15. Zhang, J.; Zhong, S.; Wang, T.; Chao, H.-C.; Wang, J. Blockchain-Based Systems and Applications: A Survey. *J. Internet Technol.* **2020**, *21*, 1–14. [CrossRef]
16. PTV Group. Available online: https://www.ptvgroup.com/en/solutions/products/ptv-vissim (accessed on 1 April 2021).
17. Shams, A.; Zlatkovic, M. Effects of capacity and transit improvements on traffic and transit operations. *Transp. Plan. Technol.* **2020**, *43*, 602–619. [CrossRef]
18. Jehad, A.E.; Ismail, A.; Borhan, M.N.; Ishak, S.Z. Modelling and optimizing of electronic toll collection (ETC) at Malaysian toll plazas using microsimulation models. *Int. J. Eng. Technol.* **2018**, *7*, 2304–2308. [CrossRef]
19. Yulianto, B. Traffic Management and Engineering Analysis of the Manahan Flyover Area by using Traffic Micro-Simulation VISSIM. *IOP Conf. Ser. Mater. Sci. Eng.* **2020**, *852*, 012005. [CrossRef]
20. Jeihani, M.; James, P.; Saka, A.A.; Ardeshiri, A. Traffic recovery time estimation under different flow regimes in traffic simulation. *J. Traffic Transp. Eng.* **2015**, *2*, 291–300. [CrossRef]
21. Golshan Khavas, R.; Hellinga, B.; Zarinbal Masouleh, A. Identifying parameters for microsimulation modeling of traffic in inclement weather. *Transp. Res. Rec.* **2017**, *2613*, 52–60. [CrossRef]
22. Li, Z.; Chitturi, M.; Zheng, D.; Bill, A.; Noyce, D. Modeling reservation-based autonomous intersection control in VISSIM. *Transp. Res. Rec.* **2013**, *2381*, 81–90. [CrossRef]

23. Camargo Pérez, J.; Carrillo, M.H.; Montoya-Torres, J.R. Multi-criteria approaches for urban passenger transport systems: A literature review. *Ann. Oper. Res.* **2015**, *226*, 69–87. [CrossRef]
24. Patil, A.; Herder, P.; Brown, K. Investment Decision Making for Alternative Fuel Public Transport Buses: The Case of Brisbane Transport. *J. Public Transp.* **2010**, *13*, 115–133. [CrossRef]
25. Mukherjee, S. Selection of Alternative Fuels for Sustainable Urban Transportation under Multi-criteria Intuitionistic Fuzzy Environment. *Fuzzy Inf. Eng.* **2017**, *9*, 117–135. [CrossRef]
26. Yedla, S.; Shrestha, R.M. Multi-criteria approach for the selection of alternative options for environmentally sustainable transport system in Delhi. *Transp. Res. Part A Policy Pract.* **2003**, *37*, 717–729. [CrossRef]
27. Pogarčić, I.; Frančić, M.; Davidović, V. Application of ahp method in traffic planning. In *16th International Symposium on Electronics in Traffic*; Meše, P., Hernavs, B., Eds.; Electrotechnical Association of Slovenia: Ljubljana, Slovenia, 2008; pp. 1–8.
28. Effat, H.A.; Hassan, O.A. Designing and evaluation of three alternatives highway routes using the Analytical Hierarchy Process and the least-cost path analysis, application in Sinai Peninsula, Egypt. *Egypt. J. Remote Sens. Space Sci.* **2013**, *16*, 141–151. [CrossRef]
29. Karunanayake, K.T.S.; Weerakoon, W.M.N.R.; Wickramasinghe, V. Use of analytic hierarchy process (AHP) for solving traffic issues due to ad-hoc urban planning. In *International Conference on Sustainable Built Environment*; University of Moratuwa, Sri Lanka: Kandy, Sri Lanka, 2012; pp. 1–13.
30. Kutlu Gündoğdu, F.; Duleba, S.; Moslem, S.; Aydın, S. Evaluating public transport service quality using picture fuzzy analytic hierarchy process and linear assignment model. *Appl. Soft Comput.* **2021**, *100*, 106920. [CrossRef]
31. Villegas Flores, N.; Cruz Salvador, L.C.; Parapinski dos Santos, A.C.; Madero, Y.S. A proposal to compare urban infrastructure using multi-criteria analysis. *Land Use Policy* **2021**, *101*, 105173. [CrossRef]
32. Shekhovtsov, A.; Kozlov, V.; Nosov, V.; Sałabun, W. Efficiency of methods for determining the relevance of criteria in sustainable transport problems: A comparative case study. *Sustainability* **2020**, *12*, 7915. [CrossRef]
33. Awasthi, A.; Chauhan, S.S. Using AHP and Dempster-Shafer theory for evaluating sustainable transport solutions. *Environ. Model. Softw.* **2011**, *26*, 787–796. [CrossRef]
34. PTV Group AG. Vissim 7 User Manual. Available online: https://www.scribd.com/document/287732324/Vissim-7-Manual-transfer-ro-25nov-ccb8e5-pdf (accessed on 15 March 2018).
35. Saaty, T.L.; Joyce, A.M. *Thinking with Models*, 1st ed.; Pergamon Press: Oxford, UK, 1981; ISBN 0-08-026475-1.
36. Saaty, T.L. A scaling method for priorities in hierarchical structures. *J. Math. Psychol.* **1977**, *15*, 234–281. [CrossRef]
37. Bartošíková, R.; Bilíková, J.; Strohmandl, J.; Šefčík, V.; Taraba, P. Modelling of decision-making in crisis management. In Proceedings of the 24th International Business Information Management Association Conference—Crafting Global Competitive Economies: 2020 Vision Strategic Planning and Smart Implementation, Milan, Italy, 6–7 November 2014; International Business Information Management Association, IBIMA: Norristown, PA, USA, 2014; pp. 1479–1483.
38. Kengpol, A.; Tuammee, S. The development of a decision support framework for a quantitative risk assessment in multimodal green logistics: An empirical study. *Int. J. Prod. Res.* **2016**, *54*, 1020–1038. [CrossRef]
39. Koohathongsumrit, N.; Meethom, W. Route selection in multimodal transportation networks: A hybrid multiple criteria decision-making approach. *J. Ind. Prod. Eng.* **2021**, *38*, 171–185. [CrossRef]
40. De Felice, F.; Petrillo, A. A new multicriteria methodology based on Analytic Hierarchy Process: The "Expert" AHP. *Int. J. Manag. Sci. Eng. Manag.* **2010**, *5*, 439–445. [CrossRef]
41. Russo, R.D.F.S.M.; Camanho, R. Criteria in AHP: A systematic review of literature. *Procedia Comput. Sci.* **2015**, *55*, 1123–1132. [CrossRef]
42. Roháčová, I.; Marková, Z. Analýza metódy AHP a jej potenciálne využitie v logistike. *Acta Montan. Slovaca* **2009**, *14*, 103–112.
43. Mikušová, M.; Čopíková, A. Using the method of multi-criteria decision making to determine the competency model of crisis manager. In *DIEM—Dubrovnik International Economic Meeting*; University of Dubrovnik: Dubrovnik, Croatia, 2015; pp. 470–486.
44. Rohrbach, B. Kreativ nach Regeln—Methode 635, eine neue Technik zum Lösen von Problemen. *Absatzwirtschaft* **1969**, *12*, 73–76.
45. Wilson, C. Using Brainwriting For Rapid Idea Generation. Available online: https://www.smashingmagazine.com/2013/12/using-brainwriting-for-rapid-idea-generation/ (accessed on 13 December 2013).
46. Moslem, S.; Çelikbilek, Y. An integrated grey AHP-MOORA model for ameliorating public transport service quality. *Eur. Transp. Res. Rev.* **2020**, *12*, 68. [CrossRef]
47. Jayasooriya, N.; Bandara, S. Calibrating and Validating VISSIM Microscopic Simulation Software for the Context of Sri Lanka. In Proceedings of the 2018 Moratuwa Engineering Research Conference (MERCon), Moratuwa, Sri Lanka, 30 May–1 June 2018.
48. Fabianova, J.; Michalik, P.; Janekova, J.; Fabian, M. Design and evaluation of a new intersection model to minimize congestions using VISSIM software. *Open Eng.* **2020**, *10*, 48–56. [CrossRef]
49. Bede, Z.; Nemeth, B.; Gaspar, P. Simulation-based analysis of mixed traffic flow using VISSIM environment. In Proceedings of the 2017 IEEE 15th International Symposium on Applied Machine Intelligence and Informatics (SAMI), Herl'any, Slovakia, 26–28 January 2017.
50. Huang, F.; Liu, P.; Yu, H.; Wang, W. Identifying if VISSIM simulation model and SSAM provide reasonable estimates for field measured traffic conflicts at signalized intersections. *Accid. Anal. Prev.* **2013**, *50*, 1014–1024. [CrossRef]
51. Wang, P.Q.; Lu, Z.G.; Zhang, H.Q.; Wang, F.J. Using Analytic Hierarchy Process to evaluate the improvement schemes of the Maoliling Tunnel based on the VISSIM simulation. In Proceedings of the 4th International Conference on Civil Engineering and Urban Planning (CEUP), Beijing, China, 25–27 July 2015.

Article

Deep Journalism and DeepJournal V1.0: A Data-Driven Deep Learning Approach to Discover Parameters for Transportation

Istiak Ahmad [1], Fahad Alqurashi [1], Ehab Abozinadah [2] and Rashid Mehmood [3,*]

1 Department of Computer Science, Faculty of Computing and Information Technology, King Abdulaziz University, Jeddah 21589, Saudi Arabia; iahmad0004@stu.kau.edu.sa (I.A.); fahad@kau.edu.sa (F.A.)
2 Department of Information Systems, Faculty of Computing and Information Technology, King Abdulaziz University, Jeddah 21589, Saudi Arabia; eabozinadah@kau.edu.sa
3 High Performance Computing Center, King Abdulaziz University, Jeddah 21589, Saudi Arabia
* Correspondence: rmehmood@kau.edu.sa

Abstract: We live in a complex world characterised by complex people, complex times, and complex social, technological, economic, and ecological environments. The broad aim of our work is to investigate the use of ICT technologies for solving pressing problems in smart cities and societies. Specifically, in this paper, we introduce the concept of deep journalism, a data-driven deep learning-based approach, to discover and analyse cross-sectional multi-perspective information to enable better decision making and develop better instruments for academic, corporate, national, and international governance. We build three datasets (a newspaper, a technology magazine, and a Web of Science dataset) and discover the academic, industrial, public, governance, and political parameters for the transportation sector as a case study to introduce deep journalism and our tool, DeepJournal (Version 1.0), that implements our proposed approach. We elaborate on 89 transportation parameters and hundreds of dimensions, reviewing 400 technical, academic, and news articles. The findings related to the multi-perspective view of transportation reported in this paper show that there are many important problems that industry and academia seem to ignore. In contrast, academia produces much broader and deeper knowledge on subjects such as pollution that are not sufficiently explored in industry. Our deep journalism approach could find the gaps in information and highlight them to the public and other stakeholders.

Keywords: natural language processing (NLP); topic modelling; BERT; transportation; newspaper; magazine; academic research; journalism; deep learning; smart cities

Citation: Ahmad, I.; AlQurashi, F.; Abozinadah, E.; Mehmood, R. Deep Journalism and DeepJournal V1.0: A Data-Driven Deep Learning Approach to Discover Parameters for Transportation. *Sustainability* **2022**, *14*, 5711. https://doi.org/10.3390/su14095711

Academic Editors: Tommi Inkinen, Tan Yigitcanlar and Mark Wilson

Received: 16 March 2022
Accepted: 2 May 2022
Published: 9 May 2022

1. Introduction

1.1. A Complex World, Governance Failures, and Deep Journalism

We live in a complex world characterised by complex people, complex times, and complex social, economic, and technological environments. Because this was not enough, our complex activities have complex effects on our ecological environment. This is not an easy time for our governments to to manage matters that affect our social, economic, and environmental sustainability. There is clear evidence that governments are failing at addressing education, healthcare, public safety, and the list goes on [1–9]. The recent COVID-19 pandemic is a major example of global governance failure both at preventing such pandemics (caused by the effects of our lifestyles, processed food we eat, and other activities that damage our planet's environment) and managing the COVID-19 pandemic [10–15]. Governments are elected by people, and in a sense, government failure is also a failure of the public. It is time that we take responsibility for both success and failure and look into ways of collaboratively improving governance.

While there are many reasons for government failures, we believe the lack of information availability is a fundamental reason that limits a government's ability to act smartly and allows a lack of transparency to creep into policies and actions, leading to corruption and failure. While the sincerity and intentions of the people involved could be a major cause of shortcomings in any institution or system, particularly large-scale public institutions and systems, practical efforts can be made to reduce silos and segmentation and bridge the gaps in the information and knowledge available to different communities through direct or indirect cross-sectional conversations and collaborations enabled through automated and autonomous technologies such as deep learning, big data analytics, and others.

To this end, this paper introduces the concept of deep journalism, a data-driven deep learning-based approach for discovering multi-perspective parameters related to a topic of interest. We examine the academic, industrial, public, governance, and political parameters for the transportation sector as a case study to introduce deep journalism and our tool, DeepJournal (Version 1.0), that implements our proposed approach. The concept of deep journalism will be illustrated in the rest of this section (and this paper) as we introduce the challenges facing the transportation sector and our work on detecting parameters for it as viewed by the public, governments, industry, and academia. The production and distribution of multi-perspective parameters are expected to provide a holistic and multi-perspective view of a sector and help bridge the knowledge and collaboration gaps that exist to reduce inefficiencies and failures.

1.2. Transportation and Challenges

Transportation is fundamental to modern societies and economies. However, transportation is a major challenge considering the many issues that this sector faces and the design parameters that need considering in developing successful policies, systems, and operations. The issues facing transportation include the safety of people and goods, rising costs, growth of megacities, long commutes for work, parking problems, damage to health and the planet, and more.

Several modes of transportation exist—road, rail, air, and marine—each with its own challenges. Road transportation is considered the backbone of modern economies although it costs over 1 million deaths and 50 million human injuries annually [16]. Rail transport requires huge capital investments and is therefore subject to monopolies, is relatively inflexible in terms of adjustments to individual passenger needs, cannot be moved around, and may be underutilised in different times and situations (such as during the recent COVID-19 pandemic). Moreover, heavily utilised trains are prone to frequent faults [17], cancellations [18,19] accidents [20–22], etc. [23]. Air transportation faces many challenges including pollution [24,25], high costs [26], high safety [27,28] and security risks [29,30], huge capital investments, fuel requirements [31], and others. Marine transportation also faces many challenges such as pollution, security risks [32–34], increasing costs, and environmental regulations [35,36]. These challenges are threatening the sustainability of our societies, economies, and our planet.

There is a need for innovative approaches based on collaborative thinking enabled through the availability of integrated information. Academia is not being used to its full potential [37]. What is possible in terms of technology and the potential of academia and people is not being matched with what is being done. Policy and action need to work together through dialogue to make information available to all bodies working in the transportation sector, the government, industry, academia, journalism, and the public. Deep Journalism could provide a solution.

1.3. Summary of the Proposed Work

In this work, we bring together a range of deep learning, big data, and other technologies to discover transportation parameters from three different perspectives using three different types of data sources, viz., a newspaper (*The Guardian*), a transportation technology magazine (*Traffic Technology International*), and academic literature on transportation (from Web of Science). The three types of data sources provide three different views of the transportation domain, that is, a view as seen by the public and governed by the political and other institutions, a second view from the transportation industry, and a third view as seen by the academics and researchers. Certainly, these views are not mutually exclusive and are to some extent affected by each other, but they do represent different perspectives with considerable differences. We call this approach Deep Journalism for two reasons. First, we call it deep in the actual sense of the word because it allows capturing and reporting a relatively deeper view of a topic (e.g., transportation) from multiple perspectives, dimensions, stakeholders, and depths. Second, we use deep learning to automatically discover multi-perspective parameters about a topic.

The newspaper dataset that we built to discover parameters for public, governance, and political aspects of transportation is collected from a UK-based newspaper, *The Guardian*. We collected all the articles from *The Guardian* newspaper that contain the word "transport" (in the title of the news, the full text of the news article, or the meta-information about the article) and found a total of 14,381 unique articles dated between September 1825 [38] and January 2022 [39]. We discovered a total of 25 parameters from *The Guardian* dataset and grouped them into 6 macro-parameters, namely Road Transport; Rail Transport; Air Transport; Crash and Safety; Disruptions and Causes; and Employment Rights, Disputes, and Strikes.

The industry and technology magazine dataset that we built to discover parameters about industrial aspects of transportation was collected from a technology-focused magazine, *Traffic Technology International (TTI)*, a popular magazine reporting the latest transport technologies and news. We collected all the articles, a total of 5193 articles dated between February 2015 [40] and January 2022 [41] from the magazine website without any filters or search queries because this magazine only covers transportation-related news. We discovered a total of 15 parameters from the *TTI* dataset and grouped them into 5 macro-parameters, namely Industry, Innovation, and Leadership; Autonomous and Connected Vehicles; Sustainability; Mobility Services; and Infrastructure.

The academic-view dataset that we built to discover parameters for the academia-focused aspects of transportation is collected from an academic database, Web of Science. We collected in aggregate 21,446 research article abstracts (with titles and keywords) in the English language only from about 20 categories of academic disciplines in Web of Science, such as transportation science and technology, engineering, environmental science, telecommunications, economics, computer science, business, and others. The collected article abstracts were limited to the publishing years 2000 [42] to 2022 [43]. We discovered 49 transportation parameters from the academic dataset and grouped them into 6 macro-parameters. These are policy, planning and sustainability; transportation modes; logistics and SCM; pollution; technologies; and modelling.

We implemented the proposed deep journalism approach into a tool called DeepJournal (Version 1.0). The tool is able to discover transportation parameters using the datasets described above. The tool comprises four software components; Data Collection and Storage, Data Preprocessing, Parameter Modelling and Discovery, and Validation and Visualisation. The three datasets were collected using web scraping and other techniques and were preprocessed to remove duplicate and irrelevant data, tokenise data, clean up the data, and lemmatise data to generate data in a form that can be processed by the deep learning processing engine. We used a pretrained BERT (bidirectional encoder representations from transformers) word embedding model [44] to capture the contextual relations within the data. The BERT model was used along with UMAP (uniform manifold approximation and projection) [45] (a dimension reduction technique), HDBSCAN (hierarchical density-based spatial clustering of applications with Noise) [46] (a clustering algorithm), and a class-based TF–IDF (term frequency–inverse document frequency) score, to automatically group documents in the datasets into document clusters.

Subsequently, we discovered transportation parameters and macro-parameters from each dataset using the document clusters along with the domain knowledge and a range of quantitative analysis methods performed on the clustered data including similarity metrics [47], hierarchical clustering [48], term score [49], keyword score [50], and inter-topic distance map [51]. A range of visualisation methods were used to elaborate on the datasets, document clusters, and the discovered parameters. These include dataset histograms [52], taxonomies, similarity matrices [53], temporal progression plots, word clouds, and others. Multiple taxonomies of transportation from public, industry, and academic views were extracted using automatic clustering of datasets. Figure 1 depicts a high-level combined multi-perspective taxonomy of transportation as viewed by the public, industry, and academia. The first and second level branches in the figure show the discovered macro-parameters and parameters, respectively.

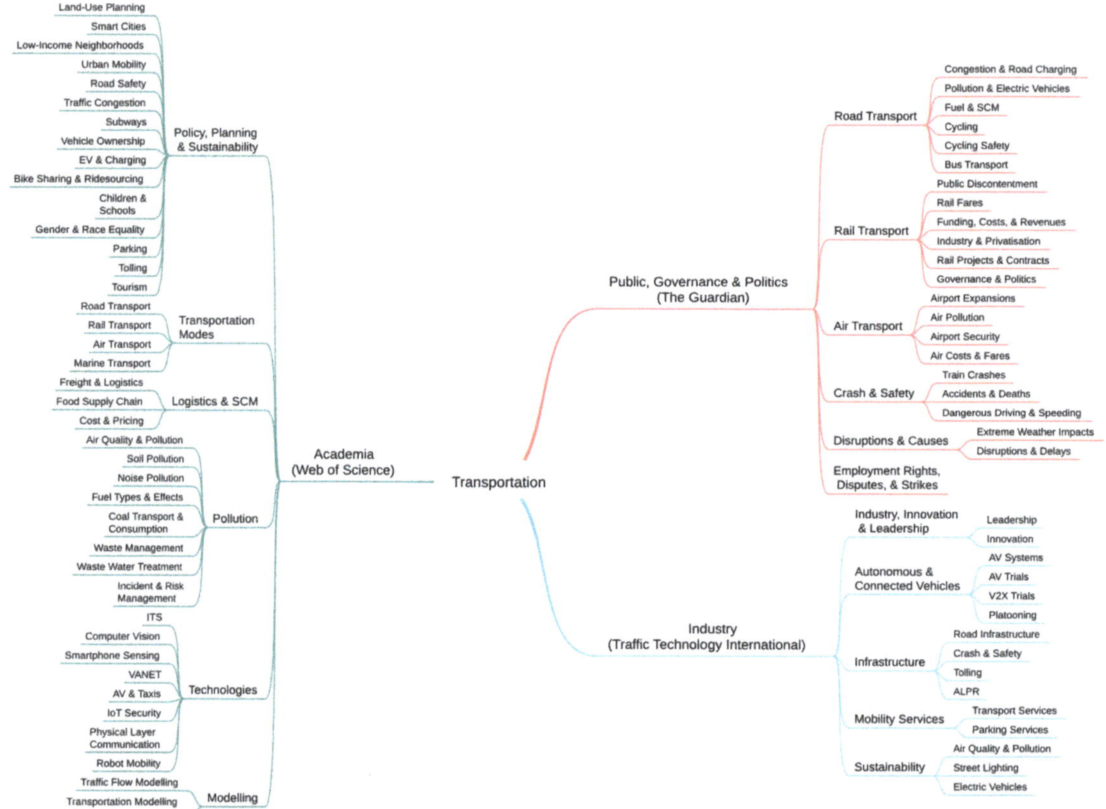

Figure 1. A multi-perspective taxonomy of transportation.

The findings related to a multi-perspective view (public, governance, political, industrial, and academic) of transportation show that there are many important problems such as transportation operations and public satisfaction that industry and academia seem to ignore, or perhaps if they do focus on them, the solutions do not achieve measurable results from the policymakers and industrialists. We can also see that academia produces much broader and deeper knowledge on the subject, while many important issues such as pollution are not publicised. Our deep journalism approach could find the gaps and highlight them for the public and other stakeholders.

The validation of our results can be considered internal or external. The internal validation is performed by investigating whether the articles and documents belonging to a certain parameter are related to the parameter. We have provided discussions on many articles in each dataset as to how those articles relate to the parameters. The external validation is performed by comparing parameters, keywords, and quantitative metrics across the three datasets (i.e., the three perspectives of transportation). The external validation is also performed by using sources other than the three dataset sources. Moreover, both the internal and external validation is performed using the depictions produced by various visualisation methods.

Further details on the methodology and design of our deep journalism approach and the DeepJournal tool are presented in Section 3.

1.4. Broader Aim, Novelty, and Contributions

The broader aim of our work is to investigate the use of ICT technologies for solving pressing problems in smart cities and societies. Specifically, in this paper, we investigate the use of deep learning and digital methods to discover and analyse cross-sectional multi-perspective information to enable better decision making and develop better instruments for academic, corporate, national, and international governance. The contributions of this paper can be summarised as follows.

1. We investigated the use of deep learning and big data analytics methods in transportation and show that important parameters related to the design and operations and the broader environmental parameters can be automatically discovered using cutting-edge technologies. The parameters can be discovered from multiple perspectives involving specific foci, stakeholders, etc. The transportation sector is used as a case study, and the approach can be applied to other sectors. Important parameters and insights are reported and explained.
2. We discovered 25 parameters and 6 macro-parameters for transportation from the public, governance, and political perspectives using a newspaper, *The Guardian.*
3. We discovered 15 parameters and 5 macro-parameters for transportation from an industrial perspective using a transportation magazine, *Traffic Technology International (TTI).*
4. We discovered 49 parameters and 6 macro-parameters for transportation from an academic perspective using the well-known database of scientific literature, Web of Science.
5. We built three datasets specifically for the work presented in this paper. These datasets will be provided openly to the community for further research and development.
6. We brought the three analytics from multiple perspectives together and introduced and investigated the novel concept of deep journalism that could be applied to any problem, sector, or domain.
7. We developed a complete big data analytics tool, DeepJournal Version 1.0, from scratch for this purpose. The tool is general and can be used on other datasets and sectors.
8. We elaborated on 89 transportation parameters and hundreds of dimensions reviewing 400 technical, academic, and news articles.

The literature review (Section 2) establishes that this work is novel in several respects: the proposed scheme of deep journalism, the developed digital methods and tools for the purpose, the use of three (or more) data sources to create a multi-perspective view of the transportation sector, the three datasets, the specific findings, and more.

1.5. Journalism, Citizen Journalism, and Deep Journalism

The public in many parts of the world is frustrated with their governments regarding the governance of social, economic, and other aspects of public lives. The public trust in governments has declined sharply with the emergence of phenomena such as kleptocracy, partisanship, and populism leading to extremism in our societies [54,55]. The main issues are related to the public's perception that the responsibility to provide impartial information and ideal governance lies with other people and not themselves. The American Press Institute defines the purpose of journalism as "to provide citizens with the information they need to make the best possible decisions about their lives, their communities, their societies, and their governments" [56]. Traditional journalism has failed at this purpose due to many reasons such as difficulty in maintaining the freedom and impartiality of media organisations and funding cuts, leading to public mistrust [57,58]. This is, for instance, highlighted by the UN News with a statement by UN Secretary-General, António Guterres, "at a time when disinformation and mistrust of the news media are growing, a free press is essential for peace, justice, sustainable development, and human rights" [59]. The distrust of traditional journalism coupled with the rise of the Internet, digital technologies, and digital and social media has given rapid rise to citizen journalism, i.e., journalism by the general public particularly using digital and social media [60–64]. Citizen journalism has also been referred to as public journalism, democratic journalism, participatory journalism, and other names.

Citizen journalism is complex due to its multifaceted, multidimensional, multilevel, and multimodal nature [65]. It is multifaceted due to its embrace of "a wide array of societal institutions, organisations, groups, and social actors at the intersection between journalism, community, and democracy". It is multidimensional due to it "embracing not only news production and creation but also news consumption and sharing, thus, generating interactive processes among news producers, consumers, and citizens". It is multilevel due to it "comprising journalists, sources, and news audiences at the individual level (micro-level), news organisations and other societal institutions at the organisational level (meso-level), and interorganisational networks in local communities and beyond (macro-level)". It is multimodal because it operates "across diverse communication platforms and channels" including radio, television, Internet, social media platforms, and more [65].

While citizen journalism solves some of the problems of traditional journalism, it comes with its own problems such as subjectivity and potentially lacking regulations, standards, quality, and responsibility [66–68]. The responsibility to maintain ideals lies with all people, and therefore, everyone has the responsibility and needs to work towards upholding honesty, sincerity, equity, freedom, and other ideals. Traditional and citizen journalism need to work together, and their problems need to be resolved collaboratively by the public.

Regarding journalism, the specific aim of this paper is to contribute to this area of journalism and help improve it through academic integrity and rigour. Academics should be in the vanguard of objective information, sincerity, impartiality, equity, and other ideals. Academics should search, pursue, propagate and defend these ideals. If the academics fail to do so, then the responsibility lies with common people to pursue and be in the vanguard of the ideals needed to maintain a free society. The idea of deep journalism is to make impartial, cross-sectional, and multiperspective information available, to bring rigour to journalism by nurturing responsibility in people by making it easy to generate information for the public benefit using deep learning, and to make tools and information available to common people so they can search and defend the ideals of freedom, including social, environmental and economic sustainability.

1.6. Software and Hardware

The work reported in this paper was developed on the Aziz supercomputer that comprises a total of 500 CPU, GPU, and Intel MIC nodes. In addition, we also used Google Colab to run some experiments. Specifically, we used an Nvidia V100 GPU with 32 GB RAM, which combines 5120 CUDA cores and 640 Tensor cores for deep learning and other HPC loads. The V100 has a double performance of 7.066 TFLOPS and a single performance of 14.13 TFLOPS. The software and platforms used in this work include Python as the programming language, along with Pandas [69], Numpy [70], BERTopic [71], NLTK [72], Scikit-Learn [73], Gensim [74], SentenceTransformer, and PyTorch [75]. The data visualisation libraries used in this work include Seaborn [76], Plotly [77], and Matplotlib [78].

1.7. Section Organisation

The rest of the paper is structured as follows. Section 2 reviews the related works and establishes the research gap. Section 3 describes the deep journalism methodology and the design of our tool. Section 4 introduces and discusses the parameters for public, governance, and politics. Sections 5 and 6 discuss the parameters for industry and academia, respectively. Section 7 provides discussion. Section 8 concludes and gives directions for future work.

2. Literature Review

We discuss in this section the works related to the proposed work in this paper. We conducted an extensive review of academic research on the use of artificial intelligence (AI) and data analytics for transportation. We did not find any work directly related to our paper. However, to place our work in the context of the overall body of work on data analytics in transportation, we review works in three areas. First, we discuss studies related to the use of AI and machine learning for transportation. Subsequently, we review research works that analyse and detect transport-related events by using social media data. Finally, we discuss works on the scientometric analysis of the general transportation literature, including scientometric analysis studies on specific areas of transportation.

Researchers have used machine learning for different problems in transportation. For example, a large body of work on the use of deep learning is in object detection, environment perception, health effect, resilience in transport, etc [79]. For example, Wang et al. [80] proposed a model MobileNetv1_yolov3lite to detect objects and speed in real-time. Zhu et al. [81] presented an overview of datasets, evaluation criteria, and future work on environment perception, i.e., vehicle tracking, scene understanding, traffic sign detection, lane and road detection, etc. for intelligent vehicles. Deep learning has also been used for many transport modelling problems, including collaborative decision making for environment perception [82], incident detection [83], disaster management [84], rapid transit systems for megacities [85], and traffic flow modelling [86]. Some other research works are on traffic flow prediction [87], autonomous vehicles [88], vehicular networking [89], automatic license plate recognition [90], crash prediction [91], and others.

Researchers have also used various social media data to analyse and detect different events to discover and solve transportation issues. For example, Alomari et al. [92] used a tool and machine learning algorithms for traffic event detection by using a total of 2,511,000 tweets and transportation-related concerns detection during the COVID-19 pandemic [93]. Later, in another study, they [16] used 33.5 million tweets for event detection and road traffic social sensing by using distributed machine learning algorithms. Their research demonstrated Twitter's efficacy in spotting major occurrences without previous information. Suma et al. used a big data tool for automatic event detection [94] from Twitter data and also used apache spark to automatically detect and validate the events [95]. Traffic incident detection is another challenge for the transportation system. Zhang [96] proposed LDA and a clustering-based algorithm to detect traffic incidents by using the Twitter dataset. They used the Carlo K-test to validate their research outcomes. There is other research on using social media datasets for various topics in transportation such as transportation planning and management, the traffic monitoring system, traffic event detection, etc. Wang et al. [97] proposed a traffic management system (i.e., traffic alert and warning) using Twitter data and the LDA topic modelling algorithm. In 2020, a BERT-based automatic traffic alert system was developed by Wan et al. [98]. The authors used Twitter data to evaluate their system. Additionally, they implemented a question-answering model to extract the location, time, and nature of the traffic events.

The works that can be considered more related to this paper are those where researchers used scientometric analysis for transportation-related topics. For example, Heilig et al. [99] used scientometric analysis to perform a study on academic research on public transportation, which offers better knowledge of articles, authors, countries, and keywords based on citation information. They used 7868 research articles with 160,132 references from 2009 to 2013. This is the only work that looked into the transportation area as a whole. All other works on scientometric analysis have focused on specific topics in transportation, and these are discussed in the following paragraph.

Das et al. [100] analysed 15,357 paper abstracts from 7 years of Transportation Research Board (TRB) yearly meetings (2008–2014) by using LDA to show the research patterns and intriguing histories of transport research. Sun and Yin [101] proposed an LDA-based topic modelling approach to find the research topics and temporal information over the last 25 years of transportation. They collected transport-related abstracts of 17,163 articles from 22 journals between 1990 and 2015 and applied LDA to discover 50 key academic research topics. In 2021, Putri et al. [102] proposed a systematic review of ITS by using the LDA and named entity recognition. They retrieved 23,823 titles and abstracts from the Scopus database between 1974 and 2020. Their research findings include the evolution of ITS development and related research areas. Some other research work has been conducted on several transport-related topics. For example, road safety is a significant component of the transportation system. Zou et al. [103] presented a scientometric analysis to reveal the core research area of road safety. The authors found that road safety studies focused mostly on driver psychological behaviour, prevention of traffic accidents, the impact of driver risk factors, and the analysis of the consequences and frequency of road crashes. In another research study, Gao et al. [104] presented a scientometric analysis on traffic safety sign research from 1990 to 2019. The authors collected 3102 articles from the Web of Science database and used Citespace to analyze and visualise the research domain. They discovered that most of the research had been conducted in the last 10 years. Their research also found that the United States is in the lead position in traffic sign research. AV is the most heavily researched topic to improve the transportation system. A scientometric analysis on autonomous vehicles was conducted by Faisal et al. [105]. They collected a total of 4645 research articles between 1998 and 2017 to perform the scientometric analysis. Their research presented the development of AV systems by analyzing the authors, affiliations, citations, and publications in AV research.

Research Gap

The literature review shows that the existing research on the use of machine learning in transportation has mainly focused on autonomous vehicles, object detection, and others. There are some works on social media data analysis for detecting events in transportation. The very few works that have focused on scientometric analysis are very different from our work. We did not find any research papers that have used newspapers, transport magazines, and academic research articles altogether. Our work is novel in several respects: the proposed scheme of deep journalism, the developed digital methods and tools for the purpose, the use of three (or more) data sources to create a multiperspective view of the transportation sector, the three datasets, the specific findings, and more.

3. DeepJournal V1.0: Methodology and Design

The proposed system methodology and design is depicted in Figure 2 to analyse contextual topics that discover the transportation issues, challenges, development, and future planning by using newspaper, magazine, and research article abstracts. The software architecture consists of four software components which are described in the following subsections. The methodology overview, including the master algorithm, is provided in Section 3.1. In this research, we use d three types of data sources named *The Guardian*, *Traffic Technology International*, and Web of Science. Sections 3.2–3.5 summarise these three data sources, discuss the data collection algorithm, and describe the datasets. Sections 3.6–3.9 cover data preprocessing, parameter modelling, parameter discovery and quantitative analysis, and validation and visualisation, respectively.

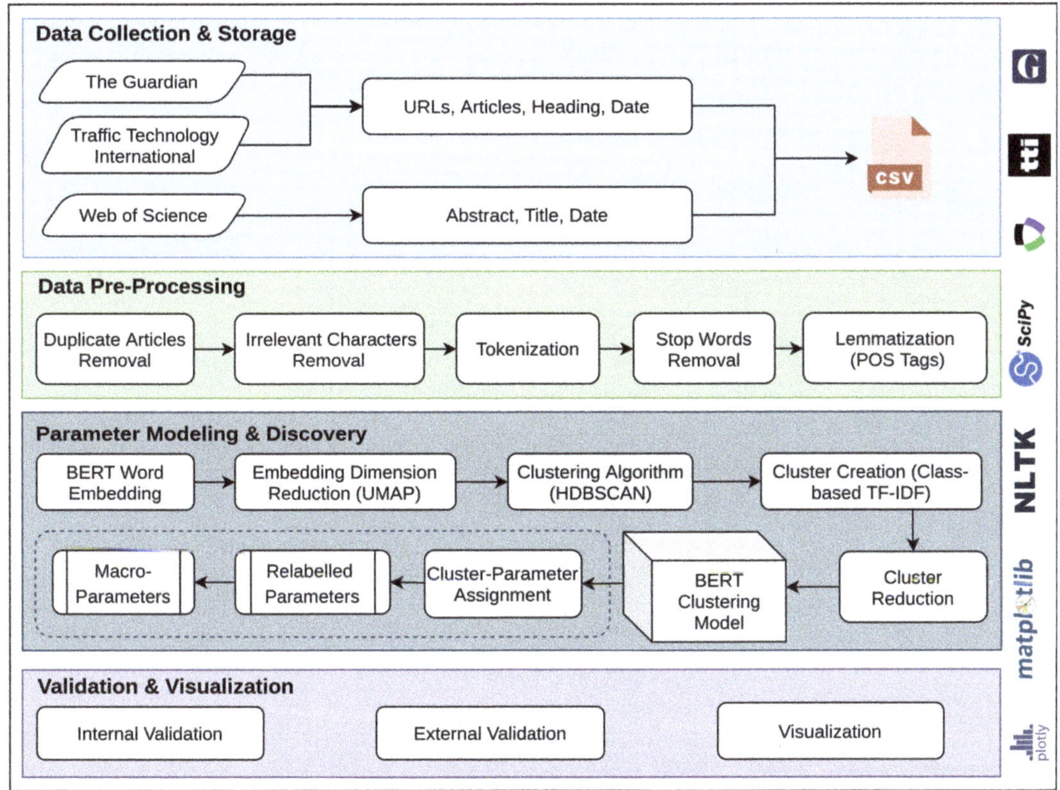

Figure 2. DeepJournal V1.0: the system architecture.

3.1. Methodology Overview

Algorithm 1 outlines the master algorithm where the inputs are the search queries and website URLs which are needed for the data collection. The dataset was collected using the web scraping technique and stored in a CSV file. Then, the CSV file was loaded into the Pandas data frame, and the articles were passed to the data preprocessing function which removes duplicate articles, accomplishes tokenisation, removes irrelevant characters and stop words from the articles, and performs lemmatisation with allowed POS (part-of-speech) tags, i.e., noun, adjective, verb, and adverb and generates cleaned tokens. After that, a pretrained BERT (bidirectional encoder representations from transformers) [44] word embedding model was used to capture the contextual relations between the words. Subsequently, we used the UMAP (uniform manifold approximation and projection) [45]/ which is a dimension reduction technique, HDBSCAN (hierarchical density-based spatial clustering of applications with noise) [46] which is a clustering algorithm and a class-based TF–IDF (term frequency–inverse document frequency) score to calculate the importance of words for each cluster. Additionally, we reduced the number of clusters by merging the most similar clusters. After that, we saved the clustering model and assigned the cluster to each article. Then, the clusters were relabelled as parameters and the parameters were grouped into macro-parameters using the domain knowledge along with a similarity matrix, hierarchical clustering, and other quantitative analysis methods. Finally, we visualised and validated the parameters against external and internal sources.

Algorithm 1 Master Algorithm

Input: *searchQuery, weblink*
Output: *article with lebelled parameter and visualization*

$CSV_file \leftarrow dataCollection(searchQuery)$
$article_DF \leftarrow read_CSV(CSV_file)$
$process_artcle \leftarrow dataPreProcessing(article_DF)$
$word_embedding \leftarrow createBERT_Embedding(process_artcle)$
$umap_embedding \leftarrow dimensionReduction(word_embedding)$
$HDBSCAN_clustering \leftarrow dimensionReduction(umap_embedding)$
$calculate_ClassTFIDF \leftarrow clustering(HDBSCAN_clustering)$
$clusters \leftarrow clusterReduction(calculate_ClassTFIDF)$
$model \leftarrow saveModel(BERTClusteringModel)$
$parameters \leftarrow relabelled(clusters)$
$parameter_visualization(parameters)$

3.2. Data Collection

We used three types of data sources in this research: *The Guardian* (newspaper), *Traffic Technology International* (magazine), and the Web of Science (academic research). We utilised web scraping techniques (i.e., Python, BeautifulSoup, Requests, and Pandas) to obtain the *The Guardian* and *TTI* datasets from their corresponding websites. We collected the Web of Science dataset from its website as it allows users to download the dataset as a CSV format. We discuss the data collection steps for *The Guardian*, *Traffic Technology International*, and Web of Science in Sections 3.3–3.5, respectively.

3.3. Dataset: Newspaper Articles (The Guardian)

The newspaper dataset was collected from the UK-based newspaper *The Guardian* from September 1825 to January 2022. We retrieved all transport-related articles from the website using the web scraping technique and collected about 14,855 articles.

Algorithm 2 shows the steps of the data collection process. Initially, we used "transport" keywords to search for the related articles on the website. After that, we passed the web link to the newspaper as a parameter in the algorithm. We divided our data collection methodology into two functions: article link collection and data collection. In the first function, after acquiring all the links from the web page content, we removed the irrelevant links and saved the links as a data frame. In the data collection function, we analysed the HTML and JavaScript code to obtain the article, date, and headline from the web page content. For each news article, we acquired the related heading and publication date. We saved the data in a data frame and finally saved the data frame into a CSV file. After retrieving the articles, we encountered a few duplicate articles. We eliminated all duplicate articles, resulting in 14,381 unique articles from *The Guardian*.

Algorithm 2 Data Collection (*The Guardian*)

Input: searchQuery, weblink
Output: CSV file

```
function ARTICLELINK(weblink)                 ▷ weblink: https://www.theguardian.com/uk/
    pages ← length(totalPage)                 ▷ total web page after searching
    for pageNumber ← 1 to pages do
        url ← weblink/transport?page=pageNumber
        content ← obtain content from URL
        links_DF ← save links as DataFrame from content     ▷ remove irrelevant links
    end for
end function
function DATACOLLECTION(links_DF)
    links ← length(links_DF)
    for pageNumber ← 1 to links do
        content ← obtain content from link
        article ← obtain article from content
        headline ← obtain headline from content
        date ← obtain publication date from content
        guardian_DF ← article, headline, date
        guardian_CSV ← guardian_DF
    end for
end function
```

Figure 3 shows the histogram of *The Guardian* news articles. The y-axis indicates the number of news articles for the increasing word count per news article. We noticed that the prevalent length of news articles was 200–500 words. The number of news articles that contained more than 800 words was relatively small. The maximum number of words in a document was 8341. For more visual understanding, the zoomed portion inside the graph is shown. The figure also shows the density against the increasing number of words per news article. The maximum density is around 0.0016.

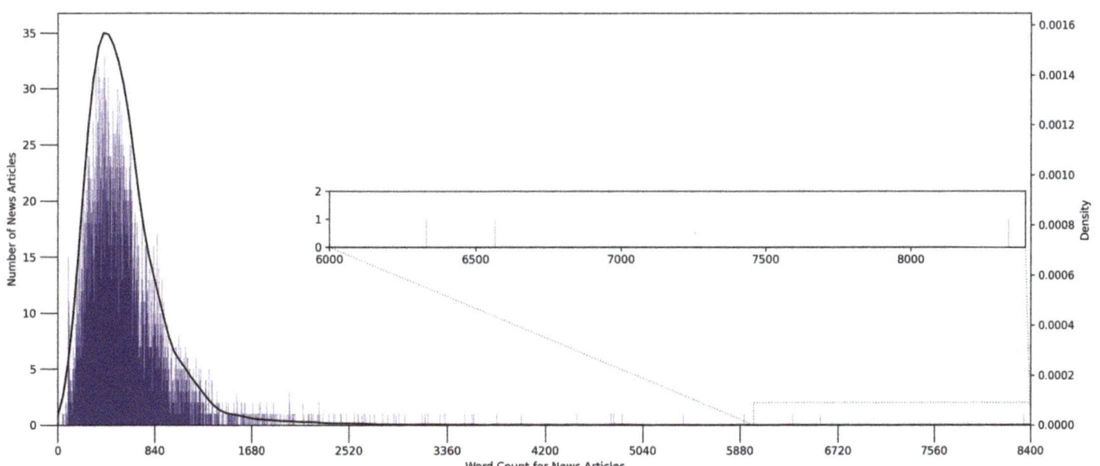

Figure 3. Histogram (*The Guardian* news articles).

3.4. Dataset: Technology Magazine Articles (Traffic Technology International)

TTI stands for *Traffic Technology International* which is a popular magazine related to the latest transport technology. From February 2015 to January 2022, we gathered 10,620 articles from all categories on the *TTI* website using the web scraping approach.

Algorithm 3 shows the steps of the data collection process. We divided the algorithm into two functions: article link collection and data collection. In the beginning, we passed the web link to the article link collection function. We used two loops, where the first loop was for the category list and the second loop was for the total web page number for each category. We used a dictionary-type variable to store the category as a key and the total web page number for that category as a value. After obtaining all the links from the web pages, we removed the irrelevant links and saved the links as a data frame. In the data collection function, we analysed the HTML and JavaScript code to obtain the article, publication date, and headline from the web page content. For each news article, we received the related heading and publication date. We saved the data in a data frame and finally saved the data frame into a CSV file. After saving the data, we found a lot of duplicate data as some articles are common in multiple categories.Therefore, we removed the duplicate articles from the dataset, and finally, we found 5193 unique articles.

Algorithm 3 Data Collection (*Traffic Technology International*)

Input: weblink
Output: CSV file

```
function ARTICLELINK(weblink) ▷ weblink: https://www.traffictechnologytoday.com/
    categoryList ← list of the category
    pagesDict ← dictionary type variable (key: category, value:pages number)
    for category ← 1 to length(categoryList) do
        for pageNumber ← 1 to pagesDict[categoryList[category]] do
            url ← weblink/news/categoryList[category]/page/pageNumber
            content ← obtain content from URL
            links_DF ← save links as DataFrame from content ▷ remove irrelevant links
        end for
    end for
end function
function DATACOLLECTION(links_DF)
    links ← length(links_DF)
    for pageNumber ← 1 to links do
        content ← obtain content from link
        article ← obtain article from content
        headline ← obtain headline from content
        date ← obtain publication date from content
        TTI_DF ← article, headline, date
        TTI_CSV ← TTI_DF
    end for
end function
```

Figure 4 depicts the histogram of the *Traffic Technology International* magazine articles. The y-axis and x-axis demonstrate the number of magazine articles and the increasing word count for magazine articles, respectively. The majority of magazine articles were between 300 and 450 words and 500 to 600 words (see graph peaks).The number of news articles that contained more than 600 words was relatively small. The maximum number of words in a document was 2323. The magnified plot inside the figure is presented for the convenience of the reader. The graph also depicts the density in relation to the increasing quantity of words per magazine article. The highest density is around 0.005.

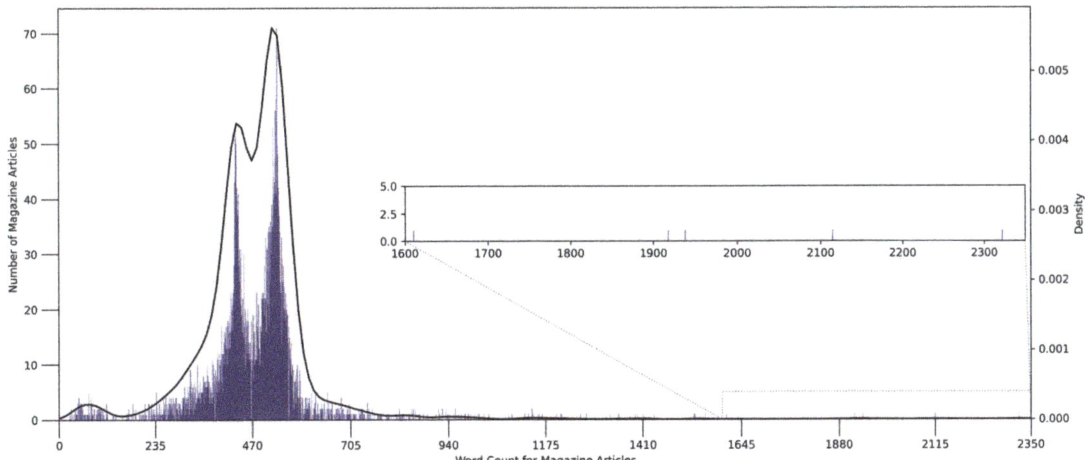

Figure 4. Histogram (*Traffic Technology International* articles).

3.5. Dataset: Academic Articles (Web of Science)

We obtained the most pertinent documents from the Web of Science, the most comprehensive database with a consistent query language and data format. Furthermore, it facilitates access to subject indexes, citation indexes, and other databases from other disciplines which assists in the discovery of relevant research and the evaluation of its findings. We collected 21,446 research articles by using "transportation" keyword from several Web of Science categories, for example, transportation science technology, engineering electrical electronics, transportation, environmental science, telecommunications, economics, computer science information system, business, etc. The document type was limited to proceedings papers, articles, and review articles. Excluded were publications produced from news items, corrections, book chapters, data papers, book reviews, letters, editorial materials, and so on. Furthermore, we narrowed our search filtering option to the English language and the publishing years 2000–2022. In addition, we utilised advanced search and selected the "topic search" option which yielded results from the title, abstract, and keywords columns.

Figure 5 illustrates the histogram of the Web of Science research article abstracts. The y-axis and x-axis show the number of article abstracts as well as the increasing word count for article abstracts. The majority of article abstracts contained between 150 and 250 words. A few article abstracts had more than 450 words. The number of article abstracts that contained more than 400 words was relatively small. The maximum number of words in an article abstract was 1132. The magnified plot inside the figure is presented for the convenience of the reader. The graph also shows the density in relation to the increasing quantity of words per article abstract. The highest density is around 0.006.

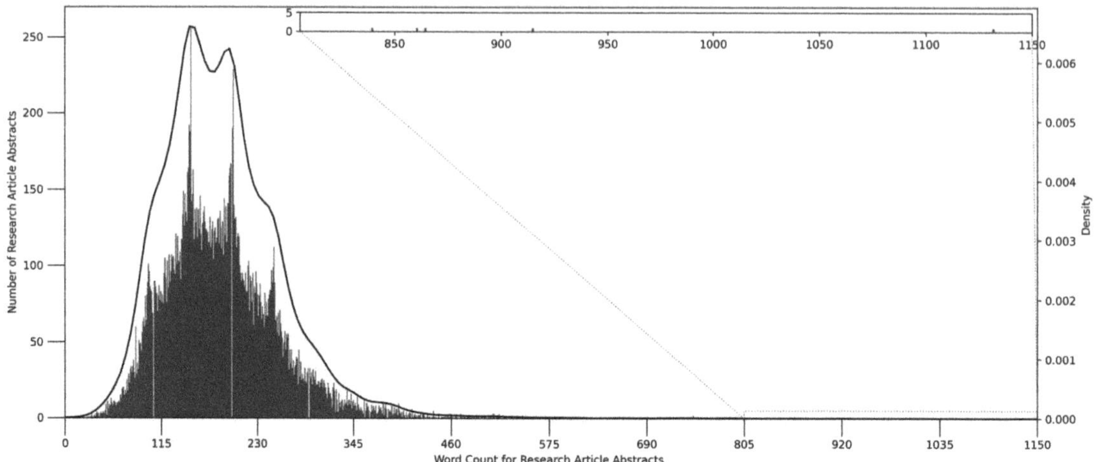

Figure 5. Histogram (Web of Science article abstracts).

3.6. Data Preprocessing

We employed the same preprocessing algorithm for the three datasets. Algorithm 4 shows the preprocessing steps as follows: (1) remove duplicate articles, (2) remove irrelevant characters, (3) tokenisation, (4) stop word removal, and (5) lemmatisation with POS tags. In the beginning, we read the CSV file using the Python package "Pandas" and saved it as a data frame (DF). In the second step, we removed all duplicate data to reduce the data redundancy, and in the third step, we removed all irrelevant characters, for example, several Unicode characters, from the texts. Furthermore, in the fourth step, we tokenised the texts using the simple_preprocess function, which is included in the Python package called "gensim". The fifth step was to remove the stop words from the article. Initially, we used the NLTK predefined stop words list for clustering and implemented the BERT parameter model. After getting the parameter from the BERT parameter model, we reviewed the corresponding keywords and explored the unnecessary keywords that were obtaining a high probability score in the parameter. After some testing, we finalised a list of stop words that did not carry significant importance for generating parameters. As a result, in our final model, we added those keywords to the stop words list and removed them from the articles. In the final step of data preprocessing, we used lemmatisation using the "Spacy" engine and allowed only four types of parts of speech tags, including noun, verb, adjective, and adverb. After the preprocessing step, we obtained the cleaned articles, which were used for parameter modelling.

Algorithm 4 Data Preprocessing

Input: *articles*
Output: *clean articles*

$article_DF \leftarrow read_CSV(CSV_file)$
$rd_DF \leftarrow removeDuplicate(article_DF)$
$ric_DF \leftarrow removeIrrelevantCharacters(rd_DF)$
$token_DF \leftarrow tokenizer(ric_DF)$
$rSW_DF \leftarrow removeStopWords(token_DF)$
$lemma_DF \leftarrow lemmatization(rSW_DF)$
$clean_DF \leftarrow cleanArticle(lemma_DF)$

3.7. Parameter Modelling

We utilised the BERT topic modeling method [71] to cluster the data and discover parameters. At the beginning of parameter modelling, we generated a word-embedding model using the BERT (Bidirectional Encoder Representations from Transformers), which is a transformer-based approach developed by Google [44]. Word embedding is a low-dimensional, dense vector representation of words, and BERT develops contextual embeddings. In this paper, we used the pretrained "distilbert-base-nli-mean-tokens" model as it maintains the balance between performance and execution time. We implemented a dimensional reduction algorithm, UMAP, to keep the maximum information in a lower dimensionality. Furthermore, we used HDBSCAN to group identical articles together that define a cluster or parameter. HDBSCAN is a density-based approach that complements UMAP effectively, considering UMAP retains a range of local structures even at lower dimensionality. Additionally, HDBSCAN does not compel data points to clusters since they are considered outliers.

Furthermore, a class-based TF–IDF (term frequency–inverse document frequency) score was used to calculate the importance of words for each parameter. By determining the frequency of a word in a given document as well as the measure of how prominent the word is in the entire corpus, TF–IDF provides a means of comparing the relevance of words between documents. However, if we consider all documents in a single group as a distinct document and then execute TF–IDF, we will achieve significance scores for words inside a cluster. This significance score is called the c TF–IDF score. The more significant the words inside a cluster, the more representative the parameter. As a result, we can obtain keyword-based descriptions for every parameter. Equation (1) [50] describes the formula of the c-TF–IDF score, where f is the frequency of each word derived for each class c and divided by the number of words w. The total number of unjoined documents (d) is then divided by the total frequency of words (f) throughout all classes (cc).

$$c - TF - IDF_c = \frac{f_c}{w_c} \times \log \frac{d}{\sum_p^{cc} f_p} \tag{1}$$

We fit all of the articles and trained the BERT parameter model after obtaining the c-TF–IDF. The number of parameters was then decreased by recalculating the articles' c-TF–IDF matrices and then iteratively merging the most often occurring parameter with the most similar one based on the respective c-TF–IDF matrices.

Finally, we assigned parameters to all the articles and saved the model. As the parameter was originally represented as an integer number, we further scrutinised the corresponding parameter articles and relabelled the parameters and aggregated them into macro-parameters using domain knowledge and quantitative analysis methods which are discussed in the next section.

3.8. Parameter Discovery and Quantitative Analysis

We discovered the parameters and macro-parameters using domain knowledge and quantitative analysis methods (i.e., term score, keyword score, intertopic distance, hierarchical clustering, and similarity matrix).

3.8.1. Term Score

A list of keywords (terms) for each parameter does not express the context of the related parameter in the same way. To find a parameter, we must first determine how many keywords are required, as well as the starting and finishing positions of significant keywords. We visualised the keywords c-TF–IDF score for each parameter by sorting them in decreasing order [71]. This term score visualisation has a significant influence on identifying the parameter.

3.8.2. Intertopic Distance Map

The intertopic distance map is a two-dimensional representation of the parameters, with the area of the parameter circles proportional to the number of words in the dictionary associated with each parameter. The circles are formed using a MinMaxScaler algorithm depending on the words they contain, with parameters closer together sharing more words [71].

3.8.3. Keyword Score

In the BERT parameter model, we obtained a set of keywords representing a parameter, where each keyword has an importance score or c-TF–IDF score (see Section 3.7) that describes the context of the parameter.

3.8.4. Hierarchical Clustering

Hierarchical clustering systematically pairs the parameters based on the cosine similarity matrices between the parameter embeddings [71]. By systematically pairing clusters, hierarchical clustering assembles a unique cluster of nested clusters. At each phase, beginning with the correlation matrix, all clusters are attempted in all possible pairs, and the pair with the greatest average inter-correlation within the experimental cluster is chosen as the new unique cluster.

3.8.5. Similarity Matrix

The similarity matrix was visualised as a heatmap using the Plotly library in Python to show the similarity between parameters based on the cosine similarity matrix [71]. We computed the similarity matrix by calculating the cosine similarity score between the parameters embedding to show the relationship between the parameters. We have used Plotly "BnGu" (green to blue) as the continuous color scale where the dark blue color represents the highest similarity relationship between parameters, while the light green represents the lowest similarity.

3.9. Validation and Visualisation

The validation of our results can be considered to be internal or external. The internal validation was performed by investigating whether the articles and documents belonging to a certain parameter are related to the parameter. We have provided discussions on many articles in each dataset as to how those articles relate to the parameters. The external validation was performed by comparing parameters, keywords, and quantitative metrics across the three datasets (i.e., the three perspectives of transportation). The external validation was also performed by using sources other than the three dataset sources. Moreover, both the internal and external validation were performed using the depictions produced by various visualisation methods.

A range of visualisation methods were used to elaborate on the datasets, document clusters, and the discovered parameters. These include dataset histograms [52], taxonomies, similarity matrices [53], temporal progression plots, word clouds, and others. For example, we visualised the temporal progression for both parameters and macro-parameters. Initially, we merged the similar representable parameters and then counted the number of articles (intensity) by grouping the parameters and article publication year. Consequently, we obtained a list of intensities for each parameter with specific years. After that, we sorted the list according to the year and plotted the intensity against the year for each parameter. We also plotted the macro-parameter temporal progression in the same way by integrating the parameters of each macro-parameter.

We used several python libraries for these visualisations including Seaborn, Plotly, and Matplotlib.

4. Public, Governance and Politics: Transportation Parameters Discovery

In this section, we discuss the parameters detected by our BERT model from *The Guardian* dataset. The parameters are grouped into six macro-parameters. We provide an overview of the parameters and macro-parameters in Section 4.1. The quantitative analysis is discussed in Section 4.2. Subsequently, we discuss each macro-parameter in separate sections, Sections 4.3–4.8. Finally, Section 4.9 discusses the temporal analysis of the parameters and macro-parameters.

4.1. Overview and Taxonomy (The Guardian)

We detected a total of 25 parameters from *The Guardian* dataset using BERT. These 25 parameters were grouped into 6 macro-parameters using the domain knowledge along with similarity matrix, hierarchical clustering, and other quantitative analysis methods. The methodology and process of the discovery of parameters and their groupings into macro-parameters have already been discussed in Section 3.

Table 1 lists the parameters and macro-parameters of *The Guardian* dataset. The parameters are categorised into 6 macro-parameters, including road transport, rail transport, air transport, crash and safety, disruptions and causes, and employment rights, disputes, and strikes (Column 1). The parameters are listed in Column 2, where some of them are merged. For example, Parameters 9 and 4 have been combined into a single parameter, rail projects and contracts. The third column indicates the cluster number. The percentage of the number of articles is recorded in the fourth column. Our BERT model labelled 50.5% of articles as the outlier clusters. The outlier clusters are more analogous to the average article compared to any of the other clusters. Consequently, we ignored these clusters, and the rest of the 49.5% of articles are listed in the fourth column. The fifth column represents the top keywords associated with each parameter.

Figure 6 provides a taxonomy of the transportation domain extracted from a newspaper-focused on public, governance and politics. The taxonomy was created using the parameters and macro-parameters discovered from *The Guardian*. The first-level branches show the macro-parameters, the second-level branches show the discovered parameters, and the third-level branches show the most representative keywords.

4.2. Quantitative Analysis (The Guardian)

This section discuss the term score, word score, intertopic distance map, hierarchical clustering, and similarity matrix.

Each parameter is represented by a group of keywords, although not all of these words describe the parameter equally. The term probability declined representation depicts how many keywords are required to describe a parameter and when the benefit of adding more keywords begins to diminish (see Section 3.8). When we assess the keywords, only the top 7 to 10 terms in each parameter accurately describe the parameter, as shown in Figure 7. Because all of the other probabilities are so close to one another, ranking them becomes more or less useless. When we analysed the top keywords per parameter to discover the parameter, we used this information to focus on the top seven or so keywords in each parameter.

Table 1. Parameter and Macro-Parameters for Transportation (Source: *The Guardian*).

Macro-Parameters	Parameters	No.	%	Keywords
Road Transport	CRC	7	2.25	road, congestion, traffic, motorway, transport, charge, scheme, car, lane, motorist, toll, government, year, city, pricing, vehicle, public, need, increase
	Pollution and EV	17	1.56	car, vehicle, emission, diesel, pollution, fuel, electric, air, petrol, clean, new, carbon, year, hydrogen, engine, power, government, manufacturer, hybrid, green
	Fuel and SCM	19	1.47	fuel, oil, price, petrol, driver, duty, tax, government, shortage, car, rise, supply, motorist, increase, cost, road, high, tanker, year
	Cycling	24	1.02	bike, cycling, cycle, cyclist, city, ride, road, bicycle, route, lane, scheme, year, car, traffic, transport, way, safe, work, street, day
	Cycling Safety	23	1.08	cyclist, cycle, road, bike, cycling, death, kill, pedestrian, traffic, ride, safety, accident, driver, year, lorry, safe, helmet, injury, lane, number
	Bus Transport	12	1.87	bus, service, routemaster, transport, route, local, public, new, passenger, year, cut, travel, run, operator, journey, city, old, work, time
Rail Transport	PD	6	2.34	train, rail, passenger, service, company, network, year, railtrack, franchise, timetable, railway, new, delay, operator, line, run, work, industry, time
	Rail Fares	11	2.06	fare, ticket, rise, rail, train, passenger, season, price, increase, inflation, travel, cost, year, commuter, cheap, pay, company, average, peak, buy
	FCR	3	2.67	rail, year, company, network, profit, bn, government, franchise, bus, train, cost, railtrack, passenger, fare, rise, increase, revenue, transport, public
	Industry and Privatisation	21	1.13	shareholder, railtrack, company, eurotunnel, byer, debt, government, tunnel, french, share, administration, creditor, investor, channel, group, year, bn, private, public
	Rail Projects and Contracts	9	2.16	franchise, rail, government, train, contract, railway, company, year, railtrack, run, network, service, line, new, transport, operator, bid, public, plan
		4	2.39	rail, project, line, transport, north, high, speed, government, new, network, plan, train, link, route, bn, cost, year, build, city
	Governance and Politics	2	2.75	government, byer, transport, labour, public, minister, private, railtrack, company, last, tory, year, plan, decision, tube, rail, political, privatisation, secretary
Air Transport	Airport Expansions	0	3.57	runway, airport, heathrow, expansion, third, government, new, aviation, decision, noise, flight, build, stanste, plan, air, capacity, climate, expand, environmental
	Air Pollution	8	2.19	emission, carbon, aviation, climate, airline, change, fuel, flight, air, environmental, government, biofuel, reduce, industry, offset, global, green, energy, scheme
	Airport Security	1	3.55	flight, airport, passenger, airline, plane, drone, security, fly, pilot, aircraft, delay, air, cancel, travel, heathrow, check, ash, staff
	Air Costs and Fares	18	1.48	airline, airport, year, flight, passenger, carrier, cost, price, baa, Easyjet, business, market, Ryanair, last, fly, profit, Heathrow, industry, air
Crash and Safety	Train Crashes	15	1.78	crash, safety, train, railtrack, rail, signal, accident, report, inquiry, Paddington, track, health, company, bar, railway, manslaughter, network, potter, disaster, prosecution
		22	1.10	train, carriage, crash, track, passenger, accident, incident, injure, scene, driver, derail, service, rail, safety, police, emergency, line, fire, injury, come
	Accidents and Deaths	16	1.72	police, crash, accident, incident, scene, woman, injure, die, man, injury, train, vehicle, driver, kill, hospital, car, family, motorway, death, cause
	DDS	14	1.79	driver, speed, limit, drive, camera, road, police, driving, mph, drink, penalty, motorist, offence, test, accident, drug, fine, safety, year, death
Disruptions and Causes	EWI	10	2.09	snow, weather, temperature, flood, road, cold, rain, heavy, condition, wind, service, warning, expect, ice, close, area, fall, cause, delay, morning
	Disruptions and Delays	13	1.82	train, service, weekend, holiday, passenger, expect, delay, busy, work, station, line, travel, day, fire, hour, weather, rail, disruption, run, traffic
ERDS	ERDS	5	2.38	strike, union, action, unite, staff, dispute, rmt, member, tube, crew, industrial, pay, talk, worker, ballot, offer, day, cabin, service, work
		20	1.24	strike, southern, rmt, train, dispute, union, service, action, member, guard, talk, driver, rail, passenger, staff, conductor, company, day, run, aslef

CRC = Congestion and Road Charging; PD = Public Discontentment; FCR = Funding, Costs, and Revenues; DDS = Dangerous Driving and Speeding; EWI = Extreme Weather Impacts; ERDS = Employment Rights, Disputes, and Strikes.

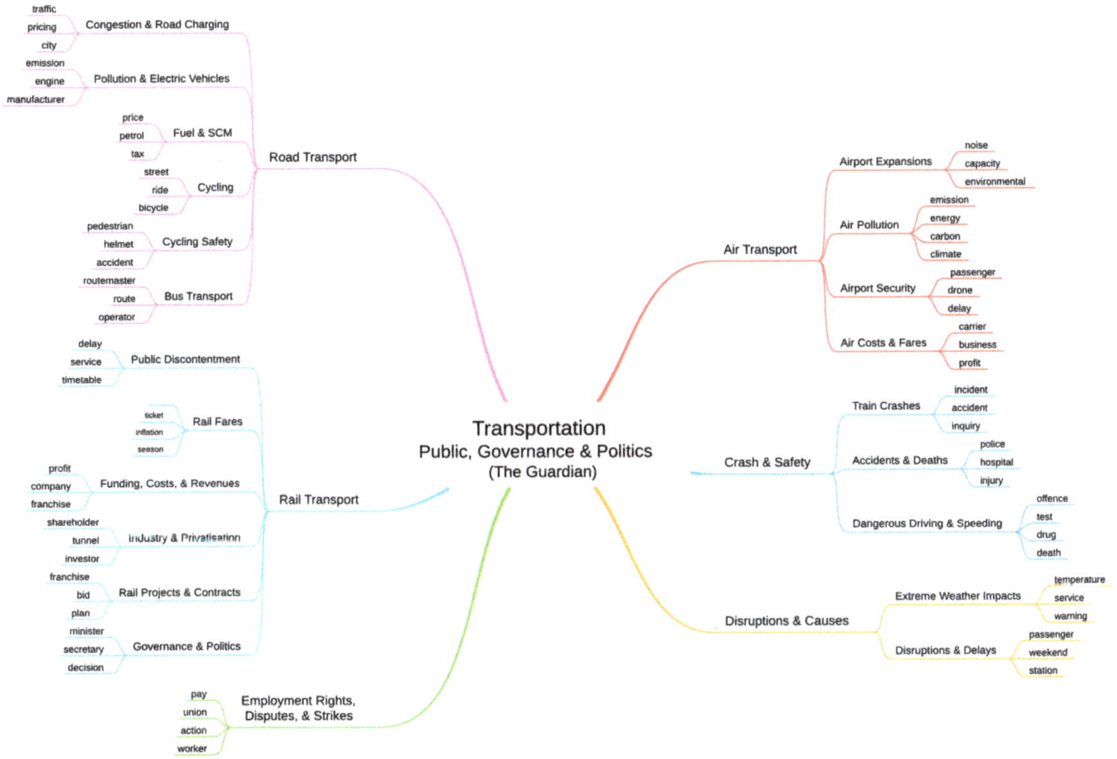

Transportation
Public, Governance & Politics
(The Guardian)
Road Transport
Congestion & Road Charging
traffic
pricing
city
Pollution & Electric Vehicles
emission
engine
manufacturer
Fuel & SCM
price
petrol
tax
Cycling
street
ride
bicycle
Cycling Safety
pedestrian
helmet
accident
Bus Transport
routemaster
route
operator
Rail Transport
Public Discontentment
delay
service
timetable
Rail Fares
ticket
inflation
season
Funding, Costs, & Revenues
profit
company
franchise
Industry & Privatisation
shareholder
tunnel
investor
Rail Projects & Contracts
franchise
bid
plan
Governance & Politics
minister
secretary
decision
Employment Rights, Disputes, & Strikes
pay
union
action
worker
Air Transport
Airport Expansions
noise
capacity
environmental
Air Pollution
emission
energy
carbon
climate
Airport Security
passenger
drone
delay
Air Costs & Fares
carrier
business
profit
Crash & Safety
Train Crashes
incident
accident
inquiry
Accidents & Deaths
police
hospital
injury
Dangerous Driving & Speeding
offence
test
drug
death
Disruptions & Causes
Extreme Weather Impacts
temperature
service
warning
Disruptions & Delays
passenger
weekend
station

Figure 6. A taxonomy of transportation extracted from *The Guardian* dataset.

c-TF-IDF score
0.1
0.08
0.06
0.04
0.02
1
3
5
7
9
11
13
15
17
19
Term Rank

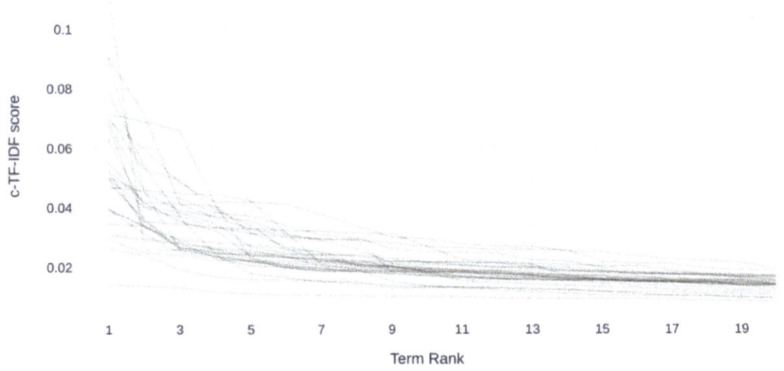

Figure 7. Term score (*The Guardian*).

Figure 8 depicts the top five keywords for each parameter. The keywords are sorted based on the importance score, or c-TF–IDF score (see Section 3.8). There are 25 subfigures, and in each subfigure, the horizontal line indicates the c-TF–IDF scores, and the vertical line indicates the keywords. For example, the first subfigure is the airport expansion parameter, which is represented by the five keywords such as runway, airport, Heathrow, expansion, and government, having 0.07, 0.05, 0.49, 0.39, and 0.26 scores, respectively.

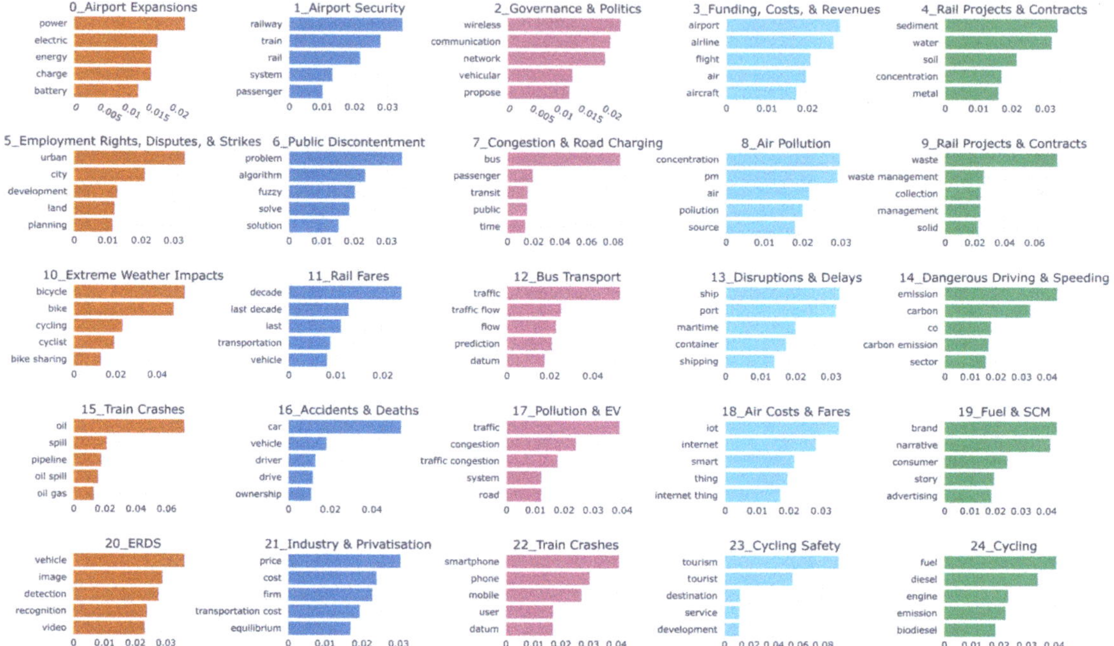

Figure 8. Newspaper article parameter with keywords c-TF–IDF score.

Figure 9 shows the intertopic distance map (see Section 3.8), where six macro-clusters are separately identified.

Figure 10 describes the hierarchical clustering of the 25 clusters and systematically pairs the clusters based on the similarity matrix (see Section 3.8). We noticed that clusters No. 6, 3, 9, 2, and 4 created a unique cluster that we labelled as the rail transport parameter.

Figure 11 visualises the similarity matrix among the parameters (see Section 3.8). Note the dark blue colour between clusters 22 and 16 which showed a high similarity score because both clusters 22 (train, carriage, and crash) and 16 (police, crash, and accidents) have high resemblance. For example, whenever a train or carriage crash happens, at that time there is a high possibility of an accident, and police might react at that time.

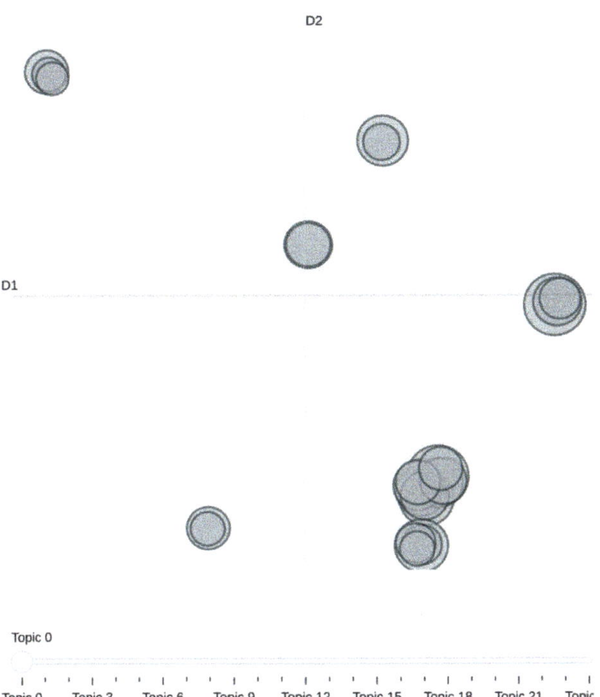

Figure 9. Intertopic distance map (*The Guardian*).

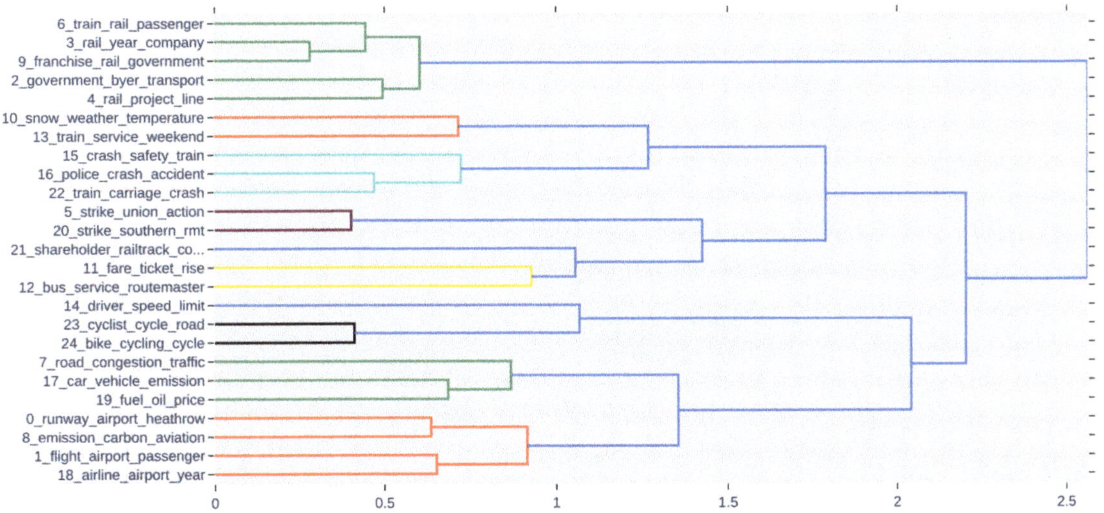

Figure 10. Hierarchical clustering (*The Guardian*).

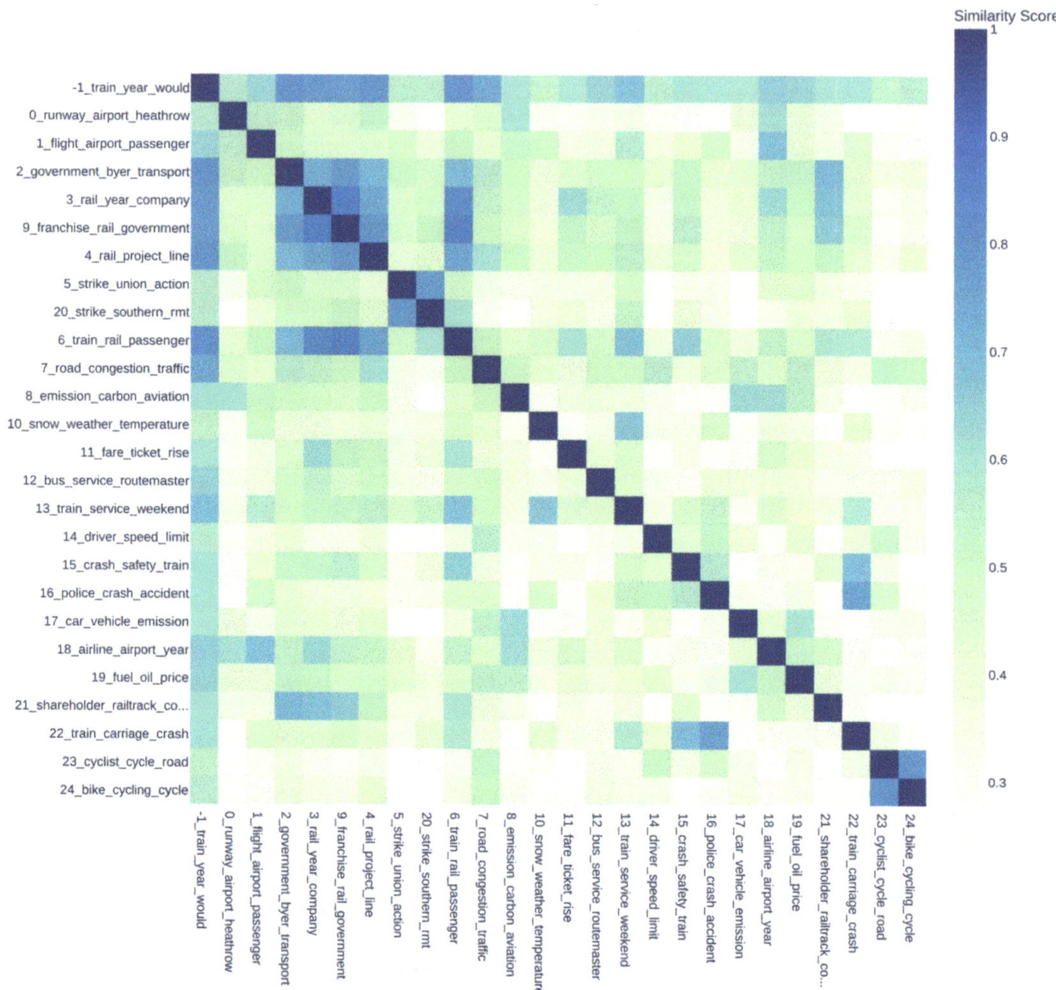

Figure 11. Similarity matrix (*The Guardian*).

4.3. Road Transport

The macro-parameter road transport includes the following parameters: congestion and road charging; pollution and electric vehicles, fuel and SCM (supply chain management), cycling, cycling safety, and bus transport.

4.3.1. Congestion and Road Charging

The congestion and road charging parameter concerns road congestion and the imposed congestion charging to address the congestion. It is represented by keywords (detected by our model) such as road, congestion, traffic, charge, scheme, car, and city. Looking at the news articles that belong to this parameter we were able to find a number of topics that capture various dimensions of this parameter. These include congestion charging [106], traffic reduction and management [107], smart roads [108], parking [109], walking [110], cycling [111], congestion charge for non-residents [112], e-scooters [113], etc. For example, Harvey reports in a *Guardian* article [114] that the traffic congestion levels in September 2020 in outer London increased above the prepandemic (COVID-19) lockdown levels in 2019. The article also shows the impact of congestion charging on outside central London and central London traffic between prepandemic, 2019, traffic and 2020.

4.3.2. Pollution and Electric Vehicles

The pollution and electric vehicles parameter captures various dimensions of transportation from *The Guardian*. These dimensions and related news include high air pollution and fears of high risks for COVID-19 infection [115], London being the worst city in Europe in terms of the damages to health due to air pollution [24], inadequacy of electric vehicle reforms alone in solving the pollution problem and the need for holistic solutions [116], the proposed increase in diesel and petrol vehicle prices to curb pollution [117], UK cities delaying creating clean air zones supposedly for COVID-19 excuses [118], the fall in air pollution levels in London by 50 percent through anti-pollution measures reported in April 2020 [119], the UK to introduce E10, a high-ethanol fuel, to cut pollution [120], the UK's plans to ban diesel and petrol vehicles by 2035 or even earlier [121], Bristol's plan (late 2019) to ban diesel vehicles [122], Oxford's plan (late 2017) to become the first zero emissions area in the world [123], charging station deserts and monopoly [124], opening of first all-electric vehicle charging station [125], Tesla struggling in the US, asking funds from government [126], a 2008 article on myths about renewable energy [127], the concerns that despite electric and hybrid car sales the emission gains are only 1% between 2011 and 2021 [128], and many more news and topics. The parameter includes the following keywords detected by our model: car, vehicle, emission, diesel, pollution, fuel, electric, petrol, carbon, and hydrogen.

4.3.3. Fuel and SCM (Supply Chain Management)

The fuel and SCM (Supply Chain Management) parameter contains keywords fuel, oil, price, petrol, driver, duty, tax, government, shortage, car, and rise. Many news articles in this parameter are about fuel prices, shortages, crises, [31,129], and costs of supply chains [130–132]. For example, a *Guardian* article [133] dated 17 November 2021 discussed the rising costs affecting all streams of businesses featuring case studies on agriculture, farming, hospitality, transport (individuals, small and large businesses), manufacturing, and retail. We found in this parameter a fascinating article on just-in-time supply chains by Kim Moody [134], Moody writes, "Decades of deregulation, privatisation and market worship have left society vulnerable to the unbidden force of "just-in-time" supply chains. No amount of government subsidies, ... will be enough to address the crises we face, from the pandemic to climate breakdown, ... Now is the time to think about not just how we make and consume things, but also how we move them ". Moody discusses how we became used to a 'just-in-time world', while not appreciating the complexity of it, including cross-continent shipping, fuel price variations, floods, closed roads, last-mile delivery people and their well-being, and the most important, the triple bottom-line effects of all of it.

4.3.4. Cycling

The cycling parameter captures the transportation issues associated with cycling and bikes, an issue that has become important due to climate and health. It includes the following keywords: bike, cycling, cycle, cyclist, city, ride, road, bicycle, route, lane, scheme, year, car, traffic, transport, way, safe, work, street, and day. The parameter discloses several important dimensions of cycling through *The Guardian* news articles, including the planned 5000 miles of dedicated cycle routes in the UK announced in June 2000 and supported by the charity Sustrans [135], barring charity cyclists from using new trains [136], the increase in the number of bikes and rise of the 'born-again bikers' in the UK, covered in February 2004 [137], ministers setting examples for bike usage in 2004 [138], a major rise of weekend cycling in the UK [139], the loss of a legal case made by cycling and rambling campaigners to prohibit vehicle driving in the Lake District [140], funding to nurture increased walking and cycling in the UK [141], cycle thefts [142], the rise of cycling holidays [143], a new 500-mile cycle route network in West Midlands, UK [144], cycling with young children [145], Uber launching electric bikes for hire in Islington borough [146], and more.

4.3.5. Cycling Safety

The parameter cycling safety is about the risks and safety of cycling due to vehicles and other hazards on the road. Our model detected the following keywords for the parameter: cyclist, cycle, road, bike, cycling, death, kill, pedestrian, traffic, ride, safety, accident, driver, year, lorry, safe, helmet, injury, lane, and number. This parameter captured some issues related to road accidents in general, from the early 2000s such as higher accident risks for children from deprived areas [147] and the use of explicit graphic accident images in ads to nurture road safety [148], but the parameter was dominated by road risks and safety for cyclists. Examples include the increase in cyclist deaths in the UK in 2020 [149]; advice on keeping safe while cycling [150]; cheaper insurance for drivers who take cycle training to improve cyclist safety [151]; concerns that pavements are being used for cycle stands and other purposes, causing problems for pedestrians [152]; the possibility that Great Britain could become a great cycling nation, but road safety is a hurdle [39]; and more.

4.3.6. Bus Transport

The parameter bus transport is represented by keywords including bus, service, route, local, public, passenger, and operator. The parameter captures bus transport issues in the UK, though most of these are applicable to other countries in one way or another. The dimensions and issues of bus transport include schemes from the government to provide cheaper bus fares for pensioners introduced in August 2000 [153]; government-hired pensioners in 2000 to go undercover and check bus service quality [154]; the proposals in late 2002 to scrap cheaper fares for pensioners and instead provide for jobless and students [155]; better employment conditions for bus drivers [156]; a bus strike in London in January 2015 and its effects on commuters [157]; an article from 2019 discussing the importance for integrated public transport across the UK while allowing city mayors to have freedom for local transport operations [158]; the launch of buses in the UK in 2020 that filter air ("air-filtering buses") while in operation around a city [159]; the need for security for bus drivers against coronavirus infection [160], a boost in electric buses in the UK with a GBP 20 million contract [161]; the behaviour of passengers towards physical distancing measures deteriorating as people get vaccinated [162]; a 2021 news report discussing the government's plans to introduce a major redesign of public transport with new bus lanes, new fare plans, and richer bus schedules [163]; changes in commuting patterns due to COVID-19 [164]; compulsory masks on tranport in London [165]; severely negative effects of privatisation of bus services outside London [166]; the effect of the COVID-19 pandemic on setting back public attitudes by two decades regarding giving preference to private cars over public transport for safety reasons [167]; an article discussing the downfall of British public transport services by bus privatisation [168]; and more.

4.4. Rail Transport

The macro-parameter rail transport defines the characteristics of the transportation sector that relate to trains and railways, as captured by the topic modelling of *The Guardian* dataset. Rail transport includes the parameters, public discontentment; rail fairs; funding, costs, and fares; industry and privatisation; rail projects and contracts; and governance and politics.

4.4.1. Public Discontentment

The public discontentment parameter is represented by keywords train, rail, passenger, service, company, network, year, railtrack, timetable, and railway. The overarching theme of the news articles in this parameter is the state of public discontentment with the rail services in the UK. The range of issues that the public is discontented with includes train delays, particularly due to ineffective train schedules. For example, *The Guardian* reported on 18 May 2019 that a new rail timetable was announced by the Rail Chief, UK, to improve the chaotic situation with the rail services in the UK due to many cancellations and delays in train services during the last year, 2018–2019 [19]. A couple of months later, people again encountered severe delays and cancellations affecting the train schedule due to overhead wire damage in July 2019 on the mainlines connecting London with Scotland, northeast England, and other regions [169]. A revised rail timetable was developed and put in place in late 2019 to enhance the rail services promised to be the biggest change in the UK train schedule, but the plans were ruined reportedly due to staff shortages [170]. These and similar incidents caused public upheaval and discussions on train delays schedules around the UK.

Other issues of discontent include poor train accessibility [171], delay in project completions [172,173], dissatisfaction with specific train service providers [174,175], discontent of company staff with their management [176], companies trying to win back customer satisfaction [177], change of management due to discontentment with services [178], and more.

4.4.2. Rail Fares

The rail fares parameter is depicted by keywords such as fare, ticket, rail, season, price, commuter, cheap, and peak. The parameter includes issues such as EasyJet in fare wars with Virgin trains [179], the withdrawal of cheaper fares amid train delays and cancellations [180], denial of compensation subsequent to the Hatfield crash in 2000 for those who did not keep their tickets [181], a planned increase in rail fares in England reported in 2020 [182], rail fares to increase by 3.8% in March 2022 reported in December 2021 [183], the launch of budget rail London–Edinburgh announced in September 2021 [184], postpandemic flexible rail season tickets [185], and more.

4.4.3. Funding, Costs, and Revenues

The train funding, costs, and fares parameter includes the keywords rail, company, network, profit, government, cost, fare, rise, revenue, public, and others. One of the news articles in this parameter, dated 5 December 2021, discusses the hefty budget cuts required by the UK government from train operators who are contracted to deliver train services for a fixed price while the revenues and risks are born by the government [186]. While the consequences of the pandemic on train travel patterns are obvious and being explored, some groups argue that it is critical to maintain services and cut costs to attract passengers and save taxpayer money. There are other political and public issues, including job cuts that harm some segments of the public. At the same time, it is necessary to reduce costs and improve public services. Other articles touch on a range of dimensions and issues of this parameter including penalties and cutting bonuses [187], bailouts [188], increase in demands and revenues [189–191], funding and funding gaps [192], and more.

4.4.4. Industry and Privatisation

The industry and privatisation parameter is characterised by keywords such as shareholder, railtrack, company, eurotunnel, buyer, debt, government, tunnel, share, and investor. The parameter captures transportation dimensions surrounding governance, privatisation, and industry, mainly for rail transportation. The earliest article [193] in this parameter dates back to 7 February 1964 about the agreement on the Channel Tunnel between the French and British governments seen as "a sound investment". The Channel Tunnel as we know was opened in 1994. We then witness a news article [194] from 1999 opposing the cabinet view on partly privatising the public transport system in the UK due to its weaknesses. We also see an article [195] from December 2000 deliberating comments from a chief executive officer of Atkins who was a major stakeholder in two London Underground bids that "the public does not appreciate the benefits brought to the railways by the private sector". These and similar issues [196] show the debates around and pros and cons of privatisation versus government-owned services. Another issue or dimensions that we can learn from our

BERT model is the legal battles between companies such as the one reported in a *Guardian* article [197] from 2001 about the company Virgin planning to sue the company Railtrack for their losses due to the Hatfield train crash [198]. The legal battles between companies also extend to the leadership of a company being offered a job by another company such as reported in December 2009 by Dan Milmo that the chief of Tube Lines was offered a position at National Express [199]. There is also an article from September 2020 reporting the former transport secretary being offered a lucrative contract by Hutchison Ports [200,201]. Other news and dimensions of this parameter include the Stagecoach offer in 2009 to its rival National Express for a merger [202], the Channel Tunnel operator Eurotunnel's hope in 2007 for "investors to back a debt-for-equity swap" to save it from bankruptcy [203], the problems with the public and private sector working together such as the London Underground public–private partnership (PPP) and East Coast Rail [204], the rail and bus company FirstGroup rejecting a takeover bid from the American company Apollo [205], the postBrexit rebranding of Eurotunnel to Getlink [206], and more.

4.4.5. Rail Projects and Contracts

The rail projects and contracts parameter was created by merging two clusters (numbers nine and four) because the two clusters contained keywords pointing to similar subjects. The parameter is represented by keywords franchise, rail, government, train, contract, railway, company, service, bid, plan, rail, project, line, transport, north, high, speed, government, plan, and route. *The Guardian* confirmed on 18 November 2021 [207] that the eastern link of HS2 connecting Leeds was abandoned by the government, and this caused fury among the affected segments of the public.

4.4.6. Governance and Politics

The governance and politics parameter represents the government's decision or plan-related keywords including government, buyer, transport, labour, public, minister, private, decision, political, privatisation, and secretary. For example, *The Guardian* reported on 14 November 2021 [23] that the government dropped the plan for HS2 and instead decided to support projects that benefit the ruling party. HS2 was reportedly promised by the prime minister during the very early days of his job. It was expected to provide a new high-speed railway link serving as the foundation of Britain's transportation network.

4.5. Air Transport

The macro-parameter air transport includes the parameters airport expansions, air pollution, airport security, and air costs and fares.

4.5.1. Airport Expansions

The airport expansions parameter is about expansions planned for airports and related facilities that are needed to meet the increasing demands for air travel [208] as well as about the opposition to expansions due to their negative impacts on climate [209,210]. This parameter includes the keywords runway, airport, Heathrow, expansion, government, aviation, decision, flight, build, and plan. For example, the matter of London Heathrow Airport expansion and construction of a third runway has remained a matter of discussion for many years. The project was approved, but climate activists challenging it, leading to the issue becoming bogged down in the courts [211]. Asthana, Laville, and Kale in a *Guardian* news item [212] discussed the Court of Appeal decision announced in March 2020 to deem the expansion unlawful due to the UK government's failure of not considering the climate impacts of the expansion. This topic has continued to remain in the news due to the airport trying to challenge the court decision [213]. In addition, Tim Crosland, the lawyer and a campaigner for environmental protection was found guilty (May 2021) by the supreme court and lost his appeal (December 2021) for disclosing the court decision before its official announcement to the public [214].

4.5.2. Air Pollution

The air pollution parameter contains the following keywords: emission, carbon, aviation, climate, airline, fuel, environmental, biofuel, and others. The parameter relates to air pollution caused by air transport. For example, Ungoed-Thomas from *The Guardian* wrote in a news item [215] about the high number of flights being taken by UK government staff (293 every day according to a report) despite the UK government's promises to protect the climate and make the government greener.

4.5.3. Airport Security

The airport security parameter is represented by keywords such as flight, airport, passenger, drone, and security. This parameter is exemplified in a news March 2008 article by Dodd and Milmo [30] reporting an incident of a breach, the second within a three-week period, where a man succeeded in reaching a runway at Heathrow airport.

4.5.4. Air Costs and Fares

The air costs and fares parameter represents the transportation characteristics connected to the costs and fees incurred by air transportation providers and consumers. The keywords include airline, airport, flight, passenger, carrier, cost, price, business, market, and profit. An example of various issues that come under this parameter is a *Guardian* news article reported by Topham and Kollewe [26] on 19 October 2021. The news is about Heathrow airport potentially increasing charges for passengers by 56 percent by 2023. Topham and Kollewe explained that Heathrow will be permitted by the CAA, the Civil Aviation Authority, to raise the landing charges considerably from the summer of 2022. This was in response to the airport organisation that asked for doubling the charges due to the massive business losses caused by the dearth of airport activity during the COVID-19 pandemic. CAA explained that the permission to increase charges was necessary for keeping the airport competitive and safe. The airlines are affected by the decision as the costs for their operations will increase. The news shows the complexity of the parameter in terms of the different stakeholders (airport management, airline operators, CAA, and consumers) and changing times and situations.

4.6. *Crash and Safety*

The macro-parameter crash and safety includes three parameters: train crashes, accidents and deaths, and dangerous driving and speeding.

4.6.1. Train Crashes

The keywords that represent the parameter train crashes are crash, safety, train, railtrack, rail, signal, and accident. The earliest *Guardian* article we found in this parameter dates back to one from 6 October 1999 about the worst crash of the decade between Great Western and Thames trains near Paddington in London [216] making safety of rail transport a major political issue [217], making the two train operators, Railtrack, and the government, to begin an inquiry into the crash [218], and government pledging GBP 1 billion for safety of rails [219]. This has further led to the possibility of manslaughter charges against Thames Trains and Railtrack [220]. Railtrack, a group of companies, owned a major part of the rail infrastructure in the UK from 1994. It was renationalised in 2002. Many other news items were found relating to train accidents such as the rail accident between two trains at Salisbury in November 2021 caused potentially by low adhesion between rail tracks and train wheels [221].

There have also been many news items from *The Guardian* in this parameter about losses to rail companies due to accidents, compensations, penalties, etc., [222]. The parameter also contained some articles related to rail suicides such as the article from November 2017 about urging commuters to indulge in small talk with people potentially attempting suicides [223]. It was reported in this article that about 273 people committed suicide on the railways in the

UK during 2016–17. The parameter and the contained news articles show the richness of information that can be extracted from our BERT-based modelling approach.

4.6.2. Accidents and Deaths

The accidents and deaths parameter is represented by keywords such as police, crash, accident, incident, scene, woman, injure, die, man, and injury. This parameter contains news articles about deaths and road accidents as opposed to the parameter train crashes where the focus of the articles is on train crashes and the various issues surrounding them such as financial, political, investigative, and industrial issues. Moreover, while this parameter mainly contains articles about roads, we also found some articles that involved trains such as a death (potentially a murder) by a woman pushing another person in front of a train [20]. Another example in this parameter showing the focus on deaths rather than the mode of transportation is news from October 2000 about the history of train accidents in the UK [224]. The article focuses on injuries and deaths rather than other details, and this is the reason we believe this article, though also related to train crashes, is mainly associated by our BERT model to the accidents and deaths parameter. Another example is an article about the death of a woman who was a staff member in a railway ticket office. She died because of COVID-19 infection that she may have caught due to a man claiming to have COVID-19 who spat and coughed on her while she was on duty [225]. The news is related to rail transport but is about a death. Other examples of articles in this parameter that involve railways and trains (or even air transport) but are mainly about deaths, road transport, and vehicles include [226–233].

The dimensions and issues connected to this parameter as seen through the news articles include the UK government strategy for road safety highlighting the gravity of the matter due to over 0.3 million road casualties in the UK every year (1 March 2000; [234]), the release of the driver of the bus that crashed and killed two and injured dozens of people (5 January 2007, [235]); death and injuries of various people in different incidents due to cold, black ice, road death traps, etc. (8 February 2009, [236], 31 March 2010, [237]); the M5 crashes in November 2011 [238] and March 2012 [239]; the M1 crash in December 2012 and its investigations [240]; the death of a man due to collision with a Nottingham tram (16 August 2016, [241]); the rescue of 60 children from a bus operated by Stagecoach after its crash (11 November 2021, [242]); a woman killed due to the collision of two buses near Victoria Station, London (10 August 2021, [243]), and many more.

One of the issues discovered from this parameter is the deaths on and the safety concerns of smart motorways in the UK [244,245]. This topic of smart motorways was also detected in Parameters 7 and 14 in relation to congestion reduction and speeding, respectively.

The discussions in this article are supported by a large number of articles for the discovered parameters. These may be seen as unnecessary or of little or no benefit. We discuss a large number and range of articles to show the complexity and breadth of the parameter topics. The knowledge gained through the parameter discovery and analysis process that is currently partly automatic and partly manual and will become increasingly automatic and autonomous will allow autonomous modelling, (exploratory, dynamic, and real-time) analysis, and optimisation of transportation and other sectors. The discussions presented in this article are also helpful in understanding the working and performance of BERT and other clustering algorithms.

4.6.3. Dangerous Driving and Speeding

The dangerous driving and speeding parameter is characterised by drunk, dozing, and other dangerous driving, speeding, speed limits, methods to measure and curb dangerous driving and their devastating effects, and penalties and legal punishments. The first article in this parameter is dated 1 March 2000 and is about the government pledging to introduce tougher measures for drunk-driving and speeding to reduce child pedestrian deaths and other injuries, while road safety and environmental protection groups show dismay, criticising the government for giving in to the motoring lobby [246].

The dimensions and issues related to this parameter include, among others, efforts by the government to intervene and improve dangerous driving behaviour [247]; government caving in to different lobbies, including motoring and alcohol lobbies [248]; dozing drivers causing deaths and their legal punishments [249]; the use of virtual reality in driving tests [250]; dangerous and drunk drivers and their legal punishments [251]; devices that would not let drunk driver start the vehicle [252]; drunk police officers [253]; uninsured drivers [254]; speed cameras and privacy [255]; the law being soft on dangerous and drunk drivers [256]; the benefits of lower speed limits to air quality and the environment [257]; penalties and jails for drivers using mobile phones [258]; illegal use of devices to deceive speed measuring equipment [259]; improvements to driving tests to improve driving behaviour of young people [260]; the benefits of autonomous cars to free us from dangerous drivers [261]; shocking driving speed violations during the COVID-19 lockdown [262]; and more.

4.7. Disruptions and Causes

The macro-parameter disruptions and causes comprises two parameters: extreme weather impacts and disruptions and delays.

4.7.1. Extreme Weather Impacts

The extreme weather impacts parameter captures the various impacts on transportation of extreme weathers such as snow, rain, floods, heat, and wind-storms. The keywords detected by our BERT model for this parameter include snow, weather, temperature, flood, road, cold, rain, heavy, condition, and wind. The issues and dimensions for this parameter as evidenced through *Guardian* news articles include, among others, ice bombs ("frozen effluent falling from planes") [263]; impact on rail transport causing delays, cancellations, accidents, deaths, injuries, financial losses, and more [264]; magic de-icer to help railways in applying timely brakes [265]; effects of snow on roads [266]; heaviest snow in 18 years and its effects [267]; government rejecting criticism over transport management during extreme weathers [268]; resignation of transport minister over snow chaos [269]; strong winds, snow, and floods beat up the country and bring it to a halt [270]; extreme weather effects on air travel [271]; weather impact on schooling [272]; travel chaos in the country [273]; storm Darcy, cold and snow to cause disruptions [274]; weather impacts on Christmas and its arrangements [275]; deaths due to storms [276]; weather impacts on rail repairs [277]; derailing of a train due to rain and landslide [17]; village evacuation due to extreme weather [278]; government advice to businesses not to penalise staff for following government snow advice [279]; travel chaos due to rain and high temperatures [280]; the inability of UK rail transport to deal with extreme climates and a call for investments [281]; damages to bridges due to flooding [282]; and more.

4.7.2. Disruptions and Delays

The disruptions and delays parameter contains the following keywords detected by our BERT model: train, service, weekend, holiday, passenger, expect, delay, busy, work, station, line, travel, day, fire, weather, rail, disruption, run, and traffic.

The earliest article in this parameter is from 19 November 1987 about a fire at King's Cross underground train station in London. This shows travel and other disruptions caused by the fire. The dimensions and issues related to travel disruptions include, among others, closure of many stations in London underground due to coronavirus [283], disruptions due to peak-hour services cancellations [18]; a warning for people to plan their travel due to expected heavy traffic from bank holiday getaway travellers amid expected fine weather [284]; bridge failure causing disruptions [285]; advice to avoid travel due to rail works [286]; rail services disrupted by lightening strikes [287] arson at a train station [288]; getaways for Easter expected to cause traffic at motorways [289]; a leaf clearing operation by Network Rail to reduce rail accident risks [290]; disruptions in Christmas Eve travels due to engineering problems and weather [291]; disruptions due to Notting Hill carnival [292]; heavy road traffic and delays due to rail closures [293]; disruptions and delays due to London 2012 Olympics [294]; disruptions due to tunnel falls in London Underground [295]; crowded airports and rail stations and congested roads due to school holidays and good weather [296]; and many more.

Considering the keywords and news articles in this parameter, we can say that this parameter is about travel disruptions, delays and their causes. The causes include accidents, fires, both bad and good weather, faults, repairs and new installations in transport infrastructures, and holidays, including bank holidays, Easter, and Christmas.

4.8. Employment Rights, Disputes, and Strikes

This macro-parameter has only one parameter which was created by merging two clusters, numbers 5 and 20. The parameter employment rights, disputes, and strikes captures information about employees, their rights, job cuts, employment conditions, disputes with the management, and union strikes and their impact on people and economy. Two document clusters were detected with similar keywords and articles, the cluster numbers 5 and 20, and therefore we merged them into a single parameter. We noted some difference in the two clusters with cluster number 5 containing more articles related to rail transport and cluster number 20 a bit inclined towards air transport. We also consider this parameter as a macro-parameter because of its vast impact on social, economic, and environmental sustainability. There are always apparent exceptions in the cluster documents such as this article [297] that is about cancellation of trains due to Covid but primarily belongs to the parameter employment rights, disputes, and strikes; however, on a close inspection, one can find the connection such as the mention of strikes in the aforementioned article.

The earliest article reported by *The Guardian* related to this parameter is on 14 October 1999 about rail guards voting to go for a strike over safety matters [298]. We see matters related to job cuts such as British Airways announcing on 7 December 2000 to cut 1000 jobs at the Gatwick airport [299]. Among the news related to employment rights, union disputes, and strikes we find articles including one about a dispute between the Amalgamated Engineering and Electrical Union (AEEU) and Virgin Atlantic reported on 29 December 2000 [300], the dispute between RMT (National Union of Rail, Maritime, and Transport Workers) and the government (precisely, TfL, Transport for London) over the work rosters threatening to go for a strike from 26 November 2021 [301], train drivers threatening a major rail strike in London rail in January 2000 subsequent to rail privatisation [302], British Airways employees' strike for disputes regarding salaries reported in June 2017 [303], a strike stretching multiple weeks during March 2019 by French customs over poor working conditions causing havoc to Eurostar trains [304], Yodel employees threatening to strike in September 2021 due to poor salaries and conditions causing potential disruptions to deliveries for major supermarkets and others [305], Stagecoach under threat of strike by drivers'low wages [306], a recent article (October 2021) on post-COVID-19 abuse of staff working at transport stations and other customer-facing staff by customers [307], and more.

These examples of different types of employment disputes and strikes reveal insights into a range of issues surrounding stakeholders, causes, and impacts of disputes and strikes.

4.9. Temporal Analysis (The Guardian)

In this section, we will analyse how the parameters have grown over time. Figure 12 displays the temporal progression of the parameters which are distributed into six subfigures. The vertical line of the graph indicates the number of articles which is defined as the intensity and the horizontal line indicates the years. Figure 12a depicts the temporal progression of the macro-parameter road transport. Fuel and SCM has a higher intensity compared to the others. Figure 12b illustrates the temporal progression of macro-parameter rail transport, where the rail projects and contracts and industry and privatisation parameter were started in 1960. After that, both parameters were highly discussed between 2000 and 2005. The intensity of articles for the macro-parameter air transport which includes four parameters is depicted in Figure 12c. Air pollution and air airport expansions both had a peak value of around 80 between 2007 and 2008. The temporal progression of the macro-parameter crash and safety which includes three parameters is shown in Figure 12d. We observed that there are more articles related to train crashes compared to others. Figure 12e displays the temporal progression of the macro-parameter disruptions and causes. The parameter extreme weather impacts was highly discussed in 2010 and had the highest peak value of 60. The temporal progression of the macro-parameters employment rights, disputes, and strikes, is shown in Figure 12f, where the highest peak value of intensity was more than 80 in 2010.

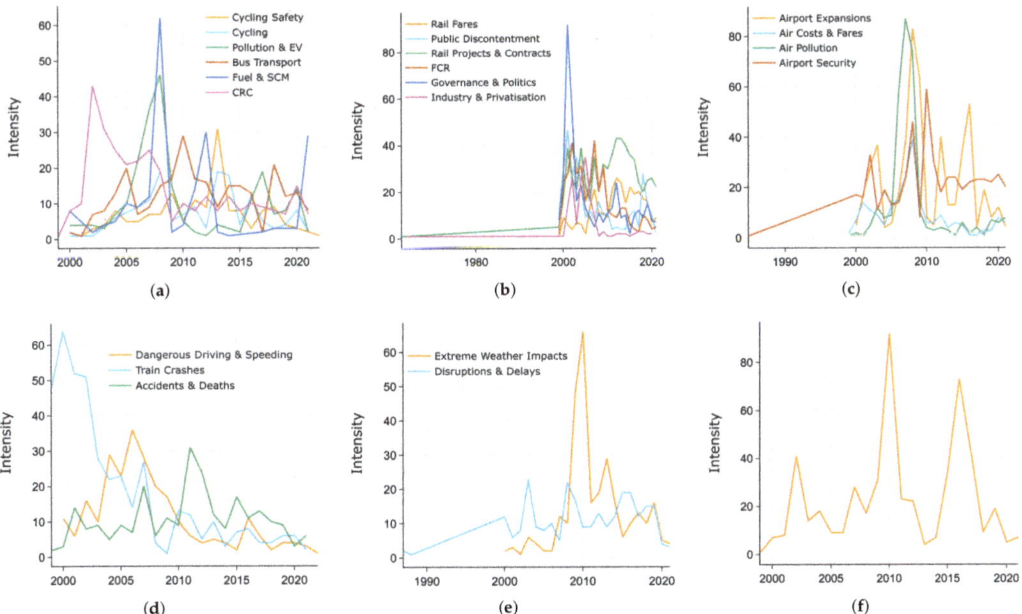

Figure 12. Temporal progression of parameters (*The Guardian*): (**a**) road transport; (**b**) rail transport; (**c**) air transport; (**d**) crash and safety; (**e**) disruptions and causes; (**f**) employment rights; disputes; and strikes.

The temporal progression of all macro-parameters is summarised in Figure 13. For the first time, rail transport was discussed in 1960. After 2000, the parameter was highly concerned and topics for discussion had the highest peak value of 225. In 2008, the macro-parameter air transport had the highest peak value. We also saw in 2020 that the macro-parameters road transport, rail transport, and air transport were equally discussed. The macro-parameters crash and safety, disruptions and causes, and employment rights, disputes, and strikes were also of equal concern in 2020.

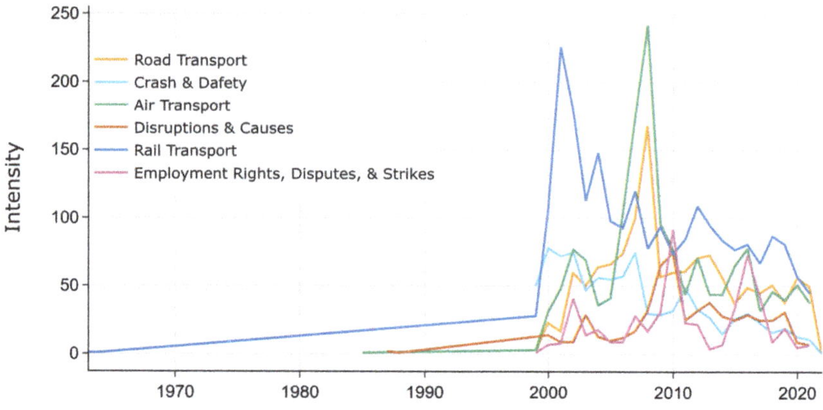

Figure 13. Aggregated macro-parameters (*The Guardian*).

5. Industry: Transportation Parameters Discovery

In this section, we discuss the parameters detected by our BERT model from the dataset acquired from the *Traffic Technology International (TTI)* magazine. The parameters are grouped into five macro-parameters. We provide an overview of the parameters and macro-parameters in Section 5.1. The quantitative analysis is discussed in Section 5.2. Subsequently, we discuss each macro-parameter in separate sections, Sections 5.3–5.7. Section 5.8 discusses the temporal analysis of the parameters and macro-parameters.

5.1. Overview and Taxonomy (Traffic Technology International Magazine)

We detected a total of 15 parameters from the *TTI* dataset using BERT. These 15 parameters were grouped into 5 macro-parameters using the domain knowledge together with a similarity matrix, hierarchical clustering, and other quantitative analysis methods. The methodology and process of the discovery of parameters and their groupings into macro-parameters have already been discussed in Section 3.

Table 2 lists the parameters and macro-parameters of the transportation magazine, *TTI*. The parameters are categorised into five macro-parameters, including industry, innovation, and leadership, autonomous and connected vehicles, sustainability, mobility services, and infrastructure (Column 1). The second and third columns list the parameters and the cluster number, respectively. The fourth column lists the proportion of the total number of articles. Our BERT model identified 58.16% of the articles as having outlier clusters. As a result of excluding this cluster, the remaining 41.84% of articles are listed in the fourth column. The top keywords related to each parameter are represented in the fifth column.

Figure 14 provides a taxonomy of the transportation domain extracted from a transportation industry-focused technical magazine. The taxonomy was created using the parameters and macro-parameters discovered from the *TTI* magazine. The first-level branches show the macro-parameters, the second-level branches show the discovered parameters, and the third-level branches show the most representative keywords.

Table 2. Parameter and Macro-Parameters for Transportation (Source: Traffic Technology International).

Macro-Parameters	Parameters	No.	%	Keywords
Industry, Innovation and Leadership	Leadership	0	8.05	technology, mobility, vehicle, transportation, event, transport, traffic, city, system, service, work, future, world, infrastructure, road, public, time, datum, smart, industry
	Innovation	4	2.60	award, project, winner, competition, transportation, traffic, win, system, tsmo, technology, road, team, category, solution, work, vehicle, industry, transport, safety, innovation
Autonomous and Connected Vehicles	AV Systems	1	4.41	vehicle, car, autonomous, system, drive, technology, datum, map, driving, driver, road, sensor, automate, company, software, autonomous vehicle, automotive, platform, self, time
	AV Trials	8	2.18	vehicle, autonomous, technology, test, drive, trial, first, driverless, autonomous vehicle, driving, testing, self, driver, road, car, system, public, company, automate, shuttle
	V2X Trials	9	1.97	australian, transport, vehicle, technology, road, government, trial, system, industry, project, infrastructure, world, safety, cohda, first, future, provide, state, automate, city
	Platooning	12	1.69	truck, driver, vehicle, platoone, technology, system, parking, drive, freight, commercial, road, truck parking, project, truck driver, trucking, company, fuel, highway, autonomous, safety
Infrastructure	Road Infrast.	2	4.10	project, lane, road, motorway, traffic, construction, bridge, improve, scheme, work, design, tunnel, junction, improvement, highway, route, mile, time, provide, reduce
	Crash and Safety	13	1.58	road, death, pedestrian, crash, safety, fatality, injury, speed, increase, report, vehicle, reduce, number, traffic, safe, kill, serious, driver, high, state
	Tolling	10	1.95	toll, system, tolling, lane, toll collection, collection, electronic, bridge, tolling system, project, transponder, contract, customer, transcore, express, road, vehicle, company, state, collection system
	ALPR	14	1.50	camera, video, system, traffic, surveillance, high, vehicle, application, technology, provide, solution, detection, plate, analytic, enforcement, range, datum, feature, capture, view
Mobility Services	Transport Services	3	3.01	system, traffic, vehicle, german, service, datum, software, mobility, company, technology, city, road, base, time, speed, transport, parking, project, driver, solution
	Parking Services	6	2.31	parking, app, system, driver, space, car, time, service, city, payment, parking space, park, available, information, smartphone, user, street, vehicle, provide, traffic
Sustainability	Air Quality and Pollution	5	2.35	air, emission, pollution, air quality, quality, vehicle, reduce, air pollution, transport, clean, city, government, local, earthsense, charge, work, help, electric, road, public
	Street Lighting	7	2.27	city, smart, smart city, energy, datum, lighting, technology, traffic, system, project, street, urban, provide, solution, sensor, light, time, help, streetlight, network
	Electric Vehicles	11	1.89	bus, passenger, system, electric, vehicle, technology, transport, public, emission, service, city, electric bus, operator, stop, route, school, help, time, bus lane, first

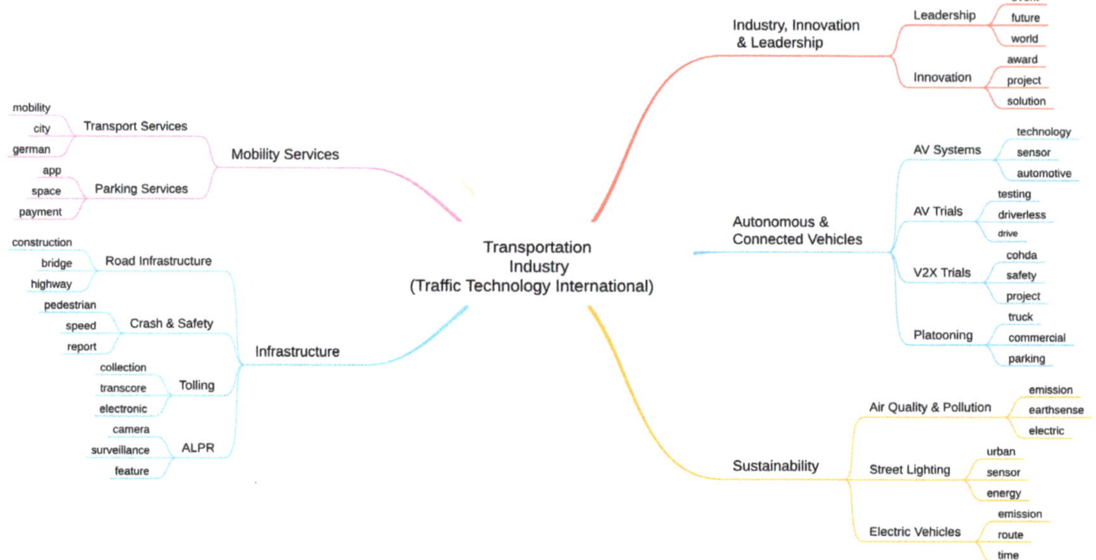

Figure 14. Taxonomy extracted from *Traffic Technology International* Magazine Dataset.

5.2. Quantitative Analysis (Traffic Technology International Magazine)

This section discuss the term score, word score, intertopic distance map, hierarchical clustering and similarity matrix.

Figure 15 shows that only the top 7 to 10 keywords in each parameter actually represent the parameter when we evaluate the keywords (see Section 3.8). Because the probabilities of all the other possibilities are so close to one another, their ranking becomes more or less meaningless. When we analysed the top keywords per parameter to discover the parameter, we used this information to focus on the top seven or so keywords in each parameter.

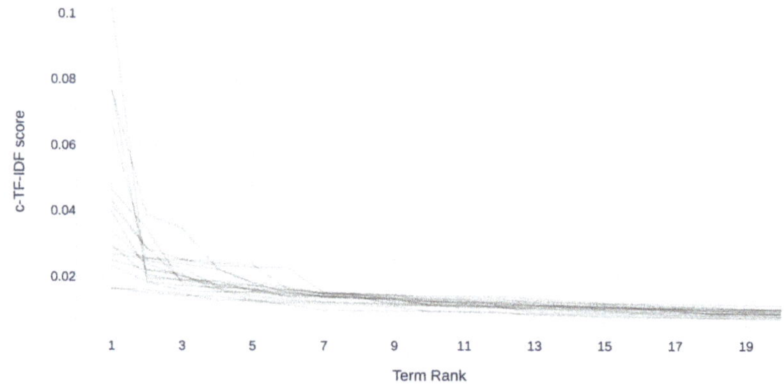

Figure 15. Term score (*Traffic Technology International* magazine).

Figure 16 depicts the top five keywords for each parameter. The importance score, or c-TF–IDF score, is used to order the keywords (see Section 3.8). There are 15 subfigures and in each subfigure, the horizontal line shows the importance score, and the vertical line shows the parameter keywords.

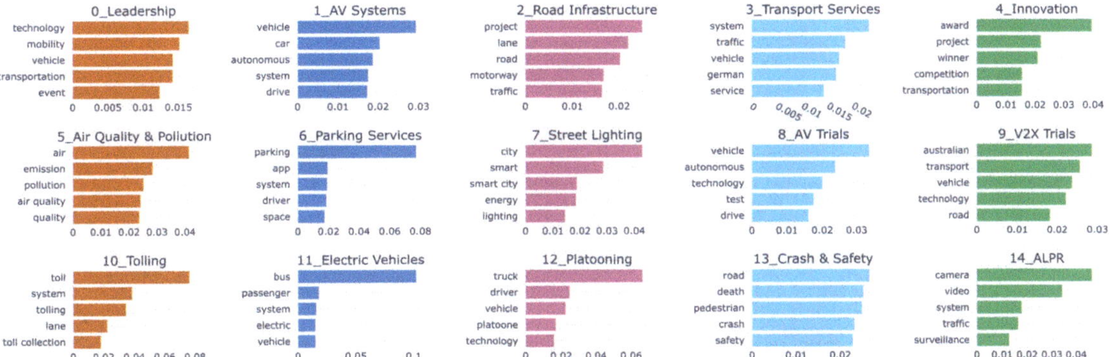

Figure 16. Magazine article parameter with keywords c-TF–IDF Score.

Figure 17 shows the intertopic distance map (see Section 3.8), where two clusters are clearly identified on the left–below corner side, and the right–upper side represents the three clusters. However, we manually tagged the parameters into five macro-parameters.

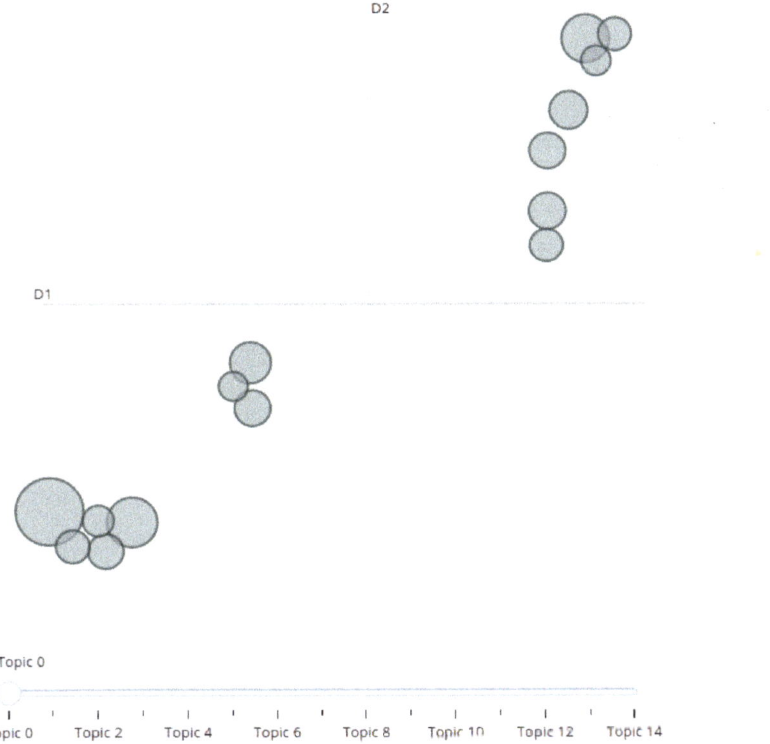

Figure 17. Intertopic distance map (*Traffic Technology International* Magazine).

Figure 18 represents the hierarchical clustering of the 15 clusters and systematically pairs the clusters based on the similarity matrix (see Section 3.8). We noticed that initially, the clusters were grouped into five clusters: (1, 8, 9, 0, 3), (13, 2, 4), (11, 5, 7), (10), and (14, 6, 12). This automated hierarchical clustering grouped the clusters correctly, with some exceptions. Furthermore, based on our knowledge and magazine articles, we manually grouped the clusters that are discussed in Table 2.

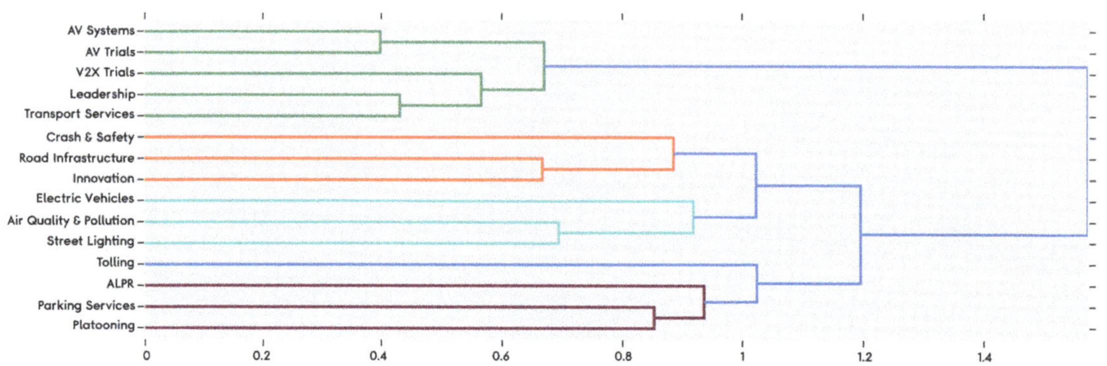

Figure 18. Hierarchical clustering (*Traffic Technology International* Magazine).

Figure 19 visualises the similarity matrix among the parameters (see Section 3.8). We used the same configuration as discussed in Figure 11. The dark blue colour represents the highest similarity relationship between parameters, while the light green represents the lowest similarity. For example, Cluster 3, labelled as transport services, and Cluster 1, labelled as AV systems, have high similarity scores as the main intention of AV systems is to improve transport services and make them smoother and more flexible. There is another high similarity between Clusters 8 and 9, which are labelled as AV trials and V2X trials, respectively.

5.3. Industry, Innovation, and Leadership

The macro-parameter industry, innovation, and leadership includes two parameters: leadership, and competitive innovation, which reveals events, appointments, innovations, and awards-related information and topics. The leadership parameter captures the transportation events and leadership news that have a high impact on transportation development. For example, the appointment of Angelos Amditis as ITS Europe chairman in 2018 [308], Laura Chace as CEO and president of ITS America in 2021 [309], and Laura Shoaf as the chair of the UK's transport group in 2021 [310], and much more news and topics. The competitive innovation parameter is about new innovations and projects related to transportation. It includes the following keywords: award, project, winner, competition, solution, and so on.We discovered an award-related announcement that occurred in Florida, and the Woolpert was awarded for operating district-wide aerial photogrammetry and various surveys [311].

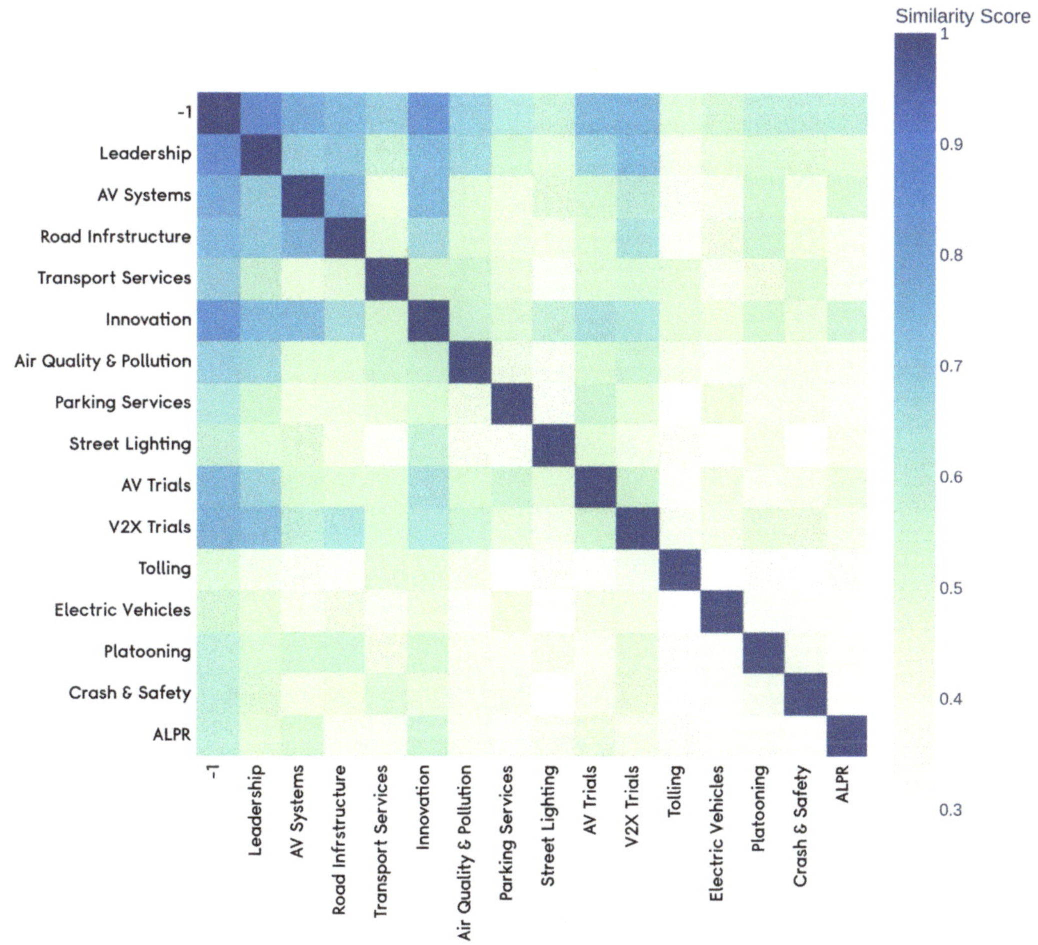

Figure 19. Similarity matrix (*Traffic Technology International* magazine).

5.4. Autonomous and Connected Vehicles Systems

The macro-parameter autonomous and connected vehicles systems includes autonomous vehicles (AV) systems, AV trials, vehicle-to-everything (V2X) trials, and platooning/truck platooning. The AV system is designed to develop an autonomous driving vehicle by using several technologies, sensors, GPS, etc. This parameter reveals several significant innovations and projects through *Traffic Technology International* magazine articles, including the first thermal sensor-equipped production for AV [312], the first use of blockchain to provide connected vehicle data by CyberCar [313], a thermal sensor technology for the AV system announced on November 5, 2019, by the Veoneer system [312], the first vehicle-to-cloud infrastructure for automated connected vehicles by SENSORIS [314], and so on. We found AV trials-related news that illustrates the AV trials by Singapore's Land Transport Authority [315].

Communication between a car and any component that can be affected by the car is referred to as V2X communication. Platooning is a technology that helps vehicles drive together and boosts road capacity by applying the automated highway system to reduce the distance between automobiles or trucks.

5.5. *Infrastructure*

The macro-parameter infrastructure includes road infrastructure, crash and safety, tolling, and ALPR (automatic licence plate recognition). The road infrastructure is represented by the following keywords: project, lane, road, motorway, traffic, construction, bridge, improvement, scheme, work, design, tunnel, junction, etc. We uncovered the following example related to this parameter: Highways England (HE) marked a turning point in road construction, encouraging better-planned roadworks and more consistent travel on motorways and key trunk routes [316]. The crash and safety is presented by road, death, pedestrian, crash, safety, fatality, injury, speed, etc. We noted that UK road deaths increased in 2016 [317]. Tolling is illustrated by toll, system, tolling, lane, toll collection, electronic, etc. We noticed that Canadas's A25 highway electronic tolling system was upgraded on 24 October 2017 [318]. ALPR is defined by the camera, video, system, traffic, surveillance, detection, plate, etc. The enforcement system employs over 120 Sicore ALPR cameras located at 80 locations along major arterial routes around the UK capital [319].

5.6. *Mobility Services*

The macro-parameter mobility services retains transport services and parking services. By applying our model, we found that the traffic enforcement system is one of the transport services [320]. The parking system is another solution to make transportation services more convenient. We found news that merged the parking payment solution and electric vehicle charging system in the UK [321].

5.7. *Sustainability / Sustainable Infrastructure*

The macro-parameter sustainability includes air pollution and quality, street lighting, and electric vehicles. Diminishing transport-sourced air pollution is one of the major concerns as the number of vehicles is dramatically increasing every day. We found the Wolverhampton project focused on diminishing air pollution and improving the air quality monitoring system [322].

Street lighting is one of the solutions for smart cities' public safety and traffic optimisation. For example, CityIQ Edge collects and processes street-level video and audio information that will enable urban areas to handle day-to-day problems [323]. To reduce carbon emissions and improve air quality, low emission electric vehicles are one of the best solutions. The following news article is an example of this parameter: the Department for Transport had contracted the UK's Transport Research Laboratory (TRL) to observe and evaluate the effectiveness and implications of low-emission buses at 13 sites around the country [324].

5.8. *Temporal Analysis (Traffic Technology International Magazine)*

In this section, we will analyze how the parameters have evolved over time. Figure 20 shows the temporal progression of the parameter which is distributed into six subfigures. The first five subfigures represent the temporal progression of five macro-parameters, whereas the last subfigure depicts the temporal progression of all macro-parameters. The vertical line of the graph indicates the number of articles which is defined as the intensity. The temporal progression of the macro-parameter industry, innovation, and leadership is depicted in Figure 20a. Leadership has a higher intensity compared to the innovation parameter. Figure 20b shows that the AV systems' intensity was increasing over time until 2017, but after that, the intensity declined. Figure 20c shows that the intensity of the road infrastructure parameter which is one of the components of the macro-parameter infrastructure was high in 2017 and then gradually decreased. The temporal progression of macro-parameter mobility services which includes two parameters, parking services and transport services is shown in Figure 20d. We observed that there are more articles related to transport services compared to parking services. The intensity of articles for the macro-parameter sustainability which includes three parameters: street lighting, air quality and pollution, and electric vehicles is depicted in Figure 20e. street lighting and air quality and pollution have both had the same peak value of 25 in 2017.

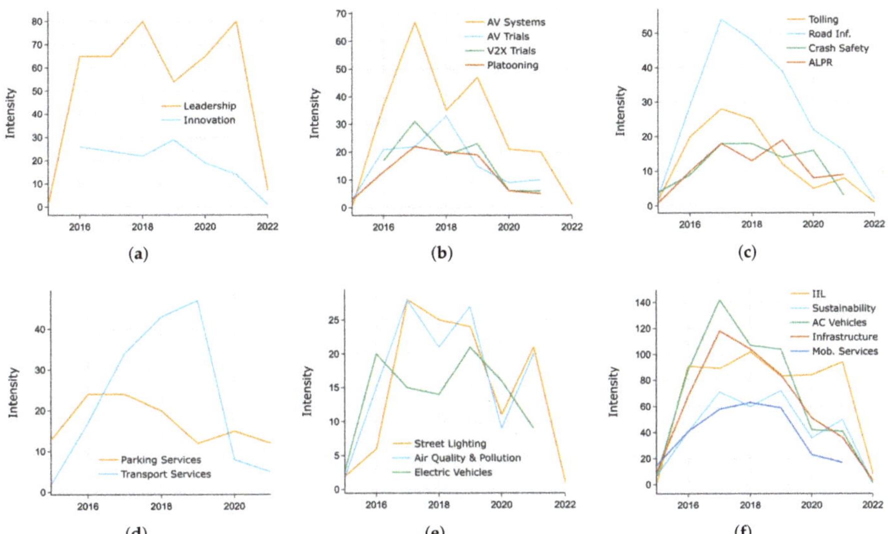

Figure 20. Temporal progression of parameters (*Traffic Technology International* Magazine): (**a**) industry, innovation, and leadership; (**b**) autonomous and connected vehicles; (**c**) infrastructure; (**d**) mobility services; (**e**) sustainability; (**f**) macro-parameters.

The temporal progression of all macro-parameters is summarised in Figure 20f. In 2017, macro-parameters autonomous and connected vehicles, infrastructure, and sustainability had the highest peak values of 140, 120, and 60, respectively.

6. Academia: Transportation Parameters Discovery

In this section, we discuss the parameters detected by our BERT model from the Web of Science. We provide an overview of the parameters and macro-parameters in Section 6.1. The quantitative analysis is discussed in Section 6.2. Subsequently, we discuss each macro-parameter in separate sections, Sections 6.3–6.8. Section 6.9 discusses the temporal analysis of the parameters and macro-parameters.

6.1. Overview and Taxonomy (Web of Science)

We detected a total of 50 parameters from *The Guardian* dataset using BERT. We skip Cluster 19 as it is related to narrative transportation [325–327], not related to general transportation. These 49 parameters were grouped into 6 macro-parameters using the domain knowledge along with a similarity matrix, hierarchical clustering, and other quantitative analysis methods. The methodology and process of the discovery of parameters and their groupings into macro-parameters have already been discussed in Section 3.

Table 3 and 4 list the parameters and macro-parameters of the academic dataset. The macro-parameters policy, planning and sustainability; transportation modes; logistics and SCM; pollution; technologies; and modelling, are listed in Column 1 with the associated parameters (Column 2). Some parameters are merged. For example, the Clusters 33 and 38, and 44 and 45 are merged as road safety and freight and logistics, respectively, in Table 3. The third column indicates the cluster number. The percentage of the number of articles is recorded in the fourth column. Our BERT model labelled 56.42% of articles as the outlier clusters. Consequently, we ignored this outlier cluster, and the rest of the 43.58% of articles are listed in the fourth column. The fifth column represents the top keywords associated with each parameter.

Table 3. Parameter and Macro-Parameters for Transportation (Source: Web of Science).

Macro-Parameters	Parameters	No.	%	Keywords
Policy, Planning and Sustainability	Land-Use Planning	5	1.47	urban, city, development, land, planning, transportation, public, sustainable, model, area, transport, spatial, system, transit, policy, economic, infrastructure, accessibility, regional, sustainability
	Smart Cities	31	0.43	smart, city, smart city, urban, system, datum, technology, citizen, transportation, service, public, concept, management, transport, energy, framework, transportation system, privacy, information, iot
	Low-Income Neighborhoods	39	0.36	job, worker, housing, commute, income, low income, low, transit, accessibility, employment, work, access, wage, job accessibility, poor, job seeker, seeker, live, area, household
	Urban Mobility	26	0.47	urban, city, growth, population, development, area, model, transportation, increase, transport, transit, demand, urban growth, network, new, land, agglomeration, spatial, build, urbanisation
	Road Safety	33	0.41	accident, road, crash, driver, safety, vehicle, collision, traffic, fatality, weather, speed, risk, traffic accident, truck, cause, system, injury, drive, road accident, death
		38	0.36	pedestrian, walk, sidewalk, walking, neighborhood, environment, walk trip, trip, build environment, household, build, walkability, walkable, behavior, model, residential, pedestrian detection, safety, crossing, street
	Traffic Congestion	17	0.89	traffic, congestion, traffic congestion, system, road, urban, time, vehicle, city, transportation, problem, propose, network, datum, base, transportation system, control, model, travel, become
	Subways	41	0.34	subway, station, subway station, underground, passenger, subway system, line, subway line, system, pm, public, passenger flow, flow, transportation, network, public transportation, platform, urban, time, city
	Vehicle Ownership	16	0.95	car, vehicle, driver, drive, ownership, model, car ownership, car follow, datum, system, transportation, base, travel, consumer, traffic, follow, household, mode, driving, information
	EV and Charging	0	5.88	power, electric, energy, charge, battery, system, electric vehicle, vehicle, voltage, grid, control, fuel, propose, high, transportation, load, motor, current, model, converter
	Bike Sharing and Ridesourcing	10	1.17	bicycle, bike, cycling, cyclist, bike sharing, sharing, share, mode, public, trip, motorcycle, city, user, station, cycle, system, travel, public bicycle, transportation, sharing system
	Children and Schools	49	0.32	school, child, student, parent, travel, active, walk, parental, school travel, travel school, mode, active travel, youth, high school, distance, walk school, choice, influence, activity, factor
	Gender and Race Equality	48	0.32	woman, gender, health, homeless, black, care, man, barrier, need, transportation, service, social, treatment, segregation, individual, community, female, child, racial, lack
	Parking	46	0.33	parking, parking lot, parking space, lot, space, parking policy, congestion, driver, policy, traffic, time, cost, car, pricing, occupancy, system, parking system, model, area, park
	Tolling	47	0.32	toll, problem, network, travel, model, travel time, time, congestion, optimal, propose, traveler, method, objective, cost, transportation, approach, path, mode, uncertainty, base
	Tourism	23	0.50	tourism, tourist, destination, service, development, travel, tourism industry, accommodation, transportation, industry, tourism development, visitor, attraction, activity, village, economic, information, tour, resource, country
Transportation Modes	Road Transport	7	1.26	bus, passenger, transit, public, time, service, stop, public transportation, route, system, model, bus stop, travel, transportation, bus service, base, transport, vehicle, datum, bus route
	Rail Transport	1	2.85	railway, train, rail, system, passenger, hsr, speed, line, high speed, transportation, operation, model, high, transport, track, traction, method, freight, propose, base
	Air Transport	3	2.19	airport, airline, flight, air, aircraft, passenger, aviation, air transportation, drone, service, model, system, market, transportation, network, air traffic, operation, datum, cost, time
	Marine Transport	13	1.08	ship, port, maritime, container, shipping, sea, vessel, ferry, model, cargo, risk, transportation, inland, system, seaport, terminal, operation, analysis, transport, base
Logistics and SCM	Freight and Logistics	44	0.33	business, logistic, company, industry, service, market, international, transportation, customer, growth, factor, sector, important, port, datum, transport, economic, investment, process, development
		45	0.33	cost, carrier, model, transportation, activity, freight, service, ltl, reduce, approach, less, analysis, transportation cost, inventory, base, process, customer, total, supply chain, implication
	Food Supply Chain	36	0.36	food, chain, product, meal, supply, supply chain, food supply, waste, food waste, temperature, production, fresh, environmental, perishable, consumer, impact, food production, store, transportation, fresh food
	Cost and Pricing	21	0.59	price, cost, firm, transportation cost, equilibrium, consumer, market, profit, auction, product, transportation, model, competition, location, valuation, segment, good, game, low, shipper

Table 4. Parameter and Macro-Parameters for Transportation (Source: Web of Science).

Macro-Parameters	Parameters	No.	%	Keywords
Pollution	Air Quality and Pollution	8	1.20	concentration, pm, air, pollution, source, emission, high, particle, pollutant, dust, winter, exposure, air pollution, contribution, summer, ozone, period, particulate, level, atmospheric
		14	1.01	emission, carbon, co, carbon emission, sector, energy, reduction, co emission, reduce, policy, transportation, climate, emission reduction, transport, low carbon, gas, greenhouse, greenhouse gas, industry, change
		25	0.47	emission, climate, climate change, change, carbon, co, sector, global, energy, environmental, co emission, warming, adaptation, global warming, carbon dioxide, dioxide, impact, transportation, policy, mitigation
		42	0.34	chinese, policy, development, city, chinese city, transportation, government, port, industry, quality, chinese government, economic, urban, air, air quality, level, emission, factor, market, pollution
	Soil Pollution	29	0.44	root, cd, plant, soil, rice, concentration, accumulation, shoot, metal, content, stress, gene, uptake, seedling, treatment, cadmium, high, decrease, increase, leave
		32	0.41	soil, metal, heavy metal, heavy, concentration, sample, plant, pollution, risk, cd, high, contamination, mining, area, water, source, environmental, dust, health, human
		4	1.59	sediment, water, soil, concentration, metal, river, high, surface, heavy metal, sample, heavy, source, pah, lake, distribution, organic, fish, erosion, area, process
	Noise Pollution	43	0.33	noise, pollution, traffic noise, noise pollution, sleep, noise level, exposure, annoyance, level, traffic, night, transportation noise, road traffic, air, health, air pollution, hour, road, cancer, noise exposure
	Fuel Types and Effects	24	0.49	fuel, diesel, engine, emission, biodiesel, production, biomass, energy, gasoline, oil, ethanol, biofuel, gas, renewable, blend, produce, cycle, diesel engine, life cycle, feedstock
	Coal Transport and Consumption	40	0.35	coal, plant, power plant, power, emission, electricity, gas, mine, mining, energy, production, fire power, coal fire, coal transportation, co, fire, transportation, fuel, thermal, natural gas
	Waste Management	9	1.19	waste, waste management, collection, management, solid, solid waste, disposal, landfill, recycling, municipal, environmental, treatment, cost, municipal solid, waste collection, collection transportation, impact, facility, material, hospital
	Waste Water Treatment	35	0.36	water, sludge, wastewater, treatment, sewer, sewage, removal, microplastic, process, pipe, environmental, wastewater treatment, desalination, concentration, environmental load, sample, reduce, high, effluent, oxygen
	Incident and Risk Management	15	0.97	oil, spill, pipeline, oil spill, oil gas, crude, crude oil, gas, petroleum, price, oil price, accident, hydrocarbon, corrosion, water, refinery, leakage, process, risk, transportation
Technologies	ITS	11	1.14	decade, last decade, last, transportation, vehicle, system, datum, past, past decade, base, technology, application, traffic, network, new, model, review, development, logistic, impact
	Computer Vision	20	0.64	vehicle, image, detection, recognition, video, plate, system, license plate, license, method, propose, object, detect, traffic, color, tracking, feature, intelligent, algorithm, base
	Smartphone Sensing	22	0.53	smartphone, phone, mobile, user, datum, mobile phone, application, sensor, information, system, mode, device, transportation, gps, time, location, base, transportation mode, service, mobility
	VANET	2	2.42	wireless, communication, network, vehicular, propose, system, vehicle, node, scheme, base, application, vanet, intelligent, mobile, protocol, sensor, intelligent transportation, access, transmission, transportation system
		28	0.44	vehicular, vehicle, application, network, system, transportation, communication, transportation system, base, protocol, intelligent, attract, vanet, intelligent transportation, researcher, propose, route, vehicular network, performance, service
		30	0.43	security, attack, privacy, vehicle, authentication, protocol, communication, network, secure, scheme, internet, propose, vanet, base, vehicular, system, iot, node, user, intelligent
	Autonomous Vehicles and Taxis	27	0.46	taxi, driver, taxi driver, passenger, taxi service, service, time, travel, share, demand, datum, taxi demand, trip, ride, autonomous taxi, autonomous, model, system, base, vehicle
	IoT Security	18	0.85	iot, internet, smart, thing, internet thing, device, security, network, application, datum, cloud, system, technology, service, thing iot, propose, intelligent, information, base, communication
	Physical Layer Communication	34	0.38	antenna, radar, frequency, system, communication, band, signal, radio, range, design, intelligent, intelligent transportation, propose, bandwidth, target, application, propose antenna, light, modulation, gain
	Robot Mobility	37	0.36	robot, mobile robot, robotic, task, system, mobile, control, approach, object, environment, problem, propose, transportation, sensor, present, base, motion, swarm, obstacle, time
Modelling	Traffic Flow Modelling	12	1.14	traffic, traffic flow, flow, prediction, datum, model, time, method, propose, network, road, system, real, forecasting, speed, traffic datum, base, intelligent, vehicle, accuracy
	Transportation Modelling Algorithms	6	1.27	problem, algorithm, fuzzy, solve, solution, propose, optimisation, method, model, objective, transportation problem, optimal, cost, genetic, programming, time, transportation, multi, network, base

Figure 21 provides a taxonomy of the transportation domain extracted from academia. The taxonomy was created using the parameters and macro-parameters discovered from the Web of Science dataset. The macro-parameters are shown on the first level of branches, the discovered parameters are shown on the second level of branches, and the most representative keywords are shown on the third level of branches.

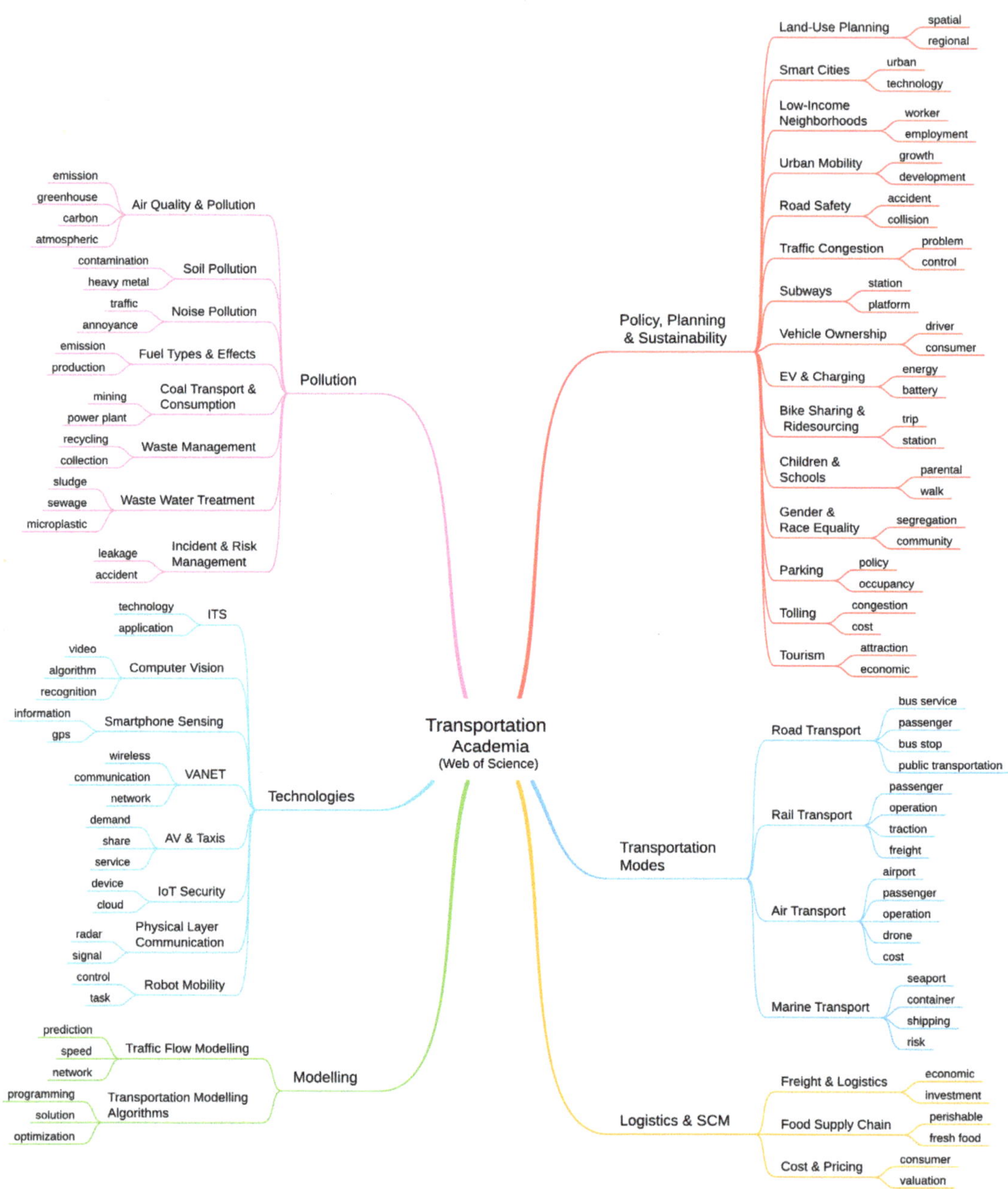

Figure 21. Taxonomy extracted from Web of Science dataset.

6.2. Quantitative Analysis (Web of Science)

This section discuss the term score, word score, intertopic distance map, hierarchical clustering, and similarity matrix.

Figure 22 depicts that only the top 13 keywords in each parameter actually represent the parameter when we evaluate the keywords (see Section 3.8). Because the probabilities of all the other possibilities are so close to one another, their ranking becomes more or less pointless. When we investigated the top keywords per parameter to label the parameter, we used this information to focus on the top 10 or so keywords in each parameter.

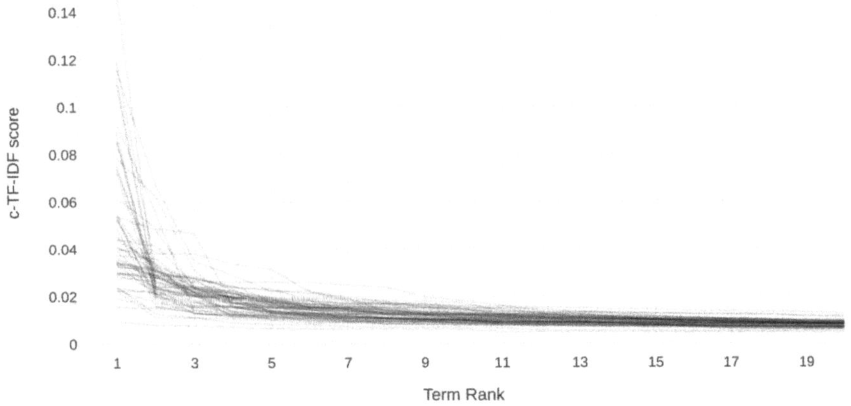

Figure 22. Term score (Web of Science).

Figure 23 shows the top five keywords for each parameter. The importance score, or c-TF–IDF score, is used to order the keywords (see Section 3.8). There are 50 subfigures and in each subfigure, the horizontal line shows the importance score, and the vertical line shows the parameter keywords.

Figure 24 shows the intertopic distance map (see Section 3.8), where nine groups of parameters are clearly identified. In the bottom left corner, one group of parameters contains more parameters than the others. There are two small-size parameters on the right side, which are comparatively small. However, we notice that the BERT model clusters are not very well clustered, so we used domain knowledge and other information to label them. Additionally, we manually labelled the parameters into six macro-parameters.

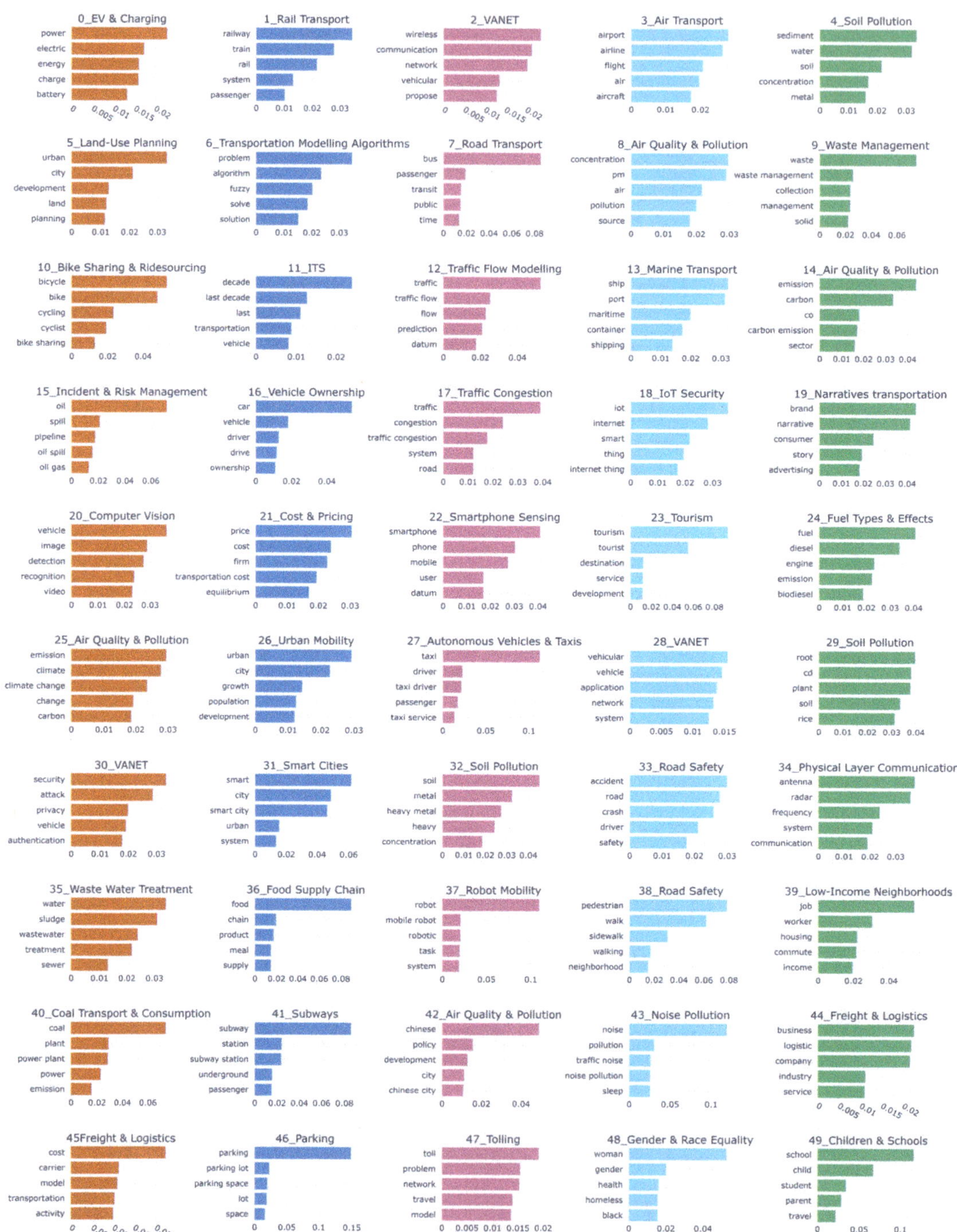

Figure 23. Academic article parameter with keywords c-TF–IDF score.

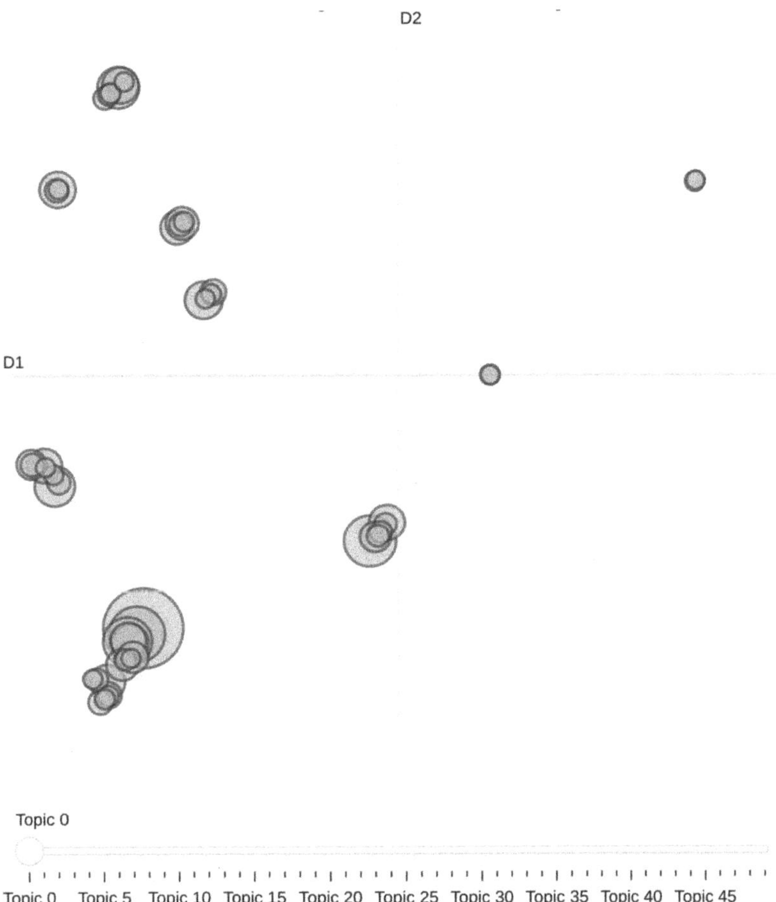

Figure 24. Intertopic distance map (Web of Science).

Figure 25 describes the hierarchical clustering of the 50 clusters and systematically pairs them based on the cosine similarity matrix (see Section 3.8). This automated hierarchical clustering grouped the clusters correctly, with some exceptions.

Figure 26 visualises the similarity matrix among the parameters (see Section 3.8). We use the same configuration as discussed in Figure 11. The dark blue colour represents the highest similarity relationship between parameters, while the light green represents the lowest similarity. For example, Cluster 12, labelled as traffic flow modelling, and Cluster 17, labelled as traffic congestion, have high similarity scores as the main focus on traffic. There is another high similarity between cCusters 2 and 30, and both are labelled as VANET.

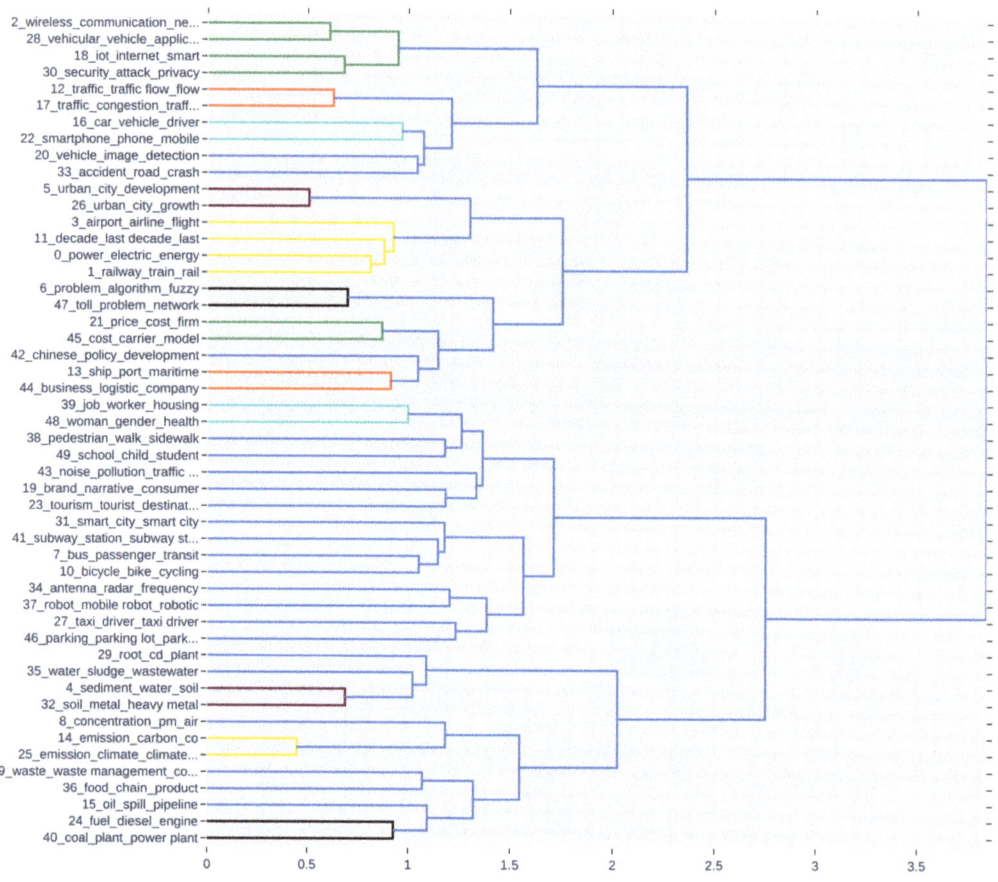

Figure 25. Hierarchical clustering (Web of Science).

6.3. Policy, Planning and Sustainability

The macro-parameter policy, planning and sustainability discusses the policy, services, planning, and development related to transportation. It includes 15 parameters, including land-use planning, smart cities, low-income neighborhoods, urban mobility, road safety, traffic congestion, subways, vehicles ownership, electric vehicles and charging, bike sharing and ridesourcing, children and schools, gender and race equality, parking, tolling, and tourism.

The land-use planning parameter discussed how to manage and map the transportation area based on several criteria. For example, analysing the usage of spatial metrics and population data to map the urban transportation areas [328] or the impact of walkability to the metro station on retail location choice [329], types of suburbanisation [330], etc. The smart cities parameter refers to urban areas that have experienced rapid growth and are technologically sophisticated. Crowd management [331], green transportation [332], and improving the quality of transportation systems [333] are the important areas of smart cities. With the progress of civilisation and rapid industrialisation, urban mobility becomes one of the key research areas in transportation. The investigation of urban mobility is the key factor for policy makers, and transport planners. The Road safety parameter is the combination of Parameters 33 and 38. This parameter is related to the research on road accidents and crashes, traffic collisions, pedestrian walking and safety [334], sidewalks, etc.

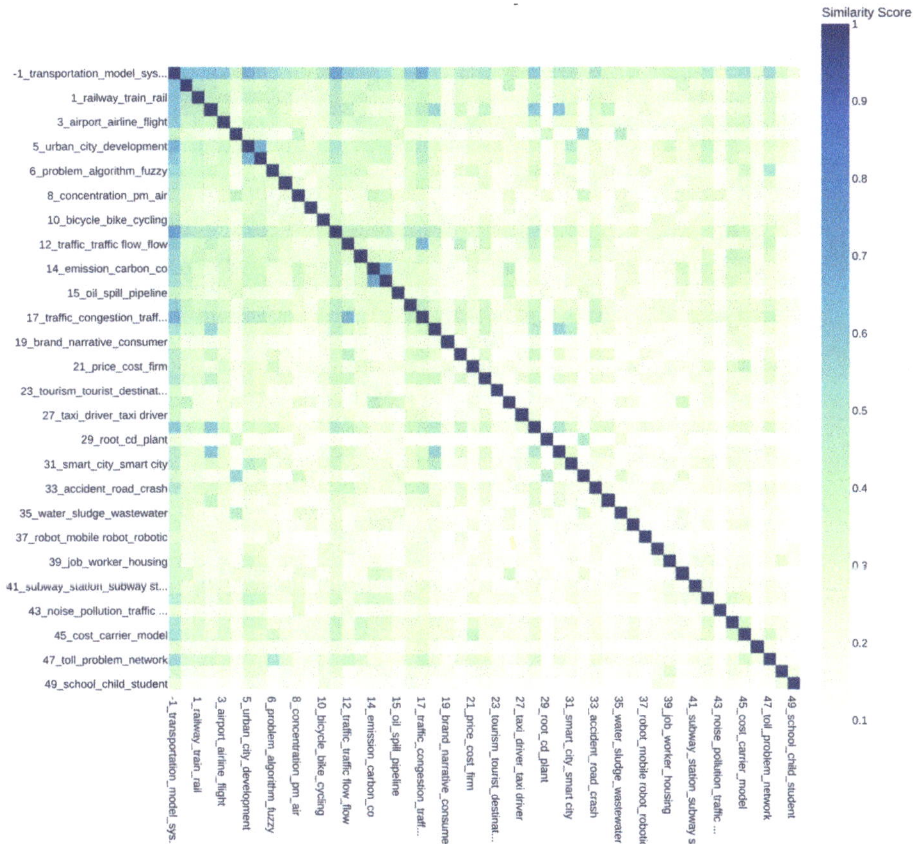

Figure 26. Similarity matrix (Web of Science).

Transportation affordability is one of the major issues in low-income neighbourhoods. Extensive research work has been conducted to analyse the issue and co-relation between transportation and low-income households. In [335], the authors examined transportation services and policies between 1960 and 2000 to determine the relationship between poverty and public transportation.In another study, the transportation affordability of low-income people in US cities was examined, and it was discovered that transit-rich neighbourhoods are more inexpensive than auto-oriented ones owing to reduced transportation costs [336]. With the advancement of civilisation and fast industrialisation, urban mobility has become one of the major research areas in transport research. For policy makers and transportation planners, investigating urban mobility is fundamental [337].

Hybrid electric vehicles (HEVs), battery electric vehicles (BEVs), plug-in electric vehicles (PEVs), and plug-in hybrid electric vehicles (PHEVs) are the most prevalent forms of electric vehicles. There are two types of charger systems: off-board and on-board, with uplink and downlink power transmission. A study of battery charger layouts, infrastructure for PHEVs, and charging voltage levels is provided by [338]. Zhang et al. [339] demonstrated efficient planning of PEVs fast-charging stations while taking into account interactions between transportation and electrical networks. In other research, Fernandez et al. [340] evaluated the impact of PEVs on distribution networks. The PHEVs batteries are charged at home from the outlet or corporate car park, which has an extra impact on the distribution grid (i.e., voltage deviations and power loss) [341].

Recently, research on bike sharing and ridesourcing are dramatically increased after COVID-19. There is a lot of ongoing research work to improve the quality of resourcing and bike-sharing systems. For example, the survey of ridesourcing research [342–344], shared autonomous vehicles [344–346], the survey of bike share [347], and traveller experience after using the ride-sharing services, such as Uber [348].

6.4. Transportation Modes

The macro-parameter transportation modes includes four parameters: road transport, rail transport, air transport, marine transport. The road transport parameter is related to public transportation. Normally, there are three types of public transportation in rural areas such as intercity buses, demand-responsive transit, and deviated fixed-route transit which is discussed in this study. One of the keys to bringing sustainable growth and a higher quality of life to urban areas is better public transit planning, control, and management [349]. Public transit uses road space more effectively than private traffic and creates fewer accidents and pollutants. Furthermore, owing to the impractical nature of public transit in many places, many choose to use private transportation rather than public transportation [350]. Much research has been conducted to improve the quality of public transportation. For example, evaluation of road transportation in Brazil [351], analysing passenger satisfaction with public transportation in Romania [352], passenger flow prediction in public transport based on the timetable in India [353], effective system design for public transport in the USA [354], etc.

The rail transport parameter shows key research areas in rail transportation such as high-speed railways [355], rail freight transportation [356], risk management [357], railway transit traffic [358], and train services [359]. The objective of high-speed rail (HSR) is to enhance mobility via inter-city conveyance and facilitate financial growth. In October 1964, Japan's first modern high-speed rail line opened, linking Tokyo and Osaka at a maximum speed of 210 km/h [360]. In 1976, British Railways established a high-speed train link between London and Bristol. Following that, France started its first high-speed rail service between Paris and Lyon in 1981. Since then, several European countries, including Spain, Germany, Italy, Belgium, and the Netherlands, have built HSR lines. South Korea built its first high-speed rail line between Daegu and Seoul in 2004 which was later extended to Busan in 2009, while Taiwan began service between Taipei and Kaohsiung in 2007. China declared in 2016 that they would have about 38,000 km of HSR tracks by 2025 [361].

The air transport parameter is represented by keywords airport, airline, flight, air, aircraft, passenger, aviation, air transportation, and others. It discusses the environmental impact of air transportation with respect to air pollution, noise, and climate change [362]. The marine transport parameter is defined by the keywords ship, port, maritime, container, shipping, sea, vessel, ferry, model, cargo, etc. Marine transportation accounts for 80–90% of worldwide trade, transporting about 10 billion tonnes of cartons and solid–liquid weighty cargo across the world's oceans each year [363]. Some research has been conducted on marine transportation trade [364,365], waterway or marine transportation systems [33,34], waterway traffic flow modelling [366], and so on,

6.5. Logistics and SCM

The macro-parameter logistics and SCM comprises three parameters, the Clusters 44 and 45 are combined as freight and logistics, food supply chain, and cost and pricing. The logistics and SCM parameter contains the following keywords, business, company, industry, service, freight, and others. The parameter relates to freight transportation, where companies or industries are using technologies to evaluate and improve their business models, logistics services, and activities. For example, marine transportation contributes 75% of all international freight [36,367], and as international freight increases, seaports need to improve their infrastructure to maintain the services and market demand. International freight transportation should be expanded to promote the growth of the logistics industry [368,369].

The food supply chain parameter is represented by keywords including food, chain, product, meal, supply, supply chain, food supply, waste, food waste, temperature, etc.

This parameter is associated with research aimed at enhancing the quality and safety of industrial food supply chain management. In[370], the authors presented an architecture for goods supply chain management with three key elements: real-time surveillance of the goods to reduce lost and perishable goods; anticipating the heat of the package; and triggering an alarm if the goods are not reserved under the acceptance conditions. Fresh food transportation is another research area in the food SCM parameter. A mathematical model was proposed by [371] to distribute the fresh food to the end customer. The distributed ledger and blockchain technologies are also used in the agri-food supply chain [372].

The cost and pricing parameter is an important research area in transportation. For example, the spatial price policy under transportation asymmetry [373], reducing transportation carrier cost [374], and analysing the global transportation system [375].

6.6. Pollution

The macro-parameter pollution contains eight parameters, namely air quality and pollution, soil pollution, noise pollution, fuel types and effects, coal transport and consumption, waste management, wastewater treatment, and incident and risk management. The transportation industry is responsible for around 30% of the greenhouse emissions that contribute to global warming [376]. In recent times, air pollution in transportation has become a major public concern. Consequently, it is crucial to enhance transport efficiency to keep urban air quality [377]. A number of studies have examined the levels of exposure to presumed air pollutants such as VOCs (volatile organic compounds) [378]. Traffic congestion also has an impact on air pollution and quality [379].

Transportation is a major contributor to soil pollution. Railway transport, for example, pollutes nearby soils with heavy metals. Heavy metal contamination has a significant detrimental influence on the natural environment, including decreased enzyme activity in soil and ecosystem destruction [380]. Noise pollution, which is mostly caused by transportation and automobile traffic [381], is a significant environmental contaminant because of its physical and mental impacts on humans.

The fuel types and effects parameter includes several types of fuels keywords such as diesel, biodiesel, gasoline, oil, biofuel, etc. Biofuel use has an impact on carbon dioxide emission in the United States transportation [382]. To determine more sustainable fuel is another area of research. In [383], the authors analyze the factors which help to choose between ethanol and gasoline for the flex-fuel vehicles. Bayramoğlu et al. [376] analyze the valve lift effects on the diesel engine.

6.7. Technologies

ITS, computer vision, smartphone sensing, VANET, autonomous vehicles and taxis, IoT security, physical layer communication, and robot mobility are the eight parameters of macro-parameter technologies.

Intelligent transportation systems (ITS) which have seen substantial advancement in the recent decade have been recognised as viable solutions to improve the transportation system's safety and reliability [384].

The computer vision parameter is the state-of-the-art research area including vehicle detection, vehicle type recognition, vehicle colour recognition [385], automatic license plate recognition [90], tracking moving vehicles, incident detection through the image processing task, vehicle counting system, traffic light detection, pedestrian detection, traffic sign recognition, etc.

The smartphone sensing parameter is another research area in the transportation field. Transport organisations actively promote public transportation to alleviate the traffic congestion produced by private vehicles. To increase the number of passengers using public transport, transport authorities can evaluate their local and premium services and market policies by analysing the travel patterns and regularity. For example, Ma et al. [386] proposed a data-mining process to identify the travel patterns in China based on the temporal and spatial features of Chinese smart card transaction data.

VANET referred to as vehicular ad-hoc networks, encloses V2V (vehicle-to-vehicle) and V2I (vehicle-to-infrastructure) communication architectures to enhance navigation, road safety, and other transport services. We noticed that VANET has been the most active research field for more than 10 years. The top three survey papers for VANET are [89,387,388] which provide an overview of VANET.

The parameter autonomous vehicles, also known as self-driving cars or driverless cars, have the ability to reduce travel time, increase road safety, improve fuel efficiency, and provide automatic parking facilities [389]. More research work is ongoing to improve the performance and quality of AV systems. For example, Haboucha et al. [390] discussed user preference concerning AV. In other research, a cost analysis of several types of AV was presented by [391].

6.8. Modelling

The macro-parameter modelling includes two parameters traffic flow modelling, and transportation modelling algorithms. Traffic flow models are important for analysing the performance of ITS applications. Extensive research work has been conducted on traffic flow models. For example, traffic flow prediction [392], traffic speed prediction [393], traffic violation detection, traffic sign detection, traffic congestion prediction, traffic data collection methods and passenger flow prediction [394]. Traffic coordination and monitoring depend on the traffic flow model. To solve the transportation modelling problems, i.e., fuzzy transportation problems where transportation source, destination, and time are represented as exponential fuzzy numbers, transportation scheduling problems, transport shipment problems, transport parameters (such as cost, capacity, speed) problem, and transportation network design problems, the researcher proposed several algorithms. For example, the genetic algorithm was used to solve the solid transportation problem [395].

6.9. Temporal Analysis (Web of Science)

In this section, we will analyse how the parameters have developed over time. Figure 27 shows the temporal progression of the parameter, which is distributed into eight subfigures. The first two subfigures represent the temporal progression of the policy, planning and sustainability parameter by showing seven and eight parameters, respectively. The next five macro-parameters show the temporal progression of transportation modes, logistics and SCM, pollution, technologies, and modelling, respectively. The last subfigure depicts the temporal progression of all macro-parameters.

The vertical line of the graph indicates the number of articles, which is defined as the intensity. The temporal progression of the macro-parameter policy, planning and sustainability is depicted in Figure 27a,b for better visualisation. The EV and Charging parameter has a higher intensity compared to the other parameters. Figure 27c shows that the rail transport intensity was increasing over time until 2017, and also more research was conducted on this parameter.

The intensity of articles for the macro-parameter logistics and SCM which includes three parameters: cost and pricing, freight and logistics and food supply chain is depicted in Figure 27d. Freight and logistics has the highest peak of about 14 in 2017. The temporal progression of macro-parameter pollution which includes eight parameters is shown in Figure 27e. We observed that there are more articles related to air quality and pollution and coal transport and consumption compared to others. Figure 27f shows the parameters of the technologies macro-parameter, where we observed that there was more research conducted on the VANET parameter. The intensity of articles for the macro-parameter modelling which includes two parameters: traffic flow modelling and transportation modelling algorithms is depicted in Figure 27g. We noticed that more research work was conducted on the transport flow modelling parameter. The temporal progression of all macro-parameters is summarised in Figure 27h. We discovered that n 2019 the macro-parameter policy, planning and sustainability had the highest peak value of 350. Pollution had the highest peak in 2021, about 250. The macro-parameter technologies and transportation modes intensity was about 150 between 2016 and 2020. There has been less research on macro-parameter modelling, logistics and SCM.

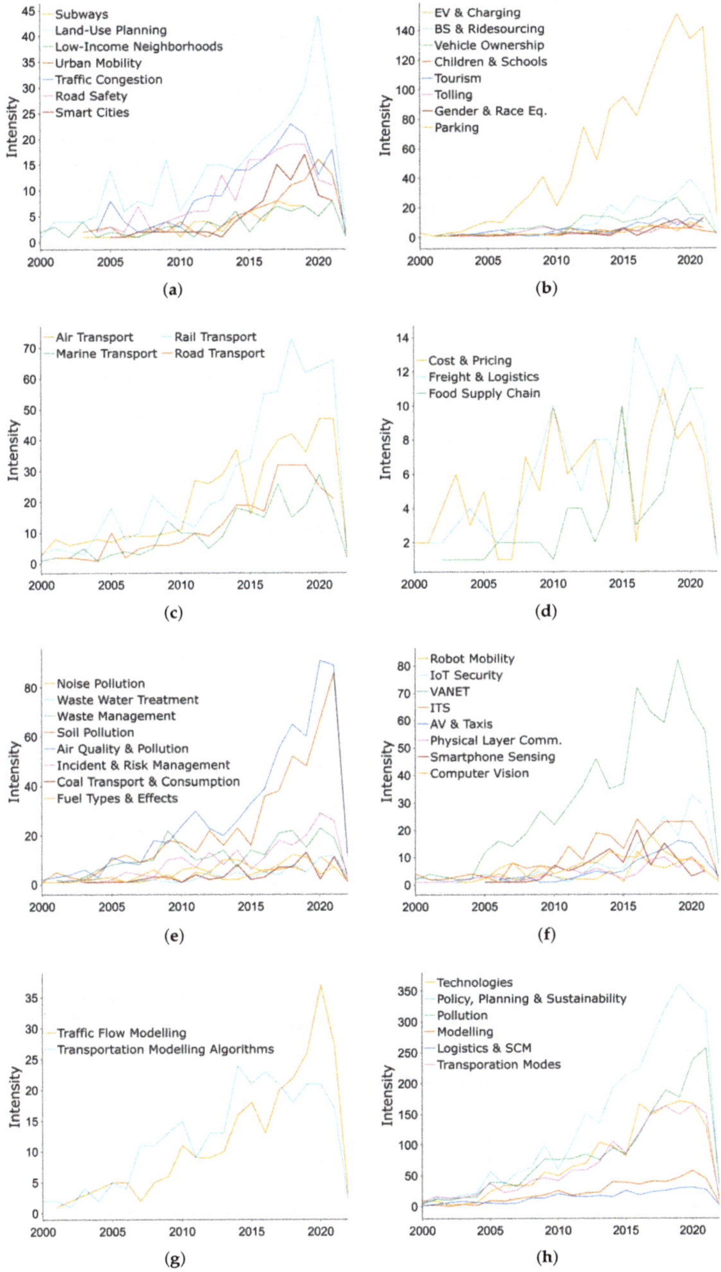

Figure 27. Temporal progression of parameters (Web of Science): (**a**) policy, planning and sustainability (i), (**b**) policy, planning and sustainability (ii); (**c**) transportation modes; (**d**) logistics and SCM; (**e**) pollution; (**f**) technologies; (**g**) modelling; (**h**) macro-parameters.

7. Discussion

The broader aim of our work is to investigate the use of ICT technologies for solving pressing problems in smart cities and societies. Specifically, in this paper, we investigate the use of digital methods to discover and analyse cross-sectional multi-perspective information to enable better decision making and develop better instruments for academic, corporate, national and international governance. We brought together a range of deep learning, big data, and other technologies to discover transportation parameters from three different perspectives using three different types of data sources, viz. a newspaper (*The Guardian*), a transportation technology magazine (*Traffic Technology International*), and academic literature on transportation (from Web of Science). These three types of transportation parameters were described in detail in Sections 4–6, respectively.

We discovered a total of 25 parameters from *The Guardian* dataset and grouped them into 6 macro-parameters, namely road transport; rail transport; air transport; crash and safety; disruptions and causes; and employment rights, disputes, and strikes. Figure 28 depicts the word cloud of the keywords of *The Guardian* parameters discovered by the BERT modelling, where the size of each keyword denotes its frequency that can be related to the importance of the keywords. The figure shows that *The Guardian* is primarily concerned with the government, trains, passengers, the general public, and accidents. As *The Guardian* is a UK-based newspaper, the train is one of the most used public transport in the UK which is clearly seen in the word cloud by the following keywords: train, rail, and railtrack. Additionally, some major transportation issues, such as accidents, delays, injuries, traffic, safety, and crashes, are also shown in this figure. Heathrow airport is the central international airport in the UK and is visible in the word cloud.

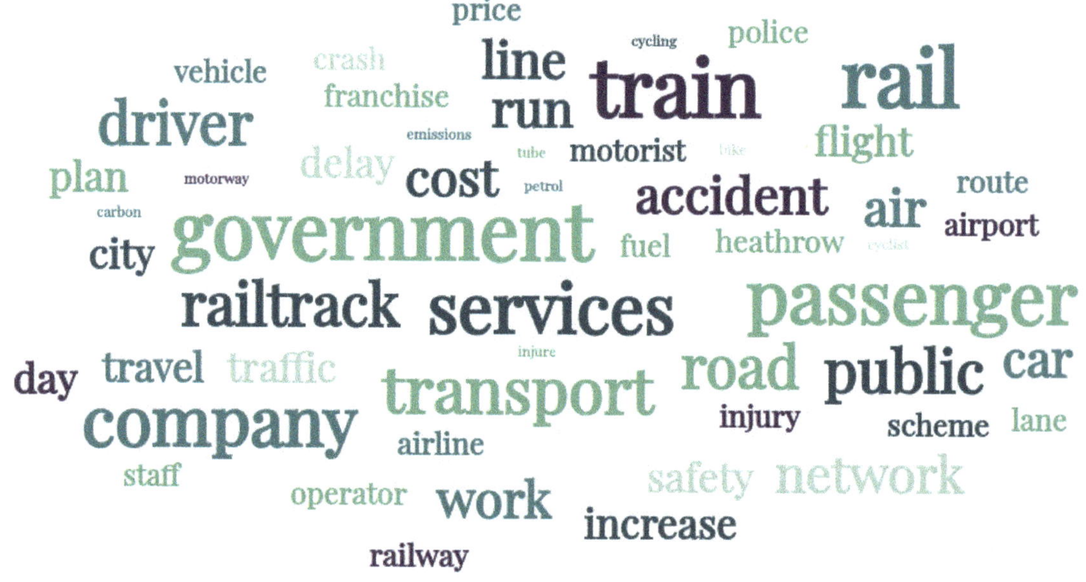

Figure 28. A word cloud generated from the transportation parameter keywords (*The Guardian*).

A total of 15 parameters were discovered from the *TTI* dataset, and we grouped them into 5 macro-parameters, namely industry, innovation, and leadership; autonomous and connected vehicles; sustainability; mobility services; and infrastructure. Figure 29 depicts the word cloud of the keywords of the *TTI* parameters discovered by the BERT modelling, where the size of each keyword denotes its frequency. The word cloud for transportation magazines is heavily dominated by the transportation industry. The keywords such as technology, project, system, traffic, solution, autonomous vehicles are highly related to the transport industry field. Datum is a well-known organisation that has been providing solutions and services in the transportation sector since 1999, especially in the UK and Europe, where the majority of transport infrastructure and projects are conducted under this organisation.

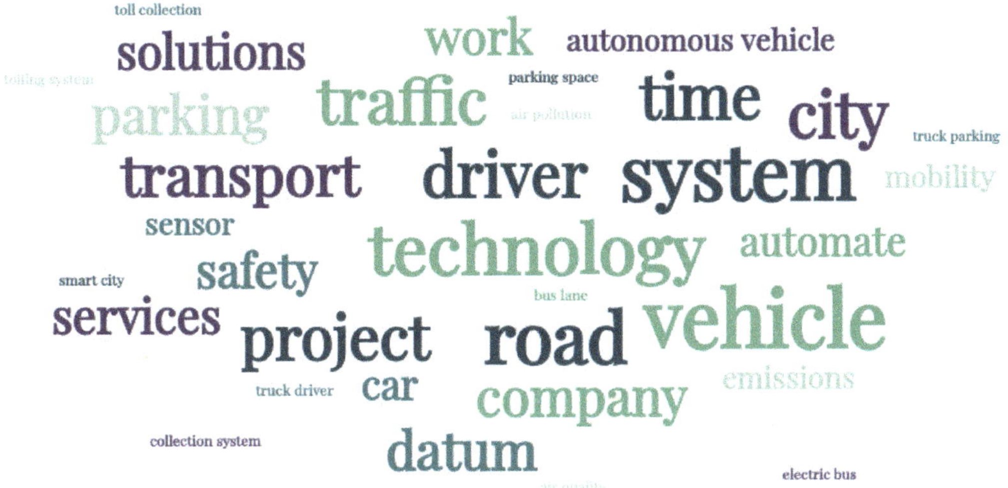

Figure 29. A word cloud generated from the transportation parameter keywords (*TTI* magazine).

We discovered 49 transportation parameters from the academic dataset and grouped them into 6 macro-parameters. These are policy, planning and sustainability; transportation modes; logistics and SCM; pollution; technologies; and modelling. Figure 30 depicts the word cloud of the keywords of the academia parameters. Academic transportation research covers a wide range of topics such as public transportation, various pollution impacts and solutions, intelligent transportation systems, traffic, supply chain management, cost, and so on.

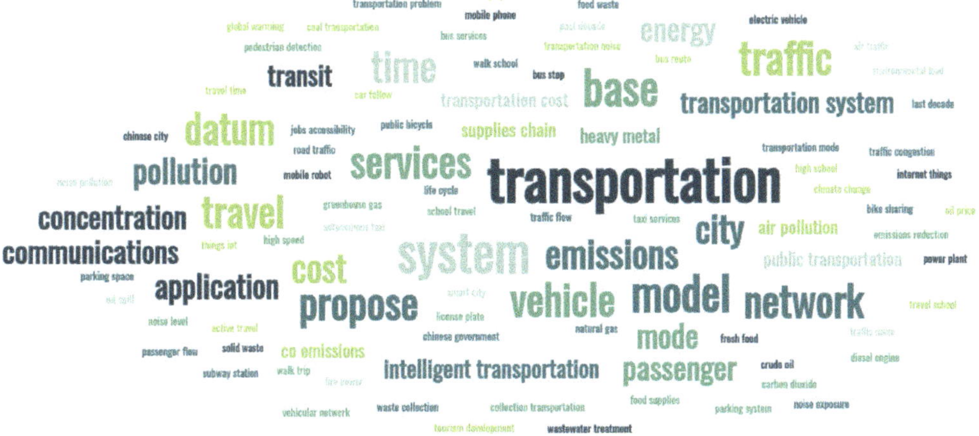

Figure 30. A word cloud generated from the transportation parameter keywords (Web of Science).

Reflecting on the multi-perspective view (public, governance, political, industrial, and academic) of transportation gained through the discovery of parameters from three different sources of datasets (see Figure 1; more detailed taxonomies and analyses are provided in the respective sections), we note that the newspaper parameters show more of a public and political view of transportation with the focus on the road, rail, and air transportation issues highlighting public discontent, accidents, cycling, pollution and effects on climate, road congestion, fuel prices, travel costs, privatisation and its impacts, dangerous driving, deaths, extreme weather impacts, and employment rights. Rail transport has a very high significance in the newspaper dataset due to the importance of the London underground and intercity rail transport in the UK. It may vary from country to country depending on the nature of transport in that country being used for intra- and intercity travel. Megacities tend to rely on rapid transit systems, but it can vary. The transport magazine parameters show an industrial view of transportation with great deal of the focus on autonomous and connected vehicles owing to several trials in progress around the world. The spotlights are also on crash and safety, tolling, parking, pollution, and electric vehicles. The Web of Science parameters show an academic view of transportation with various foci on policy, planning, sustainability, equity, children and schools, vehicle ownership, tourism, four transportation modes, freight and logistics, a wide range of pollution and its sources, modelling methods, and several technologies.

The three views show that there are many important problems such as transportation operations and public satisfaction that industry and academia seem not to place their focus on, or perhaps if they do, the solutions do not get much attention from policy makers and industrialists. For example, the fare-related parameters for public transport in *The Guardian* dataset show the importance of this topic for the public while it is not visible in the *TTI* and academic parameters. This could show the need for research in this area to model fares, provide insights, provide solutions for people for lower fares, and so on. Similarly, other parameters, issues, and contents such as the need for better employment conditions for drivers discovered from the public view (*The Guardian*) could be studied by academics and addressed by the industry, and this could set new trends where the technology-focused literature could bridge the gaps between social sciences, sustainability, and technology. We can also see that academia produces much broader and deeper knowledge on the subject, while many important issues such as a wide range of pollution affecting people and the planet do not reach the public eye. Our deep journalism approach could find the gaps and highlight them to the public and other stakeholders. We called this approach Deep Journalism because it allows capturing and reporting a relatively deeper view of a topic (e.g., transportation) from multiple perspectives, dimensions, stakeholders, and depths. Second, we use deep learning to automatically discover multi-perspective parameters about a topic.

Commenting on the relationship and contributions of this work to the specific field of journalism, we make the following observations. Traditional journalism has failed at its core purpose of providing citizens impartial information to maintain 'free' societies due to many reasons such as difficulties in maintaining the freedom and impartiality of media organisations and funding cuts, leading to public mistrust. The mistrust in traditional journalism coupled with the rise of digital technologies has given a rapid rise to citizen journalism. While citizen journalism solves some of the problems of traditional journalism, it comes with its own problems such as subjectivity and lack of regulations, standards, quality, and responsibility. In this work, we aimed to contribute to improving journalism through academic integrity and rigour. Academics should be in the vanguard of objective information, sincerity, impartiality, equity, and other ideals, and therefore the academics should search, pursue, propagate, and defend these ideals. If the academics fail to do so, then the responsibility lies on common people to pursue and be in the vanguard of the ideals needed to maintain a free society. An academic is not by profession and job but rather by action. The responsibility to maintain ideals lies with all people, and therefore everyone has the responsibility and needs to work towards upholding honesty, sincerity, equity, freedom, and other ideals. Through deep journalism, we aim to make impartial, cross-sectional, and multi-perspective information available to people, bring rigour to the journalism field by nurturing responsibility in people, make it easy for people to generate information for public benefit using deep learning, and make tools and information available to common people so they can search and defend ideals of free societies, including social, environmental, and economic sustainability. As we have explained earlier, the methods and tools developed in this work are not limited to journalism and can be used by engineers, scientists, policy makers, and others to understand the parameters for the design and operations of any sector such as healthcare, supply chain management, etc.

8. Conclusions

We live in a complex world characterised by complex people, complex times, and complex social, technological, and ecological environments. There is clear evidence that governments are failing at most public matters. The recent COVID-19 pandemic is a major example of global governance failure both at preventing such a pandemic and managing it. It is time that all of us take responsibility for both success and failure rather than criticising our governments and look into ways of collaboratively improving the governance of public matters.

While there are many reasons for government failures, we believe the lack of information availability is a fundamental reason that limits governments' abilities to act smartly and allows the lack of transparency to creep into policy and action, leading to corruption and failure. To this end, this paper introduced the concept of deep journalism, a data-driven deep learning-based approach for discovering multi-perspective parameters related to a topic of interest. We examined the academic, industrial, public, governance, and political parameters for the transportation sector as a case study to introduce deep journalism and our tool, DeepJournal (Version 1.0), that implements our proposed approach. We built three datasets specifically for the work presented in this paper. These datasets will be provided openly to the community for further research and development. We elaborated on 89 transportation parameters and hundreds of dimensions, reviewing 400 technical, academic, and news articles.

The work presented in this paper and the investigation into the proposed Deep Journalism approach is far from perfect both in terms of its definition and scope as well as its exploration in this paper such as the different types of media, the use of deep learning and other computing methods, the investigations into the specific media and sources used in this paper, and more. We work in a range of research topics including smart cities [396–400], cloud, fog, and edge computing [401–403], big data [404,405], high performance computing [406,407], education [408,409], and healthcare [410,411], and plan to benefit from these technologies and topics for extending and improving Deep Journalism in the future.

Regarding NLP (natural language processing), more research is needed to understand the clustering performance of BERT and other clustering methods. For example, some clusters have some similarities, and it is not clear why these clusters are separated by the clustering algorithms or why an article from one topic is included in another, the quality of cluster boundaries, and so on. One obvious reason is that documents are related to all clusters, some more and others less. Another reason by example is a train crash that could also be linked to a road crash because there is something about crashes and similar vocabulary there. Death and injuries can be other linked topics. Another matter related to topic modelling, NLP, and BERT performance is to investigate the effects of language used by different communities and sectors such as in the case of newspapers using a different vocabulary because dramatisation is important for them.

The aim of the work presented in this paper is partly to understand the transportation domain comprehensively and use this understanding to improve the transportation sector. Our research concerns multiple sectors and disciplines such as healthcare and smart cities, with a focus on developing interdisciplinary methods and technologies allowing for the minimisation of silos and inefficiencies through policy and action integration and the facilitation of holistic designs and optimisations. The proposed approach can also be utilised to create disciplines and curricula for teaching in schools and universities, training for staff in industry and government institutions, and more. Given ICT is penetrating all spheres of our lives, be it transport, politics, corruption investigations, legal actions, and more, capturing such details automatically is useful because it can lead to the development of automated algorithms for cross-disciplinary exploratory analysis, real-time investigations, decision making, and more.

Finally, a broader direction of investigation in the future would be to find how deep learning can further help develop multi-perspective and deeper insights into different issues, sectors, etc., related to our societies. We wish to investigate whether it is possible to develop rigorous methodologies and processes to gain deeper insights into various issues, the processes that could be automated and followed easily by the public to objectively investigate matters of civic concerns and produce impartial information that could be consumed by the public to develop informed societies. We wish to explore whether it is possible to cultivate informed societies through such processes (deep journalism), i.e., by enabling and nurturing the public to study matters of public concern through automated, rigorous, 'deep', processes and contribute to the debate through objective processes and impartial information. The definition of deep journalism proposed in this paper also entails that the scope of journalism should not be limited and should extend to any matters affecting the public and society.

We believe the possibilities for utilisation and potential impacts of the work presented in this paper and the proposed approach are significant and endless; we invite all communities interested in transportation and otherwise to investigate our proposed approach for a sustainable and joint future for all.

The bottom line is that a society is as informed as it wishes to be. This statement does not relieve oppressed governments and institutions from the responsibility for their oppressive actions against the public. It only highlights the fact that if individuals do not take responsibility for the governance of their environment, opportunists will rise and will blind them through misinformation and enslave them.

Author Contributions: Conceptualization, I.A. and R.M.; methodology, I.A. and R.M.; software, I.A.; validation, I.A. and R.M.; formal analysis, I.A. and R.M.; investigation, I.A., R.M., F.A. and E.A; resources, R.M., F.A. and E.A.; data curation, I.A.; writing—original draft preparation, I.A. and R.M.; writing—review and editing, R.M., F.A. and E.A.; visualization, I.A.; supervision, R.M. and F.A.; project administration, R.M.; funding acquisition, R.M. All authors have read and agreed to the published version of the manuscript.

Funding: The authors acknowledge with thanks the technical and financial support from the Deanship of Scientific Research (DSR) at the King Abdulaziz University (KAU), Jeddah, Saudi Arabia, under Grant No. RG-11-611-40.

Institutional Review Board Statement: Not applicable.

Informed Consent Statement: Not applicable.

Data Availability Statement: The data used in this paper will be made publicly available.

Acknowledgments: The work carried out in this paper is supported by the HPC Center at the King Abdulaziz University. The training and software development work reported in this paper was carried out on the Aziz supercomputer.

Conflicts of Interest: The authors declare no conflict of interest.

References

1. Dauvergne, P. Why is the global governance of plastic failing the oceans? *Glob. Environ. Chang.* **2018**, *51*, 22–31. [CrossRef]
2. Keping, Y. Governance and good governance: A new framework for political analysis. *Fudan J. Humanit. Soc. Sci.* **2018**, *11*, 1–8. [CrossRef]

3. Howlett, M.; Ramesh, M. The two orders of governance failure: Design mismatches and policy capacity issues in modern governance. *Policy Soc.* **2014**, *33*, 317–327. [CrossRef]
4. Keech, W.R.; Munger, M.C. The anatomy of government failure. *Public Choice* **2015**, *164*, 1–42. [CrossRef]
5. Hudson, B.; Hunter, D.; Peckham, S. Policy failure and the policy-implementation gap: Can policy support programs help? *Policy Des. Pract.* **2019**, *2*, 1–14. [CrossRef]
6. McKinnon, C. Sleepwalking into lock-in? Avoiding wrongs to future people in the governance of solar radiation management research. *Environ. Politics* **2019**, *28*, 441–459. [CrossRef]
7. Schmelzle, C.; Stollenwerk, E. Virtuous or Vicious Circle? Governance Effectiveness and Legitimacy in Areas of Limited Statehood. 2018. Available online: https://www.tandfonline.com/doi/full/10.1080/17502977.2018.1531649 (accessed on 15 February 2022).
8. Hudson, B. We Need to Talk About Policy Failure—And How to Avoid It. Available online: https://blogs.lse.ac.uk/politicsandpolicy/policy-failure-and-how-to-avoid-it/ (accessed on 15 February 2022).
9. Goldin, I. Divided Nations: Why Global Governance Is Failing, and What We Can Do about It. Available online: https://blog.politics.ox.ac.uk/divided-nations-why-global-governance-is-failing-and-what-we-can-do-about-it/ (accessed on 15 February 2022).
10. Cohen, J.; van der Meulen Rodgers, Y. Contributing factors to personal protective equipment shortages during the COVID-19 pandemic. *Prev. Med.* **2020**, *141*, 106263. [CrossRef] [PubMed]
11. Templeton, A.; Guven, S.T.; Hoerst, C.; Vestergren, S.; Davidson, L.; Ballentyne, S.; Madsen, H.; Choudhury, S. Inequalities and identity processes in crises: Recommendations for facilitating safe response to the COVID-19 pandemic. *Br. J. Soc. Psychol.* **2020**, *59*, 674–685. [CrossRef]
12. Kim, D.K.D.; Kreps, G.L. An analysis of government communication in the United States during the COVID-19 pandemic: Recommendations for effective government health risk communication. *World Med Health Policy* **2020**, *12*, 398–412. [CrossRef]
13. Williams, O.D. COVID-19 and private health: Market and governance failure. *Development* **2020**, *63*, 181–190. [CrossRef]
14. Greer, S.L.; Ruijter, A.D.; Brooks, E. The COVID-19 pandemic: Failing forward in public health. In *The Palgrave Handbook of EU Crises*; Springer: Berlin/Heidelberg, Germany, 2021; pp. 747–764.
15. Park, J.; Chung, E. Learning from past pandemic governance: Early response and Public-Private Partnerships in testing of COVID-19 in South Korea. *World Dev.* **2021**, *137*, 105198. [CrossRef] [PubMed]
16. Alomari, E.; Katib, I.; Albeshri, A.; Yigitcanlar, T.; Mehmood, R. Iktishaf+: A big data tool with automatic labeling for road traffic social sensing and event detection using distributed machine learning. *Sensors* **2021**, *21*, 2993. [CrossRef] [PubMed]
17. Topping, A.; Davies, C. Train Derails after Landslide as Heavy Rain Causes UK travel Chaos. 2016. Available online: https://www.theguardian.com/uk-news/2016/sep/16/torrential-rain-travel-chaos-england-floods (accessed on 15 February 2022).
18. Media, P. Major Disruption at Euston Station after Rush-Hour Services Cancelled. 2019. Available online: https://www.theguardian.com/uk-news/2019/oct/25/major-disruption-at-euston-station-after-rush-hour-services-cancelled (accessed on 15 February 2022).
19. Topham, G. Rail Chiefs Confident Timetable Change will Not Repeat 2018 Chaos. Available online: https://www.theguardian.com/uk-news/2019/may/18/rail-timetable-change-trains-extra-services (accessed on 18 May 2019).
20. Jones, S. Woman Held Over Tube Station Death. 2010. Available online: https://www.theguardian.com/uk/2010/oct/26/kings-cross-tube-death-arrest (accessed on 15 February 2022).
21. Topham, G. 'Railway Accidents Happen Because Someone Makes a Mistake'. 2013. Available online: https://www.theguardian.com/uk-news/2013/jul/25/railway-accidents-human-error-warning-systems (accessed on 15 February 2022).
22. Heidarysafa, M.; Kowsari, K.; Barnes, L.; Brown, D. Analysis of railway accidents' narratives using deep learning. In Proceedings of the 2018 17th IEEE International Conference on Machine Learning and Applications (ICMLA), Orlando, FL, USA, 17–20 December 2018; IEEE: Piscataway, NJ, USA, 2018; pp. 1446–1453.
23. Pidd, H. Government to Finally Drop Plan for HS2 Link to Leeds—Reports. 14 November 2021 Available online: https://www.theguardian.com/uk-news/2021/nov/14/government-to-finally-drop-plan-for-hs2-link-to-leeds-reports (accessed on 15 February 2022).
24. Valentino, S. London the Worst City in Europe for Health Costs from Air Pollution. 2020. Available online: https://www.theguardian.com/environment/2020/oct/21/london-the-worst-city-in-europe-for-health-costs-from-air-pollution (accessed on 15 February 2022).
25. Stone, T. New Report Shows Local Transport is the Main Contributor to UK Air Pollution. 2018. Available online: https://www.traffictechnologytoday.com/news/public-transit/new-report-shows-local-transport-is-the-main-contributor-to-uk-air-pollution.html (accessed on 15 February 2022).
26. Topham, G.; Kollewe, J. Heathrow Passenger Charges Could Rise by Up to 56% by 2023. 19 October 2021 Available online: https://www.theguardian.com/uk-news/2021/oct/19/heathrow-passenger-charges-rise-tickets-caa (accessed on 15 February 2022).
27. Rat, C.L. Safety Management System In Air Transportation. In *Proceedings of the International Management Conference*; Faculty of Management, Academy of Economic Studies: Bucharest, Romania, 2019; Volume 13, pp. 1051–1058.
28. Stone, T. Virginia Launches Flight Information Exchange to Improve Drone Safety. 2020. Available online: https://www.traffictechnologytoday.com/news/asset-management/virginia-launches-flight-information-exchange-to-improve-drone-safety.html (accessed on 15 February 2022).

29. Association, P. Stansted Airport Security Fault Causes Passengers to Miss Flights. 2017. Available online: https://www.theguardian.com/uk-news/2017/may/09/stansted-airport-security-fault-passengers-miss-flights (accessed on 15 February 2022).
30. Dodd, V.; Milmo, D. Runway Invader Sparks New Security Fears. 14 March 2008. Available online: https://www.theguardian.com/uk/2008/mar/14/uksecurity.transport (accessed on 15 February 2022).
31. Davies, R. Nearly Half of UK's Independent Petrol Stations Still Lack Fuel. 2021. Available online: https://www.theguardian.com/business/2021/sep/30/half-uk-independent-petrol-stations-out-of-fuel (accessed on 15 February 2022).
32. Gou, X.; Lam, J.S.L. Risk analysis of marine cargoes and major port disruptions. *Marit. Econ. Logist.* **2019**, *21*, 497–523. [CrossRef]
33. Wang, Z.; Yin, J. Risk assessment of inland waterborne transportation using data mining. *Marit. Policy Manag.* **2020**, *47*, 633–648. [CrossRef]
34. Samekto, A.A.; Kristiyanti, M. Development of Risk Management Model in Maritime Industry. *Qual.-Access Success* **2020**. Available online: https://www.proquest.com/openview/87dc6ba79b35b9df29abdeec3539e0c2/1?pq-origsite=gscholar&cbl=1046413 (accessed on 15 February 2022).
35. Stone, T. Washington Seaports to Launch Apps to Speed Up Cargo Flow, Reduce Emissions and Save Fuel. 2016. Available online: https://www.traffictechnologytoday.com/news/emissions-low-emission-zones/washington-seaports-to-launch-apps-to-speed-up-cargo-flow-reduce-emissions-and-save-fuel.html (accessed on 15 February 2022).
36. Garg, C.P.; Kashav, V. Evaluating value creating factors in greening the transportation of Global Maritime Supply Chains (GMSCs) of containerised freight. *Transp. Res. Part D Transp. Environ.* **2019**, *73*, 162–186. [CrossRef]
37. Academics Can Change the World—If They Stop Talking Only to Their Peers. 31 March 2017. Available online: https://theconversation.com/academics-can-change-the-world-if-they-stop-talking-only-to-their-peers-55713 (accessed on 15 February 2022).
38. Stockton and Darlington Railway. 24 September 1825. Available online: https://www.theguardian.com/world/1825/sep/24/transport.uk (accessed on 15 February 2022).
39. Becket, A. How Can Britain Become a 'Great Cycling Nation' When It's So Scary to Ride a Bike? 1 January 2022. Available online: https://www.theguardian.com/commentisfree/2022/jan/01/britain-cycling-scary-ride-bike (accessed on 15 February 2022).
40. Stone, T. Netherlands to Test AVs on Public Roads. 2 June 2015. Available online: https://www.traffictechnologytoday.com/news/autonomous-vehicles/netherlands-to-test-avs-on-public-roads.html (accessed on 15 February 2022).
41. Stone, T. #TRBAM 2022: South Dakota DOT Secretary Joel Jundt Named as New TRB Board Member. 11 January 2022. Available online: https://www.traffictechnologytoday.com/news/appointments/trbam-2020-south-dakota-dot-secretary-joel-jundt-named-as-new-trb-board-member.html (accessed on 15 February 2022).
42. Nagurney, A. Congested urban transportation networks and emission paradoxes. *Transp. Res. Part D Transp. Environ.* **2000**, *5*, 145–151. [CrossRef]
43. Li, X.; Hu, Z.; Zou, C. Noise annoyance and vibration perception assessment on passengers during train operation in Guangzhou Metro. *Environ. Sci. Pollut. Res.* **2022**, *29*, 4246–4259. [CrossRef]
44. Devlin, J.; Chang, M.W.; Lee, K.; Toutanova, K. Bert: pretraining of deep bidirectional transformers for language understanding. *arXiv* **2018**, arXiv:1810.04805.
45. McInnes, L.; Healy, J.; Melville, J. Umap: Uniform manifold approximation and projection for dimension reduction. *arXiv* **2018**, arXiv:1802.03426.
46. McInnes, L.; Healy, J.; Astels, S. hdbscan: Hierarchical density based clustering. *J. Open Source Softw.* **2017**, *2*, 205. [CrossRef]
47. Li, B.; Han, L. Distance weighted cosine similarity measure for text classification. In Proceedings of the International Conference on Intelligent Data Engineering and Automated Learning, Hefei, China, 20–23 October 2013; Springer: Berlin/Heidelberg, Germany, 2013; pp. 611–618.
48. Murtagh, F.; Contreras, P. Algorithms for hierarchical clustering: An overview, II. *Wiley Interdiscip. Rev. Data Min. Knowl. Discov.* **2017**, *7*, e1219. [CrossRef]
49. Tmtoolkit: Text Mining and Topic Modeling Toolkit. Available online: https://tmtoolkit.readthedocs.io/en/latest/index.html (accessed on 14 March 2022).
50. Grootendorst, M. c-TF-IDF. Available online: https://github.com/MaartenGr/cTFIDF (accessed on 14 March 2022).
51. Sievert, C.; Shirley, K. LDAvis: A method for visualizing and interpreting topics. In Proceedings of the Workshop on Interactive Language Learning, Visualization, and Interfaces, Baltimore, MD, USA, 27 June 2014; pp. 63–70.
52. Histograms. Available online: https://matplotlib.org/stable/gallery/statistics/hist.html (accessed on 14 March 2022).
53. heatmap. Available online: https://seaborn.pydata.org/generated/seaborn.heatmap.html (accessed on 14 March 2022).
54. Goodwin, M.J. *Right Response: Understanding and Countering Populist Extremism*; Chatham House: London, UK, 2011.
55. Alvares, C.; Dahlgren, P. Populism, extremism and media: Mapping an uncertain terrain. *Eur. J. Commun.* **2016**, *31*, 46–57. [CrossRef]
56. What Is the Purpose of Journalism? Available online: https://www.americanpressinstitute.org/journalism-essentials/what-is-journalism/purpose-journalism/ (accessed on 27 August 2021).
57. Carlson, M. The perpetual failure of journalism. *Journalism* **2019**, *20*, 95–97. [CrossRef]
58. Allern, S.; Pollack, E. Journalism as a public good: A Scandinavian perspective. *Journalism* **2019**, *20*, 1423–1439. [CrossRef]

59. A Free Press is 'Cornerstone' for Accountability and 'Speaking Truth to Power': Guterres. Available online: https://news.un.org/en/story/2019/05/1037741 (accessed on 28 August 2021).
60. Wall, M. Citizen Journalism. *Digit. J.* **2015**, *3*, 797–813. [CrossRef]
61. Sienkiewicz, M. Start making sense: A three-tier approach to citizen journalism. *Media Cult. Soc.* **2014**, *36*, 691–701. [CrossRef]
62. Raza, S.H.; Emenyeonu, O.C.; Yousaf, M.; Iftikhar, M. Citizen journalism practices during COVID-19 in spotlight: Influence of user-generated contents about economic policies in perceiving government performance. *Inf. Discov. Deliv.* **2022**, *50*, 142–154. [CrossRef]
63. Zeng, X.; Jain, S.; Nguyen, A.; Allan, S. New perspectives on citizen journalism. *Glob. Media China* **2019**, *4*, 3–12. [CrossRef]
64. Carmichael, V.; Adamson, G.; Sitter, K.C.; Whitley, R. Media coverage of mental illness: A comparison of citizen journalism vs. professional journalism portrayals. *J. Ment. Health* **2019**, *28*, 520–526. [CrossRef]
65. Nah, S.; Chung, D.S. *Understanding Citizen Journalism as Civic Participation*; Routledge: Philadelphia , NY, USA; 2020; p. 179. [CrossRef]
66. Al-Ghazzi, O. "Citizen Journalism" in the Syrian Uprising: Problematizing Western Narratives in a Local Context. *Commun. Theory* **2014**, *24*, 435–454. Available online: https://academic.oup.com/ct/article-pdf/24/4/435/21958510/jcomthe0435.pdf (accessed on 15 February 2022). [CrossRef]
67. Social Media Poses 'Existential Threat' to Traditional, Trustworthy News. Available online: https://news.un.org/en/story/2022/03/1113702 (accessed on 28 August 2021).
68. Mutsvairo, B.; Salgado, S. Is citizen journalism dead? An examination of recent developments in the field. *Journalism* **2022**, *23*, 354–371. [CrossRef]
69. McKinney, W. Data structures for statistical computing in python. In Proceedings of the 9th Python in Science Conference, Austin, TX, USA, 3 July–28 June 2010; Volume. 445, pp. 51–56.
70. Harris, C.R.; Millman, K.J.; van der Walt, S.J.; Gommers, R.; Virtanen, P.; Cournapeau, D.; Wieser, E.; Taylor, J.; Berg, S.; Smith, N.J.; et al. Array programming with NumPy. *Nature* **2020**, *585*, 357–362. [CrossRef] [PubMed]
71. Grootendorst, M. *BERTopic: Leveraging BERT and c-TF-IDF to Create Easily Interpretable Topics*; Cornell University: Ithaca, NY, USA, 2020 . [CrossRef]
72. Bird, S.; Klein, E.; Loper, E. *Natural Language Processing with Python: Analyzing Text with the Natural Language Toolkit*; O'Reilly Media, Inc.: Sebastopol, CA, USA, 2009.
73. Buitinck, L.; Louppe, G.; Blondel, M.; Pedregosa, F.; Mueller, A.; Grisel, O.; Niculae, V.; Prettenhofer, P.; Gramfort, A.; Grobler, J.; et al. API design for machine learning software: Experiences from the scikit-learn project. In Proceedings of the ECML PKDD Workshop: Languages for Data Mining and Machine Learning, Prague, Czech Republic, 23–27 September 2013; pp. 108–122.
74. Rehurek, R.; Sojka, P. *Gensim–Python Framework for Vector Space Modelling*; NLP Centre, Faculty of Informatics Masaryk University: Brno, Czech Republic, 2011; Volume 3. Available online: https://pypi.org/project/gensim/ (accessed on 15 Feburary 2022).
75. Paszke, A.; Gross, S.; Massa, F.; Lerer, A.; Bradbury, J.; Chanan, G.; Killeen, T.; Lin, Z.; Gimelshein, N.; Antiga, L.; et al. PyTorch: An Imperative Style, High-Performance Deep Learning Library. In *Advances in Neural Information Processing Systems 32*; Curran Associates, Inc.: Red Hook, NY, USA, 2019; pp. 8024–8035.
76. Waskom, M.; Botvinnik, O.; O'Kane, D.; Hobson, P.; Lukauskas, S.; Gemperline, D.C.; Augspurger, T.; Halchenko, Y.; Cole, J.B.; Warmenhoven, J.; et al. mwaskom/seaborn: V0.8.1 (September 2017). *Zenodo* **2017**. [CrossRef]
77. Inc., P.T. Collaborative Data Science. 2015. Available online: https://plot.ly (accessed on 15 February 2022).
78. Hunter, J.D. Matplotlib: A 2D graphics environment. *Comput. Sci. Eng.* **2007**, *9*, 90–95. [CrossRef]
79. Alam, F.; Mehmood, R.; Katib, I.; Altowaijri, S.M.; Albeshri, A. TAAWUN: A Decision Fusion and Feature Specific Road Detection Approach for Connected Autonomous Vehicles. *Mob. Netw. Appl.* **2019**, 1–17. [CrossRef]
80. Wang, N.; Chen, R.; Xu, K. A real-time object detection solution and its application in transportation. In Proceedings of the 2021 International Conference on Communications, Information System and Computer Engineering (CISCE), Beijing, China, 14–16 May 2021; IEEE: Piscataway, NJ, USA, 2021; pp. 486–491.
81. Zhu, H.; Yuen, K.V.; Mihaylova, L.; Leung, H. Overview of environment perception for intelligent vehicles. *IEEE Trans. Intell. Transp. Syst.* **2017**, *18*, 2584–2601. [CrossRef]
82. Suma, S.; Mehmood, R.; Albugami, N.; Katib, I.; Albeshri, A. Enabling Next Generation Logistics and Planning for Smarter Societies. *Procedia Comput. Sci.* **2017**, *109*, 1122–1127. [CrossRef]
83. Aqib, M.; Mehmood, R.; Alzahrani, A.; Katib, I. In-Memory Deep Learning Computations on GPUs for Prediction of Road Traffic Incidents Using Big Data Fusion. In *Smart Infrastructure and Applications*; Springer: Berlin/Heidelberg, Germany, 2020; pp. 79–114.
84. Aqib, M.; Mehmood, R.; Albeshri, A.; Alzahrani, A. Disaster Management in Smart Cities by Forecasting Traffic Plan Using Deep Learning and GPUs. In *Societies, Infrastructure, Technologies and Applications. SCITA 2017. Lecture Notes of the Institute for Computer Sciences, Social Informatics and Telecommunications Engineering (LNICST)*; Springer: Cham, Switzerland, 2017; Volume 224, pp. 139–154.
85. Aqib, M.; Mehmood, R.; Alzahrani, A.; Katib, I.; Albeshri, A.; Altowaijri, S. Rapid Transit Systems: Smarter Urban Planning Using Big Data, In-Memory Computing, Deep Learning, and GPUs. *Sustainability* **2019**, *11*, 2736. [CrossRef]
86. Aqib, M.; Mehmood, R.; Alzahrani, A.; Katib, I.; Albeshri, A.; Altowaijri, S. Smarter Traffic Prediction Using Big Data, In-Memory Computing, Deep Learning and GPUs. *Sensors* **2019**, *19*, 2206. [CrossRef]

87. Lv, Y.; Duan, Y.; Kang, W.; Li, Z.; Wang, F.Y. Traffic flow prediction with big data: A deep learning approach. *IEEE Trans. Intell. Transp. Syst.* **2014**, *16*, 865–873. [CrossRef]
88. Van Brummelen, J.; O'Brien, M.; Gruyer, D.; Najjaran, H. Autonomous vehicle perception: The technology of today and tomorrow. *Transp. Res. Part C Emerg. Technol.* **2018**, *89*, 384–406. [CrossRef]
89. Karagiannis, G.; Altintas, O.; Ekici, E.; Heijenk, G.; Jarupan, B.; Lin, K.; Weil, T. Vehicular Networking: A Survey and Tutorial on Requirements, Architectures, Challenges, Standards and Solutions. *IEEE Commun. Surv. Tutorials* **2011**, *13*, 584–616. [CrossRef]
90. Li, Y.; Liu, Y.; Wang, M. Study and realization for license plate recognition system. In Proceedings of the 2009 Asia-Pacific Conference on Information Processing, Shenzhen, China, 18–19 July 2009; IEEE: Piscataway, NJ, USA, 2009; Volume 2, pp. 501–504.
91. Theofilatos, A.; Chen, C.; Antoniou, C. Comparing machine learning and deep learning methods for real-time crash prediction. *Transp. Res. Rec.* **2019**, *2673*, 169–178. [CrossRef]
92. Alomari, E.; Katib, I.; Mehmood, R. Iktishaf: A big data road-traffic event detection tool using Twitter and spark machine learning. *Mobile Netw. Appl.* **2020**, 1–16.. [CrossRef]
93. Alomari, E.; Katib, I.; Albeshri, A.; Mehmood, R. COVID-19: Detecting government pandemic measures and public concerns from Twitter arabic data using distributed machine learning. *Int. J. Environ. Res. Public Health* **2021**, *18*, 282. [CrossRef] [PubMed]
94. Suma, S.; Mehmood, R.; Albeshri, A. Automatic event detection in smart cities using big data analytics. In Proceedings of the International Conference on Smart Cities, Infrastructure, Technologies and Applications, Jeddah, Saudi Arabia, 27–29 November 2018, Springer: Berlin/Heidelberg, Germany, 2017; pp. 111–122.
95. Suma, S.; Mehmood, R.; Albeshri, A. Automatic detection and validation of smart city events using hpc and apache spark platforms. In *Smart Infrastructure and Applications*; Springer: Berlin/Heidelberg, Germany, 2020; pp. 55–78.
96. Zhang, S. Using twitter to enhance traffic incident awareness. In Proceedings of the 2015 IEEE 18th International Conference on Intelligent Transportation Systems, Las Palmas de Gran Canaria, Spain, 15–18 September 2015, IEEE: Piscataway, NJ, USA, 2015; pp. 2941–2946.
97. Wang, D.; Al-Rubaie, A.; Clarke, S.S.; Davies, J. Real-time traffic event detection from social media. *ACM Trans. Internet Technol. (TOIT)* **2017**, *18*, 1–23. [CrossRef]
98. Wan, X.; Ghazzai, H.; Massoud, Y. Leveraging personal navigation assistant systems using automated social media traffic reporting. In Proceedings of the 2020 IEEE Technology and Engineering Management Conference (TEMSCON), Novi, MI, USA, 3–6 June 2020; IEEE: Piscataway, NJ, USA, 2020; pp. 1–6.
99. Heilig, L.; Voß, S. A scientometric analysis of public transport research. *J. Public Transp.* **2015**, *18*, 8. [CrossRef]
100. Das, S.; Sun, X.; Dutta, A. Text mining and topic modeling of compendiums of papers from transportation research board annual meetings. *Transp. Res. Rec.* **2016**, *2552*, 48–56. [CrossRef]
101. Sun, L.; Yin, Y. Discovering themes and trends in transportation research using topic modeling. *Transp. Res. Part C Emerg. Technol.* **2017**, *77*, 49–66. [CrossRef]
102. Zulkarnain; Putri, T.D. Intelligent transportation systems (ITS): A systematic review using a Natural Language Processing (NLP) approach. *Heliyon* **2021**. [CrossRef] [PubMed]
103. Zou, X.; Vu, H.L. Mapping the knowledge domain of road safety studies: A scientometric analysis. *Accid. Anal. Prev.* **2019**, *132*, 105243. [CrossRef] [PubMed]
104. Gao, J.; Wu, X.; Luo, X.; Guan, S. Scientometric Analysis of safety sign research: 1990–2019. *Int. J. Environ. Res. Public Health* **2021**, *18*, 273. [CrossRef]
105. Faisal, A.; Yigitcanlar, T.; Kamruzzaman, M.; Paz, A. Mapping two decades of autonomous vehicle research: A systematic scientometric analysis. *J. Urban Technol.* **2021**, *28*, 45–74. [CrossRef]
106. Topham, G. Sadiq Khan Considering £3.50 Daily Charge for Drivers Entering London. 2020. Available online: https://www.theguardian.com/uk-news/2020/dec/11/khan-considering-350-daily-charge-for-drivers-entering-london (accessed on 15 February 2022).
107. Walker, P. Local Councils Advised to Push Ahead with Traffic Reduction Schemes. 2020. Available online: https://www.theguardian.com/politics/2020/nov/13/local-councils-advised-to-push-ahead-with-traffic-reduction-schemes (accessed on 15 February 2022).
108. Topham, G. What Are Smart Motorways and Are They Safe? 2021 Available online: https://www.theguardian.com/world/2021/jan/19/what-are-smart-motorways-and-are-they-safe (accessed on 15 February 2022).
109. Osborne, H. London Borough of Newham tops UK Parking Fines Table. 2021. Available online: https://www.theguardian.com/society/2021/jan/18/london-borough-of-newham-tops-uk-parking-fines-table (accessed on 15 February 2022)
110. Carrington, D. Study Reveals World's Most Walkable Cities. 2020. Available online: https://www.theguardian.com/cities/2020/oct/15/study-reveals-worlds-most-walkable-cities (accessed on 15 February 2022).
111. Letters. Calm UK's Roads to Encourage Walking and Cycling. 2017. Available online: https://www.theguardian.com/politics/2017/mar/24/calm-uks-roads-to-encourage-walking-and-cycling (accessed on 15 February 2022).
112. Morris, S. Cardiff Proposes £2 Congestion Charge for Non-Residents. 2020. Available online: https://www.theguardian.com/politics/2020/jan/15/cardiff-proposes-congestion-charge-for-non-residents (accessed on 15 February 2022).
113. Connolly, K. Eco Wonder or Safety Nightmare? Germany to Vote on E-Scooters. 2019. Available online: https://www.theguardian.com/cities/2019/may/16/germany-to-vote-on-law-allowing-e-scooters-on-roads (accessed on 15 February 2022).

114. Harvey, F. Road Congestion Levels in Outer London Higher than before Lockdown. 15 September 2020. Available online: https://www.theguardian.com/environment/2020/sep/15/road-congestion-levels-in-outer-london-higher-than-before-lockdown (accessed on 15 February 2022).
115. Taylor, M. Air Pollution Roars Back in Parts of UK, Raising Covid Fears. 2020. Available online: https://www.theguardian.com/environment/2020/dec/10/air-pollution-roars-back-in-parts-of-uk-raising-covid-fears (accessed on 15 February 2022).
116. Monbiot, G. Electric Cars Won't Solve Our Pollution Problems—Britain Needs a Total Transport Rethink. 2020. Available online: https://www.theguardian.com/commentisfree/2020/sep/23/electric-cars-transport-train-companies (accessed on 15 February 2022).
117. Jolly, J. Petrol and Diesel Cars Could Cost Up to £1500 More under Proposals. 2020. Available online: https://www.theguardian.com/business/2020/sep/09/petrol-and-diesel-cars-could-cost-up-to-1500-more-under-proposals (accessed on 15 February 2022).
118. Laville, S. Manchester Becomes Latest UK City to Delay Clean Air Zone. 2020. Available online: https://www.theguardian.com/environment/2020/may/21/manchester-becomes-latest-uk-city-to-delay-clean-air-zone(accessed on 15 February 2022).
119. Taylor, M. Toxic Air Over London Falls by 50% at Busiest Traffic Spots. 2020. Available online: https://www.theguardian.com/environment/2020/apr/23/toxic-air-over-london-falls-by-50-at-busiest-traffic-spots (accessed on 15 February 2022).
120. Association, P. E10 Petrol: UK to Standardise Higher Ethanol Blend. 2020. Available online: https://www.theguardian.com/uk-news/2020/mar/04/e10-petrol-uk-to-standardise-higher-ethanol-blend (accessed on 15 February 2022).
121. Jolly, J. UK Could Ban Sale of Petrol and Diesel Cars in 12 years, Says Shapps. 2020. Available online: https://www.theguardian.com/environment/2020/feb/12/uk-ban-sale-petrol-diesel-cars-shapps-transport (accessed on 15 February 2022).
122. Murray, L. I Applaud Bristol for Banning Diesel Vehicles. But Why Not Ban All Private Cars? 2019. Available online: https://www.theguardian.com/commentisfree/2019/nov/04/bristol-ban-diesel-cars-climate-emergency (accessed on 15 February 2022).
123. Jones, H. Oxford Aims for World's First Zero Emissions Zone with Petrol Car Ban. 2017. Available online: https://www.theguardian.com/uk-news/2017/oct/11/oxford-aims-to-cut-air-pollution-car-ban-zero-emissions-zone (accessed on 15 February 2022).
124. Jolly, J. Electric Cars: UK Government Urged to Prevent 'Charging Deserts'. 2021. Available online: https://www.theguardian.com/business/2021/jul/23/electric-cars-uk-government-urged-to-prevent-charging-deserts (accessed on 15 February 2022).
125. Ambrose, J. UK's First All-Electric Car Charging Forecourt Opens in Essex. 2020. Available online: https://www.theguardian.com/environment/2020/dec/07/uk-first-all-electric-car-charging-forecourt-opens-in-essex (accessed on 15 February 2022).
126. Johnson, B. Why Obama Should Treat the Car Industry Like the Internet. 2008. Available online: https://www.theguardian.com/technology/blog/2008/dec/01/engineering-automotive (accessed on 15 February 2022).
127. Goodall, C. The 10 Big Energy Myths. 2008. Available online: https://www.theguardian.com/environment/2008/nov/27/renewableenergy-energy (accessed on 15 February 2022).
128. Syal, R. SUVs and Extra Traffic Cancelling Out Electric Car Gains in Britain. 26 February 2021. Available online: https://www.theguardian.com/environment/2021/feb/26/suvs-and-extra-traffic-cancelling-out-electric-car-gains-in-britain (accessed on 15 February 2022).
129. Editorial. *The Guardian* View on Fuel Shortages: This Crisis Needs Management. 2021. Available online: https://www.theguardian.com/commentisfree/2021/sep/28/the-guardian-view-on-fuel-shortages-this-crisis-needs-management (accessed on 15 February 2022).
130. Helm, T. Secret Report Reveals Government Fear of SCHOOLS chaos after No-Deal Brexit. 2019. Available online: https://www.theguardian.com/politics/2019/aug/03/secret-education-report-no-deal-brexit-school-chaos (accessed on 15 February 2022).
131. Weaver, M. Hundreds of KFC Shops Closed as Storage Depot Awaits Registration. 2018. Available online: https://www.theguardian.com/uk-news/2018/feb/21/hundreds-of-kfc-shops-closed-as-storage-depot-awaits-registration (accessed on 15 February 2022).
132. Weaver, M. KFC Was Warned about Switching UK Delivery Contractor, Union Says. 2018. Available online: https://www.theguardian.com/business/2018/feb/20/kfc-was-warned-about-switching-uk-delivery-contractor-union-says (accessed on 15 February 2022).
133. Jasper Jolly, Joanna Partridge, S.B.; Topham, G. There Is Nothing That Hasn't Gone Up' : How Rising Costs are Hitting UK Businesses. 2021. Available online: https://www.theguardian.com/business/2021/nov/17/how-rising-costs-are-hitting-uk-businesses-inflation-covid (accessed on 15 February 2022).
134. Moody, K. Why It's High Time to Move on from 'Just-In-Time' Supply Chains. 11 October 2021. Available online: https://www.theguardian.com/commentisfree/2021/oct/11/just-in-time-supply-chains-logistical-capitalism (accessed on 15 February 2022).
135. Addley, E. Get Out That Puncture Repair Kit…2000. Available online: https://www.theguardian.com/money/2000/jun/26/officehours.workandcareers (accessed on 15 February 2022).
136. Clark, A. Charity Cyclists Barred from New Trains. 2004. Available online: https://www.theguardian.com/uk/2004/jan/15/transport.world (accessed on 15 February 2022).
137. Jowit, J. Beware Mature Men Wobbling on Big Motorbikes. 2004. Available online: https://www.theguardian.com/uk/2004/feb/29/transport.world (accessed on 15 February 2022).
138. Staff.; agencies. On Yer Bikes, Urge Pedal-Pushing MPs. 2004. Available online: https://www.theguardian.com/politics/2004/jun/16/immigrationpolicy.transport (accessed on 15 February 2022).

139. Laker, L. Big Rise in UK Weekend Cycling Amid Calls for More Investment. 2021. Available online: https://www.theguardian.com/lifeandstyle/2021/jul/22/big-rise-in-uk-weekend-cycling-amid-calls-for-more-investment (accessed on 15 February 2022).
140. Pidd, H.; Glover, M. Campaigners Lose Legal Challenge Over Lake District 4 × 4 vehicles. 2020. Available online: https://www.theguardian.com/uk-news/2020/aug/24/campaigners-lose-legal-challenge-over-lake-district-4x4-vehicles (accessed on 15 February 2022).
141. Taylor, M. UK Increase in Cycling and Walking Must Be Nurtured, Says Minister. 2021. Available online: https://www.theguardian.com/politics/2021/mar/18/uk-increase-in-cycling-and-walking-must-be-nurtured-says-minister (accessed on 15 February 2022).
142. Hopkins, J. How to Stop Your Bicycle from Being Stolen. 2020. Available online: https://www.theguardian.com/environment/bike-blog/2020/aug/28/how-to-stop-your-bicycle-from-being-stolen (accessed on 15 February 2022).
143. Reid, C. How to Take a Cycling Holiday This Year Despite the Pandemic. 2020. Available online: https://www.theguardian.com/travel/bike-blog/2020/aug/12/how-to-take-a-cycling-holiday-this-year-despite-the-pandemic (accessed on 15 February 2022).
144. Parveen, N. West Midlands to Gain 500-Mile Cycle Network. 2020. Available online: https://www.theguardian.com/uk-news/2020/aug/11/west-midlands-to-gain-500-miles-of-cycle-routes (accessed on 15 February 2022).
145. Caird, J. How to start CYCLING with Young Children. 2020. Available online: https://www.theguardian.com/environment/bike-blog/2020/aug/04/how-to-start-cycling-with-young-children (accessed on 15 February 2022).
146. Topham, G. Uber to Launch Jump Bike Hire Scheme in Islington Borough. 2019. Available online: https://www.theguardian.com/uk-news/2019/may/24/uber-to-launch-jump-bike-hire-scheme-in-islington-borough (accessed on 15 February 2022).
147. Walters, J. Road Death Risk Higher for Deprived Children. 2002. Available online: https://www.theguardian.com/politics/2002/oct/20/uk.transport (accessed on 15 February 2022).
148. Clark, A. Shock Tactics Hailed as Road Injuries Fall. 2003. Available online: https://www.theguardian.com/uk/2003/may/02/transport.immigrationpolicy (accessed on 15 February 2022).
149. Walker, P. Cyclist Fatalities on British Roads Rose by 40% in 2020, says DfT. 2021. Available online: https://www.theguardian.com/world/2021/jun/24/cyclists-killed-on-british-roads-rose-by-40-in-2020-official-figures-show (accessed on 15 February 2022).
150. Letters. Sound Advice on Staying Safe When Cycling. 2020. Available online: https://www.theguardian.com/lifeandstyle/2020/jun/03/sound-advice-on-staying-safe-when-cycling (accessed on 15 February 2022).
151. Walker, P. Drivers Who Pass Cycle Training Scheme Could Get Cheaper Insurance. 2018. Available online: https://www.theguardian.com/uk-news/2018/nov/22/drivers-who-pass-cycle-training-scheme-could-get-cheaper-insurance (accessed on 15 February 2022).
152. Jeffries, S. Pavements Are for Everybody. Except Pedestrians. 2018. Available online: https://www.theguardian.com/commentisfree/2018/feb/27/pavements-pedestrians-cyclists-walkers-hire-bikes (accessed on 15 February 2022).
153. news, U. Pensioners to Get Cheap Bus Travel. 2000. Available online: https://www.theguardian.com/uk/2000/aug/09/transport.world (accessed on 15 February 2022).
154. Wainwright, M. Undercover Pensioners Test Bus Service Quality. 2000. Available online: https://www.theguardian.com/world/2000/mar/06/transport.uk (accessed on 15 February 2022).
155. Clark, A. Half Fares on Buses Urged for Students and Jobless. 2002. Available online: https://www.theguardian.com/uk/2002/dec/02/highereducation.transport (accessed on 15 February 2022).
156. Hill, D. London'S Bus Drivers Deserve A Better Deal. 2015. Available online: https://www.theguardian.com/uk-news/davehillblog/2015/jan/13/londons-bus-drivers-deserve-a-better-deal (accessed on 15 February 2022).
157. Tran, M. London Bus Strike Hits Commuters. 2015. Available online: https://www.theguardian.com/uk-news/2015/jan/13/london-bus-strike-hits-commuters (accessed on 15 February 2022).
158. Letters. A Case for Integrated Public Transport across the Country. 2019. Available online: https://www.theguardian.com/politics/2019/may/27/a-case-for-integrated-public-transport-across-the-country (accessed on 15 February 2022).
159. Topham, G. Air-Filtering Bus to Launch Across Six Regions in the UK. 2020. Available online: https://www.theguardian.com/global/2020/jan/24/air-filtering-bus-to-launch-across-six-regions-in-the-uk (accessed on 15 February 2022).
160. Letters. Why Aren't All Bus Drivers Being Protected? 2020. Available online: https://www.theguardian.com/world/2020/apr/07/why-arent-all-bus-drivers-being-protected (accessed on 15 February 2022).
161. Ambrose, J. UK Electric Buses Boosted by Innovative £20 m Battery Deal. 2020. Available online: https://www.theguardian.com/business/2020/jun/23/uk-electric-buses-battery-deal-zenobe-energy (accessed on 15 February 2022).
162. Blackall, M. Behaviour Is Getting Worse': The Latest from the UK Covid Frontline. 2021. Available online: https://www.theguardian.com/world/2021/mar/09/behaviour-is-getting-worse-the-latest-from-the-uk-covid-frontline (accessed on 15 February 2022).
163. Media, P. Boris Johnson to Unveil £3 bn Bus Sector Shake-Up to Drive 'Levelling Up'. 2021. Available online: https://www.theguardian.com/politics/2021/mar/15/boris-johnson-to-unveil-3bn-bus-sector-shake-up-to-drive-levelling-up (accessed on 15 February 2022).
164. Team, G.C. Tell Us: Has Your Commute Changed Due to the Pandemic? 2021. Available online: https://www.theguardian.com/cities/2021/jun/21/tell-us-has-the-pandemic-changed-your-attitude-towards-commuting (accessed on 15 February 2022).

165. Mason, R. Transport Secretary Backs London Mayor's Rule for Compulsory Masks on TfL. 2021. Available online: https://www.theguardian.com/uk-news/2021/jul/14/shapps-mask-wearing-expected-to-remain-on-public-transport (accessed on 15 February 2022).
166. Topham, G. UK Bus privatisation BREACHED Basic Rights, Says Ex-UN Rapporteur. 2021. Available online: https://www.theguardian.com/politics/2021/jul/19/uk-bus-privatisation-has-caused-poverty-and-job-losses-says-un (accessed on 15 February 2022).
167. Topham, G. Covid Set Back Attitudes to Public Transport by TWO decades, Says RAC. 2020. Available online: https://www.theguardian.com/uk-news/2020/nov/09/covid-set-back-attitudes-to-public-transport-by-two-decades-says-rac (accessed on 15 February 2022).
168. Khawaja, B. Bus Privatisation Has Destroyed a British Public Service—But There Is a Way Back. 26 July 2021. Available online: https://www.theguardian.com/commentisfree/2021/jul/26/bus-privatisation-public-service-strategy-british-private-market (accessed on 15 February 2022).
169. Association, P. East coast MAINLINE Trains Severely Disrupted by Wire Damage. 10 July 2019. Available online: https://www.theguardian.com/uk-news/2019/jul/10/east-coast-mainline-trains-severely-disrupted-by-wire-damage (accessed on 15 February 2022).
170. Walawalkar, A. Staff Shortages Hit Rail Services as New Timetable Takes Effect. 15 December 2019. Available online: https://www.theguardian.com/uk-news/2019/dec/15/staff-shortages-hit-rail-services-as-new-timetable-takes-effect (accessed on 15 February 2022).
171. Walker, A. Majority of Passengers with Disabilities Report Problems Using Trains. 2019. Available online: https://www.theguardian.com/society/2019/jul/08/majority-passengers-disabilities-report-problems-using-trains-uk (accessed on 15 February 2022).
172. Topham, G. London Seeks Further Treasury Funds for Delayed Crossrail. 2018. Available online: https://www.theguardian.com/uk-news/2018/sep/06/delays-to-london-crossrail-could-cost-tfl-20m-in-lost-revenue (accessed on 15 February 2022).
173. Topham, G. Crossrail Boss Steps Down after Project Delays. 2018. Available online: https://www.theguardian.com/uk-news/2018/nov/02/crossrail-boss-steps-down-after-project-delays (accessed on 15 February 2022).
174. Topham, G. Southern Rail Gets Lowest Satisfaction Score in Passenger Survey. 2017. Available online: https://www.theguardian.com/uk-news/2017/jul/25/southern-rail-gets-lowest-satisfaction-score-in-passenger-survey (accessed on 15 February 2022).
175. Branigan, T. Big Train Firms Fare Worst for Satisfaction. 2007. Available online: https://www.theguardian.com/uk/2007/jun/05/transport.politics (accessed on 15 February 2022).
176. Cash, M. The Southern Scandal Has Gone on Long Enough. GTR Must Lose Its Contract. 2016. Available online: https://www.theguardian.com/commentisfree/2016/jul/20/southern-scandal-gtr-rail-services (accessed on 15 February 2022).
177. Walters, J.; Ahmed, K. Railtrack Goes Half Steam Ahead. 2000. Available online: https://www.theguardian.com/uk/2000/nov/26/transport.world (accessed on 15 February 2022).
178. Topham, G. 'What Matters Is That It Works': Peter Hendy on TfL, Network Rail and Mayors. 2015. Available online: https://www.theguardian.com/business/2015/jul/03/ex-london-transport-chief-completes-move-national-rail (accessed on 15 February 2022).
179. Kelso, P. Branson Derailed in Easyjet Head to Head. 1999. Available online: https://www.theguardian.com/world/1999/dec/11/transport.uk (accessed on 15 February 2022).
180. Harper, K.; Wilson, J. Apex Fares Cancelled by Train Companies. 2000. Available online: https://www.theguardian.com/uk/2000/oct/28/transport.hatfield2 (accessed on 15 February 2022).
181. The Guardian. Most passengers miss out on rail compensation. 2000. Available online: https://www.theguardian.com/world/2000/nov/21/transport.uk (accessed on 15 February 2022).
182. Topham, G. Rail Fares in England to Rise above Inflation for First Time Since 2013. 2020. Available online: https://www.theguardian.com/money/2020/dec/16/rail-fares-in-england-to-rise-above-inflation-for-first-time-since-2013 (accessed on 15 February 2022).
183. Topham, G. Rail Fares to Increase by 3.8% in March. 2021. Available online: https://www.theguardian.com/money/2021/dec/17/rail-fares-increase-march-inflation (accessed on 15 February 2022).
184. Topham, G. FirstGroup to Launch Budget London to Edinburgh Rail Service Next Month. 2021. Available online: https://www.theguardian.com/uk-news/2021/sep/07/firstgroup-to-launch-budget-london-to-edinburgh-rail-service-next-month (accessed on 15 February 2022).
185. Topham, G.; Makortoff, K. Flexible Rail Season Tickets in England Criticised over Savings Claims. 21 June 2021. Available online: https://www.theguardian.com/money/2021/jun/21/flexible-rail-season-tickets-on-sale--england-covid-19-pandemic (accessed on 15 February 2022).
186. Topham, G. Back to the Bad Old Days': Swingeing Rail Cuts Set Alarm Bells Ringing. 5 December 2021. Available online: https://www.theguardian.com/business/2021/dec/05/back-bad-old-days-swingeing-rail-cuts-alarm-bells-ringing (accessed on 15 February 2022).
187. Milmo, D. Network Rail Cuts Bonuses over Crash. 2007. Available online: https://www.theguardian.com/business/2007/may/24/transportintheuk1 (accessed on 15 February 2022).

188. Milmo, D. Downturn Is Shunting Railway Industry towards Bail-Out. 2009. Available online: https://www.theguardian.com/business/2009/feb/01/transport-rail-industry-uk-bail-out (accessed on 15 February 2022).
189. Milmo, D. Go-Ahead 'Surprised' by Public Transport Boom. 2008. Available online: https://www.theguardian.com/business/2008/sep/06/goaheadgroup.transport (accessed on 15 February 2022).
190. Milmo, D. Drivers Taking the Bus Help FirstGroup Revenues. 2007. Available online: https://www.theguardian.com/business/2007/nov/08/transportintheuk (accessed on 15 February 2022).
191. The Guardian. Government 'Underestimates' Air Travel Boom. 2006. Available online: https://www.theguardian.com/travel/2006/nov/20/travelnews.uknews.transportintheuk1 (accessed on 15 February 2022).
192. Milmo, D. Tube Station Work Halted after Discovery of £400 m Funding Gap. 2009. Available online: https://www.theguardian.com/uk/2009/apr/01/london-transport-tube-travel (accessed on 15 February 2022).
193. Greenfield, E. Channel Tunnel is Final Choice—'a sound investment'. 1964. Available online: https://www.theguardian.com/world/1964/feb/07/france.transport (accessed on 15 February 2022).
194. Kramer, S. Bonds: They'll Work Here. 1999. Available online: https://www.theguardian.com/world/1999/nov/24/transport.uk1 (accessed on 15 February 2022).
195. Bowers, S. Tube Bidder Lauds Private Sector Role. 2000. Available online: https://www.theguardian.com/business/2000/dec/01/transportintheuk (accessed on 15 February 2022).
196. The Guardian. Transport Commissioner Attacks Labour's Tube Plans. 2001. Available online: https://www.theguardian.com/world/2001/jul/06/transport.uk (accessed on 15 February 2022).
197. Walters, J. Virgin to Sue Railtrack. 2001. Available online: https://www.theguardian.com/business/2001/apr/15/transportintheuk.transportpolicy (accessed on 15 February 2022).
198. The Guardian. Four Die as Train Derails. 2000. Available online: https://www.theguardian.com/uk/2000/oct/17/hatfield.transport (accessed on 15 February 2022).
199. Milmo, D. Tube Lines Chief Offered National Express Job. 2009. Available online: https://www.theguardian.com/business/2009/dec/07/tube-lines-finch-national-express (accessed on 15 February 2022).
200. Syal, R. Chris Grayling Lands £100 k Job Advising Some of UK's Top Ports. 2020. Available online: https://www.theguardian.com/politics/2020/sep/17/chris-grayling-lands-100k-job-advising-some-of-uks-top-ports (accessed on 15 February 2022).
201. Greenfield, P. No-Deal Brexit Ferry Company Owns No Ships and Has Never Run Channel Service. 2018. Available online: https://www.theguardian.com/politics/2018/dec/30/no-deal-brexit-ferry-company-owns-no-ships-and-has-never-run-ferry-service (accessed on 15 February 2022).
202. Whey, S. Stagecoach Makes £1.7 bn Merger Offer for Rival National Express. 2009. Available online: https://www.theguardian.com/uk/2009/oct/19/national-express-stagecoach-merger (accessed on 15 February 2022).
203. Milmo, D. Eurotunnel Shareholders Cast Decisive Vote. 2007. Available online: https://www.theguardian.com/business/2007/may/21/france.transportintheuk (accessed on 15 February 2022).
204. Inman, P. When Private Meets Public Sector: The History of a Tangled Relationship. 2018. Available online: https://www.theguardian.com/politics/2018/jan/13/private-public-sector-history-tangled-relationship (accessed on 15 February 2022).
205. Greenfield, P. FirstGroup Rejects Takeover Bid from US Equity Group Apollo. 2018. Available online: https://www.theguardian.com/business/2018/apr/11/firstgroup-rejects-takeover-bid-from-us-equity-group-apollo (accessed on 15 February 2022).
206. Topham, G. Eurotunnel Renamed Getlink in Preparation for Post-Brexit Era. 20 November 2017. Available online: https://www.theguardian.com/business/2017/nov/20/eurotunnel-rebrand-getlink-brexit-channel-tunnel (accessed on 15 February 2022).
207. Topham, G. HS2 Rail Leg to Leeds Scrapped, Grant Shapps Confirms. 18 November 2021. Available online: https://www.theguardian.com/uk-news/2021/nov/18/hs2-rail-leg-to-leeds-scrapped-grant-shapps-confirms (accessed on 15 February 2022).
208. Heathrow Airport Expansion. 2016. Available online: https://www.gov.uk/government/collections/heathrow-airport-expansion (accessed on 15 February 2022).
209. Airport Expansions. Available online: https://friendsoftheearth.uk/climate/airport-expansions (accessed on 15 February 2022).
210. Taylor, M. Campaigners Say UK Airport Expansion Plans Must Be Suspended Amid New Climate Goals. 2021. Available online: https://www.theguardian.com/world/2021/may/10/campaigners-say-uk-airport-expansion-plans-must-be-suspended-amid-new-climate-goals (accessed on 15 February 2022).
211. Third Runway. Available online: https://www.theguardian.com/environment/heathrow-third-runway (accessed on 15 February 2022).
212. Asthana, A.; Laville, S.; Kale, S. Grounded: Why Heathrow's Third Runway May Never Happen. 6 March 2020. Available online: https://www.theguardian.com/news/audio/2020/mar/06/grounded-why-heathrows-third-runway-may-never-happen (accessed on 15 February 2022).
213. Topham, G. Heathrow to Challenge Third Runway Verdict Using Climate Pledge. 2020. Available online: https://www.theguardian.com/environment/2020/oct/06/heathrow-to-challenge-third-runway-verdict-using-climate-pledge (accessed on 15 February 2022).
214. Media, P. Climate Lawyer Loses Supreme Court Appeal over Heathrow Leak. 20 December 2021. Available online: https://www.theguardian.com/environment/2021/dec/20/climate-lawyer-loses-supreme-court-appeal-over-heathrow-leak (accessed on 15 February 2022).

215. Ungoed-Thomas, J. Nearly 300 Flights within UK Taken by Government Staff... Every Day. 7 November 2021. Available online: https://www.theguardian.com/environment/2021/nov/07/nearly-300-flights-within-uk-taken-by-government-staff-every-day (accessed on 15 February 2022).
216. Harper, K. Inquiry Likely to Focus on Driver Error. 1999. Available online: https://www.theguardian.com/uk/1999/oct/06/ladbrokegrove.transport1 (accessed on 15 February 2022).
217. Carnage at Paddington. 1999. Available online: https://www.theguardian.com/uk/1999/oct/06/ladbrokegrove.transport9 (accessed on 15 February 2022).
218. Statement by Railtrack, Thames Trains, First Great Western. 1999. Available online: https://www.theguardian.com/uk/1999/oct/06/ladbrokegrove.transport5 (accessed on 15 February 2022).
219. Harper, K.; Maguire, K. Prescott Pledges £1 bn for Rail Safety. 1999. Available online: https://www.theguardian.com/uk/1999/oct/07/ladbrokegrove.transport10 (accessed on 15 February 2022).
220. Harper, K.; Hall, S. Lamentable Failures that Claimed 31 Lives. 2001. Available online: https://www.theguardian.com/uk/2001/jun/20/ladbrokegrove.transport2 (accessed on 15 February 2022).
221. Topham, G.; Morris, S. Salisbury Train Crash May Have been Caused by Leaves on Line. 2 November 2021. Available online: https://www.theguardian.com/uk-news/2021/nov/02/salisbury-train-crash-may-have-been-caused-by-leaves-on-line (accessed on 15 February 2022).
222. The Guardian. Railtrack Faces £600 m Losses. 2001. Available online: https://www.theguardian.com/world/2001/jan/15/transport.uk (accessed on 15 February 2022).
223. Khomami, N. Commuters Urged to Make Small Talk to Help Prevent Railway Suicides. 2017. Available online: https://www.theguardian.com/uk-news/2017/nov/15/commuters-urged-to-make-small-talk-to-help-prevent-railway-suicides (accessed on 15 February 2022).
224. Britain's Rail Disaster History. 2000. Available online: https://www.theguardian.com/uk/2000/oct/17/ladbrokegrove.transport (accessed on 15 February 2022).
225. Weaver, M.; Dodd, V. UK Rail Worker Dies of Coronavirus after Being Spat at While on Duty. 2020. Available online: https://www.theguardian.com/uk-news/2020/may/12/uk-rail-worker-dies-coronavirus-spat-belly-mujinga (accessed on 15 February 2022).
226. Wainwright, M. Selby Rail CRASH Motorist blames 'Fate'. 2011. Available online: https://www.theguardian.com/uk/2011/feb/28/selby-rail-crash-driver-fate (accessed on 15 February 2022).
227. Malik, S.; Agencies. Woman and Child Killed by Rush-Hour Train in Surrey. 2013. Available online: https://www.theguardian.com/uk/2013/mar/22/woman-child-killed-train-surrey (accessed on 15 February 2022).
228. Quinn, B. Woman Gets Electric Shock after Climbing on to Freight Train. 2014. Available online: https://www.theguardian.com/uk-news/2014/feb/01/woman-train-hackney-hospital (accessed on 15 February 2022).
229. Siddique, H. Heathrow Airport: Engineer Killed in Vehicle Crash on Airfield. 2018. Available online: https://www.theguardian.com/uk-news/2018/feb/14/heathrow-airport-serious-accident-on-airfield-disrupts-flights (accessed on 15 February 2022).
230. Association, P. Woman Killed on Train May Have been Leaning Out of Window, Say Police. 2018. Available online: https://www.theguardian.com/uk-news/2018/dec/02/woman-killed-on-train-may-have-been-leaning-out-of-window-say-police (accessed on 15 February 2022).
231. Campbell, L. Man Jailed for Driving a Car Half a Mile on Railway Track in Birmingham. 2021. Available online: https://www.theguardian.com/uk-news/2021/sep/07/man-jailed-for-driving-a-car-half-a-mile-on-railway-track-in-birmingham (accessed on 15 February 2022).
232. Morris, S. Woman Hit by Branch When Leaning Out Train Window Near Bath, Inquest Hears. 2021. Available online: https://www.theguardian.com/uk-news/2021/may/26/woman-hit-by-branch-leaning-out-train-window-inquest-bethan-roper (accessed on 15 February 2022).
233. Marsh, S. Missing 12-Year-Old Girls Found after Night Locked Inside Train. 2021. Available online: https://www.theguardian.com/uk-news/2021/mar/22/missing-12-year-old-girls-found-after-night-locked-inside-train (accessed on 15 February 2022).
234. The Government's Road Safety Strategy: Text of Tony Blair's Speech. 2000. Available online: https://www.theguardian.com/world/2000/mar/01/transport.uk1 (accessed on 15 February 2022).
235. Association, P. Bail for Driver in Fatal Coach Crash. 2007. Available online: https://www.theguardian.com/uk/2007/jan/05/transport.world2 (accessed on 15 February 2022).
236. Davies, C. Britain's Freeze Claims a New Victim as Boy Falls through Ice. 2009. Available online: https://www.theguardian.com/uk/2009/feb/08/snow-deaths-ice (accessed on 15 February 2022).
237. Carrell, S. Girl DIES in school Bus Crash. 2010. Available online: https://www.theguardian.com/uk/2010/mar/31/school-bus-crash-kills-girl (accessed on 15 February 2022).
238. Association, P. M5 Crash: Somerset Residents Describe Tragic Scene. 2011. Available online: https://www.theguardian.com/uk/2011/nov/05/m5-crash-somerset-residents-scene (accessed on 15 February 2022).
239. Neild, B. M5 Coach CRASH leaves One Dead as Police Arrest Driver. 2012. Available online: https://www.theguardian.com/uk/2012/mar/24/m5-coach-crash-lorry-motorway (accessed on 15 February 2022).
240. Morris, S. Police Investigating Claims of 'BMW Race' in Lead-Up to M1 Crash. 2012. Available online: https://www.theguardian.com/uk/2012/dec/03/police-investigating-bmw-race-crash (accessed on 15 February 2022).

241. Perraudin, F. Man Dies in Collision with Nottingham Tram. 2016. Available online: https://www.theguardian.com/uk-news/2016/aug/16/man-dies-in-collision-with-nottingham-tram (accessed on 15 February 2022).
242. Wolfe-Robinson, M. Sixty Schoolchildren Rescued from Doubledecker in Lincoln after Crash. 11 November 2021. Available online: https://www.theguardian.com/uk-news/2021/nov/11/sixty-schoolchildren-rescued-from-doubledecker-in-lincoln-after-crash (accessed on 15 February 2022).
243. Slawson, N. Woman Killed as Buses Collide at Victoria Station in London. 2021. Available online: https://www.theguardian.com/uk-news/2021/aug/10/woman-killed-buses-collide-victoria-station-london (accessed on 15 February 2022).
244. Wolfe-Robinson, M. Smart Motorways Present Ongoing Risk of Death, Says Coroner. 2021. Available online: https://www.theguardian.com/world/2021/jan/18/smart-motorways-present-ongoing-risk-of-death-says-coroner (accessed on 15 February 2022).
245. Topham, G. Highways England's Smart Motorways Policy Killed Drivers, Says Ex-Minister. 2020. Available online: https://www.theguardian.com/politics/2020/jan/27/highways-englands-smart-motorways-policy-killed-drivers-exminister (accessed on 15 February 2022).
246. The Guardian. Government Accused of 'Wimping out' on Road Safety. 2000. Available online: https://www.theguardian.com/world/2000/mar/01/transport.uk (accessed on 15 February 2022).
247. Gibbs, G. Fake Crash Aimed to Deter Joyriding. 2000. Available online: https://www.theguardian.com/world/2000/mar/10/transport.uk (accessed on 15 February 2022).
248. Harper, K. Drink-Driving Cave-in to Alcohol Lobby. 2000. Available online: https://www.theguardian.com/uk/2000/mar/27/transport.world (accessed on 15 February 2022).
249. Wainwright, M. Dozing Driver Who Caused 10 Deaths Gets Five Years. 2002. Available online: https://www.theguardian.com/uk/2002/jan/12/transport.selby (accessed on 15 February 2022).
250. Carter, H. Driving test enters VIRTUAL Reality. 2002. Available online: https://www.theguardian.com/uk/2002/sep/20/transport.world (accessed on 15 February 2022).
251. Ahmed, K. Reckless Drivers Who Kill Face 14 Years in Jail. 2003. Available online: https://www.theguardian.com/politics/2003/may/11/transport.ukcrime (accessed on 15 February 2022).
252. Jowit, J. Device That Won't Let Drinkers Drive. 2004. Available online: https://www.theguardian.com/politics/2004/feb/29/uk.transport (accessed on 15 February 2022).
253. Muir, H. Met 'Soft' on Drink and Drive Officers. 2004. Available online: https://www.theguardian.com/uk/2004/jun/22/drugsandalcohol.ukcrime (accessed on 15 February 2022).
254. Jones, R. Car Menaces Still at Large. 2004. Available online: https://www.theguardian.com/uk/2004/jul/17/motoring.lifeandhealth (accessed on 15 February 2022).
255. Sturcke, J. Privacy CONCERNS over police ROAD Camera Plans. 2005. Available online: https://www.theguardian.com/uk/2005/mar/23/ukcrime.transport (accessed on 15 February 2022).
256. Morris, S. Judge Gives Discharge to 159mph PC. 2006. Available online: https://www.theguardian.com/uk/2006/aug/26/transport.ukcrime (accessed on 15 February 2022).
257. Osborne, H. Speed Limit Enforcement Cuts Pollution, Report Shows. 2006. Available online: https://www.theguardian.com/environment/2006/oct/23/travelsenvironmentalimpact.transportintheuk (accessed on 15 February 2022).
258. Dyer, C. Motorists Risk Jail for Using Phones in Car. 2006. Available online: https://www.theguardian.com/uk/2006/dec/14/ukcrime.mobilephones (accessed on 15 February 2022).
259. Wainwright, M. Driver Who Deflected Speed Guns Guilty of Perverting Justice. 2007. Available online: https://www.theguardian.com/uk/2007/aug/31/transport.world1 (accessed on 15 February 2022).
260. Milmo, D. Driving Test Changes Planned to Cut Deaths of Young Motorists. 2009. Available online: https://www.theguardian.com/uk/2009/apr/21/driving-test-changes (accessed on 15 February 2022).
261. Bennett, C. Driverless cars WILL Geld the Dangerous Stallions of the Road. 2016. Available online: https://www.theguardian.com/commentisfree/2016/dec/04/driverless-cars-boris-johnson-stallion-road-safety (accessed on 15 February 2022).
262. O'Carroll, L. RAC Hits Out at 'Truly Shocking' Lockdown Speeding Offences. 2 June 2020. Available online: https://www.theguardian.com/uk-news/2020/jun/02/rac-hits-out-at-truly-shocking-lockdown-speeding-offences (accessed on 15 February 2022).
263. The Guardian. Beware of 'Ice Bombs' from Airliners, Says MP. 2000. Available online: https://www.theguardian.com/uk/2000/nov/16/transport.world (accessed on 15 February 2022).
264. Patrick Greenfield, P.W.; Wolfe-Robinson, M. Bad Weather Causes Delays on Train Routes to Glasgow Cop26 Talks. 2021. Available online: https://www.theguardian.com/uk-news/2021/oct/31/bad-weather-causes-delays-on-train-route-to-glasgow-cop26-talks (accessed on 15 February 2022).
265. Tran, M. Network Rail Unveils 'Magic' De-Icer. 2005. Available online: https://www.theguardian.com/business/2005/nov/24/transportintheuk.money (accessed on 15 February 2022).
266. Balakrishnan, A. Peril on Roads after Heavy Snow in Northern England and Scotland. 2008. Available online: https://www.theguardian.com/uk/2008/dec/04/snow-roads-northern-england-scotland (accessed on 15 February 2022).
267. Jones, S.; Gillan, A. The Day the Snow Came—And Britain Stopped. 2009. Available online: https://www.theguardian.com/uk/2009/feb/02/snow-brings-britain-travel-chaos (accessed on 15 February 2022).

268. Mulholland, H. London Snowfall: Boris Transport Chief Rejects Livingstone Criticism. 2009. Available online: https://www.theguardian.com/politics/2009/feb/03/boris-johnson-ken-livingstone-snow-london (accessed on 15 February 2022).
269. The Guardian. Transport Minister Resigns over Scotland's Snow Chaos. 2010. Available online: https://www.theguardian.com/uk/2010/dec/11/scottish-transport-minister-stewart-stevenson-resigns-snow (accessed on 15 February 2022).
270. Williams, R. Floods, Snow and Gales Batter Britain. 2009. Available online: https://www.theguardian.com/uk/2009/feb/10/flood-snow-weather-uk (accessed on 15 February 2022).
271. Milmo, D.; Gabbatt, A. Snow Strands Passengers at Gatwick and Luton. 2010. Available online: https://www.theguardian.com/uk/2009/dec/18/snow-england-north-east-south (accessed on 15 February 2022).
272. Walker, P. Schools Reopen as UK Temperatures Rise. 2010. Available online: https://www.theguardian.com/uk/2010/jan/11/schools-reopen-uk-temperatures-rise (accessed on 15 February 2022).
273. Wainwright, M.; Milmo, D. UK Snow-Readiness Audit Ordered as Travel Chaos Takes over Country. 2010. Available online: https://www.theguardian.com/uk/2010/dec/01/snow-readiness-audit-travel-chaos (accessed on 15 February 2022).
274. Mistlin, A. UK Weather: Storm Darcy to Cause Further Disruption across UK. 8 February 2021. Available online: https://www.theguardian.com/uk-news/2021/feb/08/storm-darcy-snow-ice-further-disruption-across-uk (accessed on 15 February 2022).
275. Topham, G.; Ratcliffe, R. Smooth Christmas Getaway in Prospect with UK Transport Links Working Well. 2011. Available online: https://www.theguardian.com/uk/2011/dec/16/christmas-getaway-rail-roads-weather-good (accessed on 15 February 2022).
276. Peter Walker, B.Q.; Urquhart, C. Storms across Britain Leave Five Dead and Christmas Travel in Chaos. 2013. Available online: https://www.theguardian.com/uk-news/2013/dec/24/storms-britain-dead-christmas-travel-chaos (accessed on 15 February 2022).
277. Gwyn Topham, T.C. UK Storms and Flooding Set Back Rail Repairs. 2014. Available online: https://www.theguardian.com/environment/2014/feb/17/uk-storms-flooding-rail-repairs (accessed on 15 February 2022).
278. Davies, C.; Gayle, D. Villagers Evacuated as Snow, High Winds and Flooding Bring Havoc to UK. 2017. Available online: https://www.theguardian.com/uk-news/2017/jan/12/storms-bring-snow-high-winds-and-flood-warnings-to-britain (accessed on 15 February 2022).
279. Association, P. Scottish Firms Warned Not to Punish Staff Who Followed Snow Advice. 2018. Available online: https://www.theguardian.com/uk-news/2018/mar/04/scottish-employers-warned-penalising-staff-snow-advice (accessed on 15 February 2022).
280. Busby, M. UK Weather: Heavy Rain and High Temperatures Bring Travel Chaos. 2019. Available online: https://www.theguardian.com/uk-news/2019/jul/27/uk-weather-warning-heavy-rain-high-temperatures-met-office (accessed on 15 February 2022).
281. Topham, G. UK Railways Cannot Cope with Climate Crisis, Says Rail Boss. 2019. Available online: https://www.theguardian.com/uk-news/2019/nov/07/uk-railways-cannot-cope-with-effects-climate-change-warns-rail-boss (accessed on 15 February 2022).
282. Campbell, L. Two Bridges Swept away Amid Flooding in Scottish Borders. 29 October 2021. Available online: https://www.theguardian.com/uk-news/2021/oct/29/two-bridges-swept-away-flooding-scottish-borders (accessed on 15 February 2022).
283. Rawlinson, K. London Tube to Close Many Stations Because of Coronavirus. 2020. Available online: https://www.theguardian.com/uk-news/2020/mar/19/london-tube-to-close-many-stations-because-of-coronavirus (accessed on 15 February 2022).
284. Hall, R. Bank Holiday Weekend Travel Warning as Fine Weather Forecast for UK. 27 August 2021. Available online: https://www.theguardian.com/uk-news/2021/aug/27/bank-holiday-weekend-travel-warning-as-fine-weather-forecast-for-uk (accessed on 15 February 2022).
285. Campbell, L. Tower Bridge Stuck in Open Position Overnight after 'Technical Failure'. 2021. Available online: https://www.theguardian.com/uk-news/2021/aug/09/disruption-as-tower-bridge-stuck-open-after-technical-failure (accessed on 15 February 2022).
286. Press Association. London Waterloo Commuters Told Not to Travel Due to Engineering Delays. 2018. Available online: https://www.theguardian.com/uk-news/2018/nov/19/london-waterloo-commuters-told-not-to-travel-due-to-engineering-delays (accessed on 15 February 2022).
287. Pidd, H.; Walker, A. Lightning Strikes Disrupt rail Services in Northern England. 2018. Available online: https://www.theguardian.com/uk-news/2018/jul/27/lightning-strikes-disrupt-rail-services-in-northern-england (accessed on 15 February 2022).
288. Slawson, N.; Grierson, J. Nottingham Train Station Fire Was Arson, Police Believe. 2018. Available online: https://www.theguardian.com/uk-news/2018/jan/12/nottingham-train-station-evacuated-after-large-fire (accessed on 15 February 2022).
289. Jones, H. Easter Getaway: M6 and M4 to Bear Brunt of Bank Holiday Traffic. 2017. Available online: https://www.theguardian.com/uk-news/2017/apr/13/easter-getaway-busy-roads-and-mild-weather-forecast (accessed on 15 February 2022).
290. Clark, A. Leaf Blitz Clears Rail Lines. 2003. Available online: https://www.theguardian.com/uk/2003/nov/04/transport.world (accessed on 15 February 2022).
291. Rawlinson, K. Christmas Eve Journeys Hit by Rail Disruption across UK.2015. Available online: https://www.theguardian.com/uk-news/2015/dec/24/christmas-eve-journeys-rail-train-travellers (accessed on 15 February 2022).

292. Gayle, D. Notting Hill Carnival: Bank Holiday Rain Threatens to Dampen Spirits. 2015. Available online: https://www.theguardian.com/uk-news/2015/aug/30/bank-holiday-weather-thundery-downpours-loom-for-southern-england (accessed on 15 February 2022).
293. Topham, G. Easter Road Traffic Set to Hit 2015 Peak as Major Rail Lines Close. 2015. Available online: https://www.theguardian.com/world/2015/apr/02/easter-road-traffic-set-to-hit-2015-peak-as-major-rail-lines-close (accessed on 15 February 2022).
294. Waldram, H. How to Get to the Olympics: Alternative Travel Routes. 2012. Available online: https://www.theguardian.com/uk/london-2012-olympics-blog/2012/aug/03/how-to-get-to-olympics-alternative-routes (accessed on 15 February 2022).
295. Association, P. Bakerloo Line Hit by Delays as Part of Tunnel Falls on Track. 2012. Available online: https://www.theguardian.com/uk/2012/apr/26/bakerloo-line-delays-tunnel-track (accessed on 15 February 2022).
296. Harris, P. Britons Swelter—In Airports and Traffic Jams. 2002. Available online: https://www.theguardian.com/uk/2002/jul/28/transport.world (accessed on 15 February 2022).
297. Moloney, C. Southern Cancels London Victoria Trains for Two Weeks over Covid. 2021. Available online: https://www.theguardian.com/uk-news/2021/dec/30/southern-cancels-london-victoria-trains-for-two-weeks-over-covid (accessed on 15 February 2022).
298. Taylor, R. Rail Guards Vote to Strike over Train Safety. 1999. Available online: https://www.theguardian.com/uk/1999/oct/14/ladbrokegrove.transport1 (accessed on 15 February 2022).
299. Bowers, S. BA to Shed 1000 Jobs at Gatwick. 2000. Available online: https://www.theguardian.com/business/2000/dec/07/transportintheuk1 (accessed on 15 February 2022).
300. Gow, D. Cabin Crew Union Takes on Branson. 2000. Available online: https://www.theguardian.com/uk/2000/dec/29/transport.world (accessed on 15 February 2022).
301. Davies, C. London Underground Strikes Threaten to Disrupt Festive Season. 25 November 2021. Available online: https://www.theguardian.com/uk-news/2021/nov/25/london-underground-strikes-threaten-to-disrupt-festive-season (accessed on 15 February 2022).
302. Harper, K. London commuters FACE Rail Strikes. 2000. Available online: https://www.theguardian.com/world/2000/jan/15/transport.uk (accessed on 15 February 2022).
303. Topham, G. British Airways Cabin Crew to Hold Four-Day Strike Over Pay Dispute. 2017. Available online: https://www.theguardian.com/business/2017/jun/02/ba-shutdown-caused-by-contractor-who-switched-off-power-reports-claim (accessed on 15 February 2022).
304. Chrisafis, A. French Customs Strike Continues to Cause Cross-Channel Travel Chaos. 2019. Available online: https://www.theguardian.com/world/2019/mar/20/french-customs-strike-continues-to-cause-cross-channel-travel-chaos (accessed on 15 February 2022).
305. Butler, S. Yodel Workers to Strike, Threatening MS, Aldi and Very Deliveries. 2021. Available online: https://www.theguardian.com/business/2021/sep/16/yodel-workers-strike-marks-spencer-aldi-very-deliveries (accessed on 15 February 2022).
306. Weaver, M. Bus Network in Britain Facing Strikes over Drivers' Low Pay. 2021. Available online: https://www.theguardian.com/uk-news/2021/oct/14/bus-network-in-britain-facing-strikes-over-drivers-low-pay (accessed on 15 February 2022).
307. Tapper, J. Nurses and Shop Staff in UK Face Tide of Abuse since End of Lockdowns. 2021. Available online: https://www.theguardian.com/money/2021/oct/23/nurses-and-shop-staff-in-uk-face-tide-of-abuse-since-end-of-lockdowns (accessed on 15 February 2022).
308. Stone, T. ERTICO—ITS Europe Elects Angelos Amditis as New Board Chairman. 2018. Available online: https://www.traffictechnologytoday.com/news/its/ertico-its-europe-elects-angelos-amditis-as-new-board-chairman.html (accessed on 15 February 2022).
309. Stone, T. ITS America Picks Laura Chace as New President and CEO. 2021. Available online: https://www.traffictechnologytoday.com/news/appointments/its-america-picks-laura-chace-as-new-president-and-ceo.html (accessed on 15 February 2022).
310. Stone, T. Laura Shoaf Appointed New Chair of the UK's Urban Transport Group. 2021. Available online: https://www.traffictechnologytoday.com/news/appointments/laura-shoaf-appointed-new-chair-of-the-uks-urban-transport-group.html (accessed on 15 February 2022).
311. Stone, T. Woolpert Awarded District-Wide Aerial Photogrammetry Survey Project in Florida. 4 May 2018. Available online: https://www.traffictechnologytoday.com/news/data/woolpert-awarded-district-wide-aerial-photogrammetry-survey-project-in-florida.html (accessed on 15 February 2022).
312. Frost, A. First Thermal Sensor-Equipped Production AV Set for 2021. 2019. Available online: https://www.traffictechnologytoday.com/news/autonomous-vehicles/first-thermal-sensor-equipped-production-av-set-for-2021.html (accessed on 15 February 2022).
313. Stone, T. CyberCar Successfully Demonstrates First Use of Blockchain to Deliver Connected Car Data. 2017. Available online: https://www.traffictechnologytoday.com/news/connected-vehicles-infrastructure/cybercar-successfully-demonstrates-first-use-of-blockchain-to-deliver-connected-car-data.html (accessed on 15 February 2022).
314. Stone, T. SENSORIS Project Releases First Vehicle-to-Cloud Data Standard for CAVs. 2018. Available online: https://www.traffictechnologytoday.com/news/autonomous-vehicles/sensoris-project-releases-first-vehicle-to-cloud-data-standard-for-cavs.html (accessed on 15 February 2022).

315. Stone, T. Singapore LTA Announces Second Partner to Trial Autonomous Mobility-on-Demand Service. 3 August 2016. Available online: https://www.traffictechnologytoday.com/news/autonomous-vehicles/singapore-lta-announces-second-partner-to-trial-autonomous-mobility-on-demand-service.html (accessed on 15 February 2022).
316. Stone, T. Highways England Awards Contracts for New Regional Delivery Partnership Scheme. 7 November 2018. Available online: https://www.traffictechnologytoday.com/news/traffic-management/highways-england-awards-contracts-for-new-regional-delivery-partnership-scheme.html (accessed on 15 February 2022).
317. Stone, T. Organizations Call for New Measures as UK Road Deaths Rise in 2016. 29 September 2017. Available online: https://www.traffictechnologytoday.com/news/safety/organizations-call-for-new-measures-as-uk-road-deaths-rise-in-2016.html (accessed on 15 February 2022).
318. Stone, T. Emovis Delivers Upgraded Electronic Tolling System for Canada's A25 Highway. 24 October 2017. Available online: https://www.traffictechnologytoday.com/news/tolling/emovis-delivers-upgraded-electronic-tolling-system-for-canada%c2%92s-a25-highway.html (accessed on 15 February 2022).
319. Stone, T. Siemens Completes Roll-Out of UK's Largest Permanent Average Speed Enforcement System. 19 July 2017. Available online: https://www.traffictechnologytoday.com/news/enforcement/siemens-completes-roll-out-of-uks-largest-permanent-average-speed-enforcement-system.html (accessed on 15 February 2022).
320. Allen, J. Vitronic's Speed Enforcement Trailer Gets Type Approval in The Netherlands. 10 April 2018. Available online: https://www.traffictechnologytoday.com/news/enforcement/vitronics-speed-enforcement-trailer-gets-type-approval-in-the-netherlands.html (accessed on 15 February 2022).
321. Stone, T. New UK Partnership Combines Parking and EV Charging Payments. 23 November 2021. Available online: https://www.traffictechnologytoday.com/news/electric-vehicles-ev-infrastructure/new-uk-partnership-combines-parking-and-ev-charging-payments.html (accessed on 15 February 2022).
322. Frost, A. Wolverhampton Project Aims to Reduce Transport-Sourced Air Pollution. 24 October 2019. Available online: https://www.traffictechnologytoday.com/news/emissions-low-emission-zones/wolverhampton-project-aims-to-reduce-transport-sourced-air-pollution.html (accessed on 15 February 2022).
323. Stone, T. Ubicquia Acquires AI Platform to Strengthen Its Smart City, Traffic Optimisation Services. 4 May 2020. Available online: https://www.traffictechnologytoday.com/news/acquisitions-mergers/24001.html (accessed on 15 February 2022).
324. Stone, T. UK Department for Transport Appoints TRL to Study the Impact of Low Emission Buses. 22 August 2017. Available online: https://www.traffictechnologytoday.com/news/safety/uk-department-for-transport-appoints-trl-to-study-the-impact-of-low-emission-buses.html (accessed on 15 February 2022).
325. To, R.N.; Patrick, V.M. How the eyes connect to the heart: The influence of eye gaze direction on advertising effectiveness. *J. Consum. Res.* **2021**, *48*, 123–146. [CrossRef]
326. Anaza, N.A.; Kemp, E.; Briggs, E.; Borders, A.L. Tell me a story: The role of narrative transportation and the C-suite in B2B advertising. *Ind. Mark. Manag.* **2020**, *89*, 605–618. [CrossRef]
327. Bandyopadhyay, A.; Septianto, F.; Nallaperuma, K. Mixed feelings enhance the effectiveness of luxury advertising. *Australas. Mark. J.* **2022**, *30*, 28–34. [CrossRef]
328. Nong, D.H.; Fox, J.M.; Saksena, S.; Lepczyk, C.A. The Use of Spatial Metrics and Population Data in Mapping the Rural-Urban Transition and Exploring Models of Urban Growth in Hanoi, Vietnam. *Environ. Urban. ASIA* **2021**, *12*, 156–168. [CrossRef]
329. Tinessa, F.; Pagliara, F.; Biggiero, L.; Veneri, G.D. Walkability, accessibility to metro stations and retail location choice: Some evidence from the case study of Naples. *Res. Transp. Bus. Manag.* **2021**, *40*, 100549. [CrossRef]
330. Mantey, D.; Sudra, P. Types of suburbs in post-socialist Poland and their potential for creating public spaces. *Cities* **2019**, *88*, 209–221. [CrossRef]
331. Boukerche, A.; Coutinho, R.W.L. Crowd Management: The Overlooked Component of Smart Transportation Systems. *IEEE Commun. Mag.* **2019**, *57*, 48–53. [CrossRef]
332. Tao, H.; Wang, J.; Yaseen, Z.M.; Mohammed, M.N.; Zain, J.M. Shrewd vehicle framework model with a streamlined informed approach for green transportation in smart cities. *Environ. Impact Assess. Rev.* **2021**, *87*, 106542. [CrossRef]
333. Kim, S.J.; Kim, T.H. A study on the use of the mobile construction CALS service for the construction of a smart city. In Proceedings of the 2016 6th International Conference on Information Communication and Management (ICICM), Hertfordshire, UK, 29–31 October 2016; IEEE: Piscataway, NJ, USA, 2016; pp. 280–284.
334. Moeini, S.M. Attitudes to urban walking in Tehran. *Environ. Plan. B Plan. Des.* **2012**, *39*, 344–359. [CrossRef]
335. Sanchez, T.W. Poverty, policy, and public transportation. *Transp. Res. Part A Policy Pract.* **2008**, *42*, 833–841. [CrossRef]
336. Singer, M.E. How affordable are accessible locations? Neighborhood affordability in US urban areas with intra-urban rail service. *Cities* **2021**, *116*, 103295. [CrossRef]
337. Aljoufie, M.; Brussel, M.; Zuidgeest, M.; Van Delden, H.; van Maarseveen, M. Integrated analysis of land-use and transport policy interventions. *Transp. Plan. Technol.* **2016**, *39*, 329–357. [CrossRef]
338. Yilmaz, M.; Krein, P.T. Review of Battery Charger Topologies, Charging Power Levels, and Infrastructure for Plug-In Electric and Hybrid Vehicles. *IEEE Trans. Power Electron.* **2013**, *28*, 2151–2169. [CrossRef]
339. Zhang, H.; Moura, S.J.; Hu, Z.; Song, Y. PEV Fast-Charging Station Siting and Sizing on Coupled Transportation and Power Networks. *IEEE Trans. Smart Grid* **2018**, *9*, 2595–2605. [CrossRef]

340. Pieltain Fernández, L.; Gomez San Roman, T.; Cossent, R.; Mateo Domingo, C.; Frías, P. Assessment of the Impact of Plug-in Electric Vehicles on Distribution Networks. *IEEE Trans. Power Syst.* **2011**, *26*, 206–213. [CrossRef]
341. Clement-Nyns, K.; Haesen, E.; Driesen, J. The Impact of Charging Plug-In Hybrid Electric Vehicles on a Residential Distribution Grid. *IEEE Trans. Power Syst.* **2010**, *25*, 371–380. [CrossRef]
342. Rayle, L.; Dai, D.; Chan, N.; Cervero, R.; Shaheen, S. Just a better taxi? A survey-based comparison of taxis, transit, and ridesourcing services in San Francisco. *Transp. Policy* **2016**, *45*, 168–178. [CrossRef]
343. Dias, F.F.; Lavieri, P.S.; Garikapati, V.M.; Astroza, S.; Pendyala, R.M.; Bhat, C.R. A behavioral choice model of the use of car-sharing and ride-sourcing services. *Transportation* **2017**, *44*, 1307–1323. [CrossRef]
344. Hall, J.D.; Palsson, C.; Price, J. Is Uber a substitute or complement for public transit? *J. Urban Econ.* **2018**, *108*, 36–50. [CrossRef]
345. Fagnant, D.J.; Kockelman, K.M. The travel and environmental implications of shared autonomous vehicles, using agent-based model scenarios. *Transp. Res. Part C Emerg. Technol.* **2014**, *40*, 1–13. [CrossRef]
346. Cramer, J.; Krueger, A.B. Disruptive change in the taxi business: The case of Uber. *Am. Econ. Rev.* **2016**, *106*, 177–182. [CrossRef]
347. Fishman, E. Bikeshare: A review of recent literature. *Transp. Rev.* **2016**, *36*, 92–113. [CrossRef]
348. Alemi, F.; Circella, G.; Handy, S.; Mokhtarian, P. What influences travelers to use Uber? Exploring the factors affecting the adoption of on-demand ride services in California. *Travel Behav. Soc.* **2018**, *13*, 88–104. [CrossRef]
349. Yang, H.; Cherry, C.R. Use characteristics and demographics of rural transit riders: A case study in Tennessee. *Transp. Plan. Technol.* **2017**, *40*, 213–227. [CrossRef]
350. Liu, Y.; Liu, C.; Yuan, N.J.; Duan, L.; Fu, Y.; Xiong, H.; Xu, S.; Wu, J. Intelligent bus routing with heterogeneous human mobility patterns. *Knowl. Inf. Syst.* **2017**, *50*, 383–415. [CrossRef]
351. Freitas, A.L.P. Assessing the quality of intercity road transportation of passengers: An exploratory study in Brazil. *Transp. Res. Part A Policy Pract.* **2013**, *49*, 379–392. [CrossRef]
352. Dinulescu, R.; Bugheanu, A.M. Improving users'satisfaction by implementing the analytic hierarchy process in the public transportation system. *Environ. Eng. Manag. J. (EEMJ)* **2020**, *19*. Available online: https://eds.p.ebscohost.com/eds/pdfviewer/pdfviewer?vid=0&sid=28a579a2-2525-4f6a-97be-807b8a530bea%40redis (accessed on 15 February 2022).
353. Gummadi, R.; Edara, S.R. Analysis of passenger flow prediction of transit buses along a route based on time series. In *Information and Decision Sciences*; Springer: Berlin/Heidelberg, Germany, 2018; pp. 31–37.
354. Circella, G.; Hunt, J.D.; Stefan, K.J.; Brownlee, A.T.; McCoy, M. Simplified model of local transit services. *Eur. J. Transp. Infrastruct. Res.* **2014**, *14*. [CrossRef]
355. Ali, W.; Wang, J.; Zhu, H.; Wang, J. An Optimized Fast Handover Scheme Based on Distributed Antenna System for High-Speed Railway. In Proceedings of the 2017 IEEE 86th Vehicular Technology Conference (VTC-Fall), Toronto, ON, Canada, 24–27 September 2017; IEEE: Piscataway, NJ, USA, 2017; pp. 1–5.
356. Yijing, Y.; Xu, W. Optimization of Delay Time of Freight Trains Based on Coordinated Wagon-Flow Allocation Operation between Technical Station. In Proceedings of the 2020 5th International Conference on Electromechanical Control Technology and Transportation (ICECTT), Nanchang, China, 15–17 May 2020; IEEE: Piscataway, NJ, USA, 2020; pp. 632–635.
357. Mou, R.; Yue, Z. Risk Analysis of LNG Tanker for Railway Transportation. In *ICTE 2019*; American Society of Civil Engineers: Reston, VA, USA, 2020; pp. 462–469.
358. Kurhan, M.B.; Kurhan, D.M. The Effectiveness Evaluation of International Railway Transportation in the Direction of "Ukraine–European Union", Kaunas University of Technology. 2018. Available online: http://eadnurt.diit.edu.ua/jspui/handle/123456789/10781 (accessed on accessed on 15 Feburary 2022).
359. Khoshniyat, F.; Peterson, A. Improving train service reliability by applying an effective timetable robustness strategy. *J. Intell. Transp. Syst.* **2017**, *21*, 525–543. [CrossRef]
360. Givoni, M. Development and impact of the modern high-speed train: A review. *Transp. Rev.* **2006**, *26*, 593–611. [CrossRef]
361. Zhang, A.; Wan, Y.; Yang, H. Impacts of high-speed rail on airlines, airports and regional economies: A survey of recent research. *Transp. Policy* **2019**, *81*, A1–A19. [CrossRef]
362. Schäfer, A.W.; Waitz, I.A. Air transportation and the environment. *Transp. Policy* **2014**, *34*, 1–4. [CrossRef]
363. Schnurr, R.; Walker, T.R. Marine transportation and energy use. In *Reference Module in Earth Systems and Environmental Sciences*; Elsevier: Amsterdam, The Netherlands, 2019.
364. Mahmoodjanloo, M.; Chen, G.; Asian, S.; Iranmanesh, S.H.; Tavakkoli-Moghaddam, R. In-port multi-ship routing and scheduling problem with draft limits. *Marit. Policy Manag.* **2021**, *48*, 966–987. [CrossRef]
365. Oztanriseven, F.; Nachtmann, H. Modeling dynamic behavior of navigable inland waterways. *Marit. Econ. Logist.* **2020**, *22*, 173–195. [CrossRef]
366. Aivazidou, E.; Politis, I. Transfer function models for forecasting maritime passenger traffic in Greece under an economic crisis environment. *Transp. Lett.* **2021**, *13*, 591–607. [CrossRef]
367. Bou-Harb, E.; Kaisar, E.I.; Austin, M. On the impact of empirical attack models targeting marine transportation. In Proceedings of the 2017 5th IEEE International Conference on Models and Technologies for Intelligent Transportation Systems (MT-ITS), Naples, Italy, 26–28 June 2017; IEEE: Piscataway, NJ, USA, 2017; pp. 200–205.
368. DO, Q.H. Critical Factors Affecting the Choice of Logistics Service Provider: An Empirical Study in Vietnam. *J. Asian Financ. Econ. Bus.* **2021**, *8*, 145–150.

369. Shariff, D.; Jayaraman, K.; Shah, M.Z.; Saharuddin, A. Service quality determinants of the container haulage industry: An empirical study in Malaysia. *Int. J. Shipp. Transp. Logist.* **2016**, *8*, 66–80. [CrossRef]
370. Nechifor, S.; Târnaucă, B.; Sasu, L.; Puiu, D.; Petrescu, A.; Teutsch, J.; Waterfeld, W.; Moldoveanu, F. Autonomic monitoring approach based on cep and ml for logistic of sensitive goods. In Proceedings of the IEEE 18th International Conference on Intelligent Engineering Systems INES 2014, Tihany, Hungary, 3–5 July 2014; IEEE: Piscataway, NJ, USA, 2014; pp. 67–72.
371. Ghomi, S.F.; Asgarian, B. Development of metaheuristics to solve a transportation inventory location routing problem considering lost sale for perishable goods. *J. Model. Manag.* **2019**, *14*, 175–198. [CrossRef]
372. Domínguez, J.P.; Roseiro, P. Blockchain: A brief review of Agri-Food Supply Chain Solutions and Opportunities. *ADCAIJ Adv. Distrib. Comput. Artif. Intell. J.* **2020**, *9*, 95–106. [CrossRef]
373. Yao, J.T. The impact of transportation asymmetry on the choice of a spatial price policy. *Asia-Pac. J. Reg. Sci.* **2019**, *3*, 793–811. [CrossRef]
374. Lyu, K.; Chen, H.; Che, A. Combinatorial Auction for Truckload Transportation Service Procurement with Auctioneer-Generated Supplementary Bundles of Requests. In Proceedings of the 2020 IEEE 23rd International Conference on Intelligent Transportation Systems (ITSC), Rhodes, Greece, 20–23 September 2020; pp. 1–6. [CrossRef]
375. Galińska, B. Multiple criteria evaluation of global transportation systems-analysis of case study. In *Scientific And Technical Conference Transport Systems Theory And Practice*; Springer: Berlin/Heidelberg, Germany, 2017; pp. 155–171.
376. Bayramoğlu, K.; Yilmaz, S.; Kaya, K.D. Numerical investigation of valve lifts effects on performance and emissions in diesel engine. *Int. J. Glob. Warm.* **2019**, *18*, 287–303. [CrossRef]
377. Canh, N.P.; Hao, W.; Wongchoti, U. The impact of economic and financial activities on air quality: A Chinese city perspective. *Environ. Sci. Pollut. Res.* **2021**, *28*, 8662–8680. [CrossRef] [PubMed]
378. Wang, P.; Ying, Q.; Zhang, H.; Hu, J.; Lin, Y.; Mao, H. Source apportionment of secondary organic aerosol in China using a regional source-oriented chemical transport model and two emission inventories. *Environ. Pollut.* **2018**, *237*, 756–766. [CrossRef]
379. Macedo, J.; Kokkinogenis, Z.; Soares, G.; Perrotta, D.; Rossetti, R.J. A HLA-based multi-resolution approach to simulating electric vehicles in simulink and SUMO. In Proceedings of the 16th International IEEE Conference on Intelligent Transportation Systems, ITSC, The Hague, The Netherlands, 6–9 October 2013; IEEE: Piscataway, NJ, USA, 2013; pp. 2367–2372.
380. Meng, X.; Ai, Y.; Li, R.; Zhang, W. Effects of heavy metal pollution on enzyme activities in railway cut slope soils. *Environ. Monit. Assess.* **2018**, *190*, 1–12. [CrossRef]
381. Yang, W.; He, J.; He, C.; Cai, M. Evaluation of urban traffic noise pollution based on noise maps. *Transp. Res. Part D Transp. Environ.* **2020**, *87*, 102516. [CrossRef]
382. Suh, D.H. Interfuel substitution effects of biofuel use on carbon dioxide emissions: Evidence from the transportation sector. *Appl. Econ.* **2019**, *51*, 3413–3422. [CrossRef]
383. Andersson, L.; Ek, K.; Kastensson, Å.; Wårell, L. Transition towards sustainable transportation–What determines fuel choice? *Transp. Policy* **2020**, *90*, 31–38. [CrossRef]
384. Noori, H. Modeling the impact of vanet-enabled traffic lights control on the response time of emergency vehicles in realistic large-scale urban area. In Proceedings of the 2013 IEEE International Conference on Communications Workshops (ICC), Budapest, Hungary, 9–13 June 2013; IEEE: Piscataway, NJ, USA, 2013; pp. 526–531.
385. Fu, H.; Ma, H.; Wang, G.; Zhang, X.; Zhang, Y. MCFF-CNN: Multiscale comprehensive feature fusion convolutional neural network for vehicle color recognition based on residual learning. *Neurocomputing* **2020**, *395*, 178–187. [CrossRef]
386. Ma, X.; Wu, Y.J.; Wang, Y.; Chen, F.; Liu, J. Mining smart card data for transit riders' travel patterns. *Transp. Res. Part C Emerg. Technol.* **2013**, *36*, 1–12. [CrossRef]
387. Al-Sultan, S.; Al-Doori, M.M.; Al-Bayatti, A.H.; Zedan, H. A comprehensive survey on vehicular ad hoc network. *J. Netw. Comput. Appl.* **2014**, *37*, 380–392. [CrossRef]
388. Hartenstein, H.; Laberteaux, L. A tutorial survey on vehicular ad hoc networks. *IEEE Commun. Mag.* **2008**, *46*, 164–171. [CrossRef]
389. Fagnant, D.J.; Kockelman, K. Preparing a nation for autonomous vehicles: Opportunities, barriers and policy recommendations. *Transp. Res. Part A Policy Pract.* **2015**, *77*, 167–181. [CrossRef]
390. Haboucha, C.J.; Ishaq, R.; Shiftan, Y. User preferences regarding autonomous vehicles. *Transp. Res. Part C Emerg. Technol.* **2017**, *78*, 37–49. [CrossRef]
391. Bösch, P.M.; Becker, F.; Becker, H.; Axhausen, K.W. Cost-based analysis of autonomous mobility services. *Transp. Policy* **2018**, *64*, 76–91. [CrossRef]
392. Zhang, Y.; Huang, G. Traffic flow prediction model based on deep belief network and genetic algorithm. *IET Intell. Transp. Syst.* **2018**, *12*, 533–541. [CrossRef]
393. James, J. Citywide traffic speed prediction: A geometric deep learning approach. *Knowl.-Based Syst.* **2021**, *212*, 106592.
394. Zheng, Z.h.; Ling, X.m.; Wang, P.; Xiao, J.; Zhang, F. Hybrid model for predicting anomalous large passenger flow in urban metros. *IET Intell. Transp. Syst.* **2021**, *14*, 1987–1996. [CrossRef]
395. Tao, Z.; Xu, J. A class of rough multiple objective programming and its application to solid transportation problem. *Inf. Sci.* **2012**, *188*, 215–235. [CrossRef]
396. Yigitcanlar, T.; Kankanamge, N.; Regona, M.; Maldonado, A.R.; Rowan, B.; Ryu, A.; Desouza, K.C.; Corchado, J.M.; Mehmood, R.; Li, R.Y.M. Artificial Intelligence Technologies and Related Urban Planning and Development Concepts: How Are They Perceived and Utilized in Australia? *J. Open Innov. Technol. Mark. Complex.* **2020**, *6*, 187. [CrossRef]

397. Yigitcanlar, T.; Butler, L.W.E.D.K.M.R.C.J. Can Building "Artificially Intelligent Cities" Safeguard Humanity from Natural Disasters, Pandemics, and Other Catastrophes? An Urban Scholar's Perspective. *Sensors* **2020**, *20*, 2988. [CrossRef]
398. Yigitcanlar, T.; Mehmood, R.; Corchado, J.M. Green Artificial Intelligence: Towards an Efficient, Sustainable and Equitable Technology for Smart Cities and Futures. *Sustainability* **2021**, *13*, 8952. [CrossRef]
399. Yigitcanlar, T.; Corchado, J.M.; Mehmood, R.; Li, R.Y.M.; Mossberger, K.; Desouza, K. Responsible Urban Innovation with Local Government Artificial Intelligence (AI): A Conceptual Framework and Research Agenda. *J. Open Innov. Technol. Mark. Complex.* **2021**, *7*, 71. [CrossRef]
400. Yigitcanlar, T.; Regona, M.; Kankanamge, N.; Mehmood, R.; D'Costa, J.; Lindsay, S.; Nelson, S.; Brhane, A. Detecting Natural Hazard-Related Disaster Impacts with Social Media Analytics: The Case of Australian States and Territories. *Sustainability* **2022**, *14*, 810. [CrossRef]
401. Janbi, N.; Mehmood, R.; Katib, I.; Albeshri, A.; Corchado, J.M.; Yigitcanlar, T. Imtidad: A Reference Architecture and a Case Study on Developing Distributed AI Services for Skin Disease Diagnosis over Cloud, Fog and Edge. *Sensors* **2022**, *22*, 1854. [CrossRef]
402. Mohammed, T.; Albeshri, A.; Katib, I.; Mehmood, R. UbiPriSEQ—Deep Reinforcement Learning to Manage Privacy, Security, Energy, and QoS in 5G IoT HetNets. *Appl. Sci.* **2020**, *10*, 7120. [CrossRef]
403. Janbi, N.; Katib, I.; Albeshri, A.; Mehmood, R. Distributed Artificial Intelligence-as-a-Service (DAIaaS) for Smarter IoE and 6G Environments. *Sensors* **2020**, *20*, 5796. [CrossRef]
404. Arfat, Y.; Suma, S.; Mehmood, R.; Albeshri, A. Parallel Shortest Path Big Data Graph Computations of US Road Network Using Apache Spark: Survey, Architecture, and Evaluation. In *EAI/Springer Innovations in Communication and Computing*; Springer: Cham, Switzerland, 2020; pp. 185–214. [CrossRef]
405. Usman, S.; Mehmood, R.; Katib, I. Big Data and HPC Convergence for Smart Infrastructures: A Review and Proposed Architecture. In *EAI/Springer Innovations in Communication and Computing*; Springer: Cham, Switzerland, 2020; pp. 561–586. [CrossRef]
406. AlAhmadi, S.; Mohammed, T.; Albeshri, A.; Katib, I.; Mehmood, R. Performance Analysis of Sparse Matrix-Vector Multiplication (SpMV) on Graphics Processing Units (GPUs). *Electronics* **2020**, *9*, 1675. [CrossRef]
407. Alyahya, H.; Mehmood, R.; Katib, I. *Parallel Iterative Solution of Large Sparse Linear Equation Systems on the Intel MIC Architecture*; Springer: Cham, Switzerland, 2020; pp. 377–407. [CrossRef]
408. Mehmood, R.; Alam, F.; Albogami, N.N.; Katib, I.; Albeshri, A.; Altowaijri, S.M. UTiLearn: A Personalised Ubiquitous Teaching and Learning System for Smart Societies. *IEEE Access* **2017**, *5*, 2615–2635. [CrossRef]
409. Alswedani, S.; Katib, I.; Abozinadah, E.; Mehmood, R. Discovering Urban Governance Parameters for Online Learning in Saudi Arabia During COVID-19 Using Topic Modeling of Twitter Data. *Front. Sustain. Cities* **2022**, *4*. [CrossRef]
410. Alahmari, N.; Alswedani, S.; Alzahrani, A.; Katib, I.; Albeshri, A.; Mehmood, R. Musawah: A Data-Driven AI Approach and Tool to Co-Create Healthcare Services with a Case Study on Cancer Disease in Saudi Arabia. *Sustainability* **2022**, *14*, 3313. [CrossRef]
411. Alotaibi, S.; Mehmood, R.; Katib, I.; Rana, O.; Albeshri, A. Sehaa: A Big Data Analytics Tool for Healthcare Symptoms and Diseases Detection Using Twitter, Apache Spark, and Machine Learning. *Appl. Sci.* **2020**, *10*, 1398. [CrossRef]

Article

Transport Mode Choice for Residents in a Tourist Destination: The Long Road to Sustainability (the Case of Mallorca, Spain)

Maurici Ruiz-Pérez [1,2,*] and Joana Maria Seguí-Pons [1]

1 Geography Department, University of the Balearic Islands, 07122 Palma, Spain; joana.segui-pons@uib.es
2 GIS & Remote Sensing Service, University of the Balearic Islands, 07122 Palma, Spain
* Correspondence: maurici.ruiz@uib.es; Tel.: +34-971-173006

Received: 29 October 2020; Accepted: 11 November 2020; Published: 14 November 2020

Abstract: Sustainable mobility policies may encounter social, economic, and cultural barriers to successful implementation that need to be assessed. In this sense, knowledge of the population's mobility habits and their relationship with transport modes is particularly essential. Along these lines, a study was carried out of the patterns of transport modes chosen concerning various social and territorial variables on the island of Mallorca based on the most recent mobility surveys. The study shows that the choice of mode is influenced by a wide range of factors, such as gender, age group, motive for the trip, occupation, region of residence, duration of the trip, and proximity to Palma, the capital of the island. The results indicate that private vehicles are the most often chosen mode of transport. Private vehicles are mainly used by working men between 30 and 44 years old for journeys between home and work, which do not exceed 30 min and are preferably in areas close to Palma. Sustainable modes are little used, although they are mainly used by women, young people, and retired people for work purposes and for access to educational and health centers. The demand for transport generated by the resident population and tourist activity and the negative externalities generated by mobility in private vehicles are closely related on a municipal level (Pearson's coefficient 0.84, $p = 0.00$). However, the modal distribution does not seem to be directly related to these factors. Instead, it develops a more conditioned distribution by access to rail transport infrastructures and other geographical factors. In recent years, the Balearic Islands' public administration launched the Balearic Islands Sectorial Mobility Plan 2019–2026, which aims to promote sustainable modes and reduce the use of private vehicles. This plan represents a considerable economic investment, but will also require great institutional coordination and cultural changes in the population's perception of mobility. The study shows that the implementation of sustainable modes on the island requires a global vision of mobility issues that integrates urban planning and tourism planning to make the land-use model more sustainable.

Keywords: modal choice; sustainable mobility; Mallorca mobility; logistic regression; transport planning

---

1. Introduction

The path towards implementing a sustainable transport model in a territory is a long and complicated process that involves a broad series of changes and transformations that must be carried out in a harmonized way over time and space [1–3]. Some of these transformations are structural and require large-scale economic investments, such as construction of or improvements in sustainable transport infrastructures (roads, highways, railways, trams, subways, bike lanes, etc.) [4], or the construction of energy infrastructures for the use of transport (gas pipelines, construction of facilities for alternative energy, charging systems for electric vehicles, etc.) and communications infrastructures to facilitate the implementation and development of sharing mobility [5] (Santos, 2018). Other transformations have to do with the change in the model of land use. In this case, more sustainable land occupation scenarios should be considered, in which the population's activities require the smallest number of trips by motorized modes [6]. The implementation of sustainable modes requires updating urban planning towards more sustainable scenarios and deploying new approaches to the optimized location of public amenities [7]. Other changes have more to do with transport and mobility governance, such as developing transportation planning policies, plans, and programs oriented toward promoting and regulating sustainable modes. This is where the development of sustainable urban mobility plans (SUMPs) would be located at different geographical scales: national, regional, and municipal. In addition, transport service management systems' transformation towards public models that guarantee environmental justice is required. It is also essential to transform the population's attitude towards the acceptance and preferential use of sustainable modes from the perspectives of environmental, social, and economic improvement and the improvement of health. Finally, it should be noted that the path to a knowledge society has a direct effect on the development of sharing mobility that can help optimize modes of transport management, diversify the offer, and move towards a more sustainable model [8].

The modal choice of transport analysis is a topic of great relevance for a holistic territorial vision on sustainability, since it integrates social, economic, environmental, and governance aspects. Knowledge of the distribution of the modes of transport of the resident population in daily or long-distance trips is key to diagnosing the degree of sustainability of transport in a community. Such knowledge guides planners and managers regarding the types of actions to be carried out in its improvement [9]. There are many factors involved in the modal choice of transport by the population; some have to do with the availability of infrastructures and services. Others are of more of a social nature and are related to the demographic structure, economic activity, or the model of settlement in the territory. The mode choice process is also influenced by psychological and cultural factors that include gender or the motivation for the trip [10,11].

Usually, sustainable modes of transport are related to active and healthy modes (pedestrian or bike). The collective modes of public transport (train, metro, bus) are included equally in the sustainable group, but are not healthy. Environmentally friendly modes are also discussed, preferably referring to sustainable modes [12]. Private motorized vehicles, such as cars, vans, trucks, and motorcycles, are usually included in the group of unsustainable modes, although the use of motorcycles can sometimes be considered sustainable [13]. Electric vehicles are also considered sustainable from the perspective of fossil fuel emissions and consumption, but not for congestion or parking problems. The use of motor vehicles increases fossil fuel consumption and the emission of gases that contribute to the increase in the greenhouse effect and the degradation of air quality that accelerates climate change [14–16].

Mobility plays a fundamental role in the development of tourist destinations [17]. In this sense, it is important to maximize the efficiency of infrastructures, the access to services, and the conditions of intrinsic mobility in a tourist destination. A touristic area that is well endowed, both in infrastructure and transportation services, can increase its competitiveness [18]. In this sense, it is recommended that a tourist destination prioritize sustainable modes [19], as, in addition to assuming an environmental, social, and economic improvement, it constitutes a marketing strategy that provides an image of environmental quality, which constitutes an element of attraction for the tourist [20]. Access to accommodation, tourist resources, and equipment through sustainable modes is a guarantee of quality and has become a priority for planners and tourism managers.

With this approach, the first step towards transport sustainability in tourism zones consists of encouraging the host population to mainly use sustainable modes. This makes the seasonal dynamics of tourism less stressful for transport infrastructure and services, and leads to a reduction in the negative externalities derived from the massive use of private vehicles. In Mediterranean island environments, whose main economic activity is tourism, the negative externalities derived from unsustainable modes are unacceptable. Road congestion and lack of public transport present a harmful image for tourists, with adverse effects in the medium and long terms [21]. Therefore, the improvement in a tourist destination should not only concern the update of accommodation, leisure, and recreation infrastructure, but also, to a large extent, support mobility infrastructures and services and promote reductions in the use of private vehicles.

In the EU's political framework that promotes territorial development to ensure a model of social and economic cohesion, the massive use of private vehicles in island environments with tourist vocations can unbalance regional economies by jeopardizing the attractiveness of the destinations. In this sense, Mediterranean islands are susceptible to the harmful effects of congestion derived from traffic increases. Therefore, the main objective is for urban mobility to facilitate the economic development of the territory, the quality of life of inhabitants, and the protection of the environment [22]. Therefore, the high sensitivity of tourist destinations to mobility actions advises basing decisions on comprehensive mobility issues.

The hypothesis proposed in this work is that initiatives to promote sustainable mobility should be established and assimilated appropriately by the resident population to set an example for visitors and tourists and to contribute to building a more sustainable tourist destination. Therefore, knowledge of the key factors in the choice of transport modes by the resident population of a tourist destination is essential to diagnose the degree of sustainability of its transport system and deploy an appropriate sustainable mobility policy that includes residents and tourists. Having information on the different modes at the municipal level and knowing the reasons for their choice can help the regional planner base investments in and approaches to sustainable mobility.

To test this hypothesis, an analysis of mobility habits was carried out on the island of Mallorca (Balearics, Spain); its relationship with various social and territorial factors was observed. The study develops a geographic approach and analyzes the relationships between modal selection at the municipal, regional, and island levels.

One of the paper's main contributions to the international literature is the description of a modal choice case study in a mature tourist destination such as Mallorca. In addition, it shows the social inertia that hinders the deployment of sustainable modes and the difficulty of replacing private vehicles if essential infrastructures are not developed. The paper also describes the use of some indicators and specific analysis of mobility data, which can help rationalize investment in transport infrastructure and new facilities' locations.

The paper is structured in four sections: Firstly, an analysis of the scientific literature on the subject of the article is carried out. Then, the methodology includes a description of the case study and the statistical tools used; next, the main results achieved are described and their implications are discussed, and finally, the main conclusions are drawn from the study carried out.

2. Literature Review

The analysis of the factors determining the choice of transport mode is a subject that has aroused great interest in the international scientific community in recent decades. There are many contributions in this area, in which the influence of various factors on modal choice and their relations are detected and analyzed [23]. In the scientific literature, several variables have been identified in this process that can give rise to a multidimensional model in which the following components are distinguished: spatial (physical), social, economic, psychological, cultural, and environmental. Table 1 represents the set of dimensions and factors highlighted by different authors as playing a role in the modal choice process.

Table 1. Dimensions and factors of the modal choice of transport.

DIMENSION/ASPECTS	Transport Mode	Factors	References
The type of trip	Everyone	– Objective of the trip – Time waiting/travel – Distance travelled – Time of the trip – Month of the trip – Complexity of the trip – Possibility of developing activities during the trip – Comfort	[24–26]
The means of transport	Everyone	– Availability of public transportation – Cost of the trip – Discounts – Security – Demand – Independence	[27–30]
Environment	Bicycle Pedestrian	– Climate (temperature, precipitation)	[31–35]
Economy	Private vehicle	– Income level – House size – Car ownership	[36]
Urban design/Built environment Neighborhood spatial patterns	Pedestrian Public transport Private vehicle	– Existence of trees – Size of sidewalks – Mixed uses – Number of intersections of the road network – Urban structure – Density – Diversity – Design – Accessibility	[37–43]
Facilities	Private vehicle Bicycle Pedestrian	– Availability of parking at the destination – Internet availability for the user – Proximity – Dimensions of the facility	[9,40,44–48]
Population structure	Private vehicle	– Gender – Age of the population – Size of family travelling – Number of children/adults travelling – Education level – Nationality of the population	[49–54]
Socio-psychological		– Lifestyle – Perceptions – Attitude preferences – Residential dissonance	[11,55]

Note: For the sake of simplicity, the indicated references have been assigned to one factor, but could include multiple factors.

There is a broad debate about the role played by the more objective factors (environment, physical characteristics) or the more subjective aspects (attitudes or lifestyles) regarding the modal choice [43]. Some authors divide the factors into two groups: macroscopic, referring to characteristics of the environment and society, and microscopic, related to the intrinsic characteristics of the traveler and the attributes of the trip [56].

Regarding the tourists' modal choice, Le-Klähn emphasizes that the main factors of a visitor's/tourist's use of public transport have to do with stopping driving and avoiding congestion or problems when purchasing a private vehicle [20]. Likewise, Lumsdon analyzes the factors that should be considered for bus services designed for tourism, highlighting that the tourist seeks an experiential process rather than a proper displacement [56,57]. Nutsugbodo points out that the availability, safety, and comfort of public transport and sociodemographic characteristics are determining factors in the choice of a tourist destination [18].

In summary, the modal choice is a complex process in terms of the number of influential factors; it is very dynamic and very sensitive to spatial, socioeconomic, and psychological aspects of the type of trip, geographical environment, or people's habits. In this context, some authors have proposed the term mobility as an integrating concept of the set of factors of modal selection [58]. Mobility refers to the potential to be mobile, regardless of whether physical displacement has occurred. This intrinsic complexity implies that the success of proposed modal shifts towards sustainable modes is not a generalizable or straightforward process for any geographical area or target population.

The bibliographic study carried out shows the absence of specific works on the modal choices made by the population living in tourist destinations. This circumstance is not trivial, since these populations receive intense pressure from tourist activity in the high season, which makes them susceptible to negative externalities in terms of mobility. In most cases, the host population experiences an excess of private vehicle use generalizable to tourists (rental), as well as an overload of public transport [59].

In general, in tourist destinations, there is a unanimous agreement (social and political) that sustainable modes need to be developed; however, there is significant resistance to their deployment. These weaknesses are evident in the modal study, as they are the ballast that prevents a correct evolution towards modal sustainability.

Research on mobility in island environments is scarce. However, the problems detected are common to different authors: excessive use of private vehicles, the need for quality public transport, and the importance of public participation in the mobility planning process [60–63].

The use of sustainable mobility indicators is considered a key instrument for developing transport policies at the international, regional, and local levels [64–68]. Indicators have become a tool for monitoring progress towards sustainable development. They are widely used in the framework of sustainable urban mobility plans (SUMPs) [69]. There are numerous proposals for sustainable mobility indicators that refer to different aspects of mobility concerning the main components of sustainability: environmental, social, economic, and governance [9].

One of the most widely used indicators of sustainable mobility concerns the division of modal choice. The percentage of each mode of transport in relation to the total is a primary mobility indicator used in many studies. The modal split (the quantitative relationship between the various modes) provides essential information on the level of sustainability of the geographical area for which it has been analyzed. Some authors propose an indicator for the evaluation of private vehicle reduction [70]. It should be noted that the modal choice is usually integrated within the group of indicators that would refer to observed mobility [71].

3. Methods

3.1. Case Study

Mallorca is the island with the largest surface area (3640 km^2) of the Balearic Islands archipelago, which is located in the Mediterranean basin (Figure 1a). Mallorca, together with the islands of Minorca, Ibiza, and Formentera, is one of the seventeen autonomous communities that make up the Spanish state. The island has 53 municipalities deployed in seven regions or counties (Figure 1c).

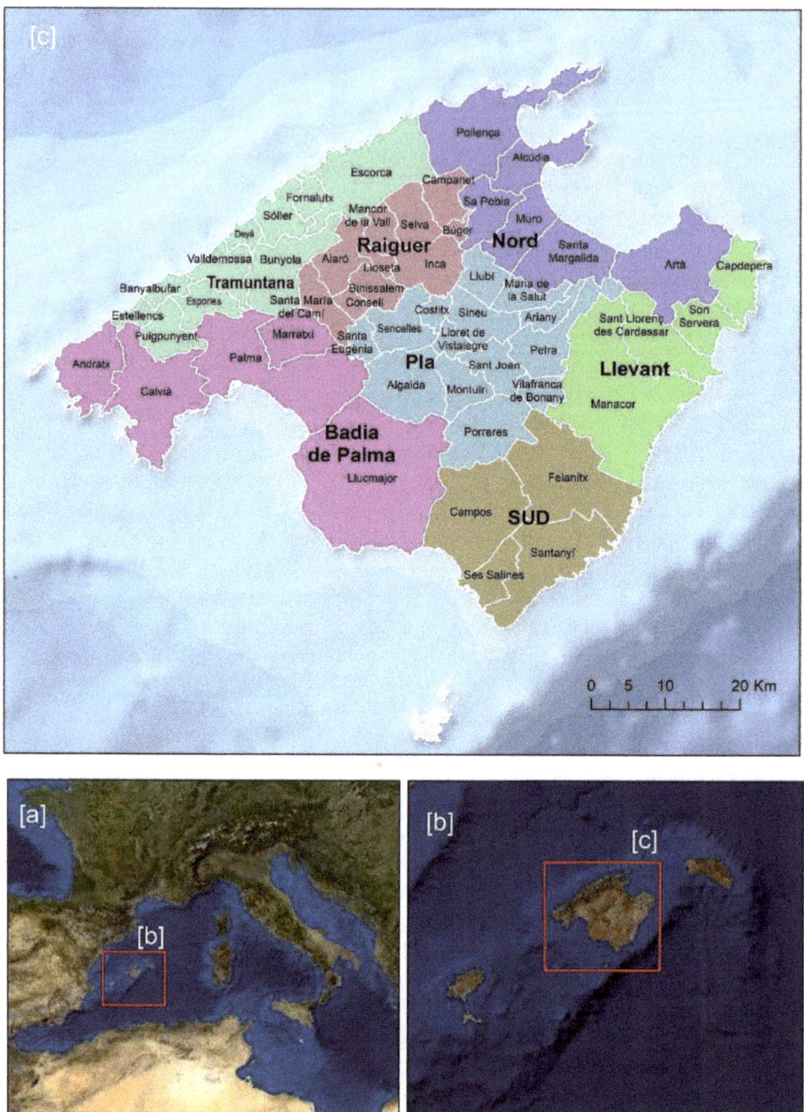

Figure 1. [**a**] Location of Mallorca in the Mediterranean basin, [**b**] location of Mallorca in the Balearic Islands Archipelago, and [**c**] municipalities and counties of Mallorca.

The population of Mallorca amounts to 896,038 people [72] with a heterogeneous territorial distribution. Palma's municipality, the autonomous community's capital, concentrates 48% of its population, followed by the municipalities of Calvia, Manacor, and Inca (Figure 2). The island combines a population model concentrated in traditional centers with a dispersed model of residential constructions on agricultural, rural, or natural land, many of which are used for tourism. This process of diffuse artificialization is known as "rururbanization" [73]. In recent decades, this dynamic has become so widespread that it constitutes a global transformation process, giving rise to an "ISLAND-CITY" model [74] that has crucial consequences on the mobility of its population.

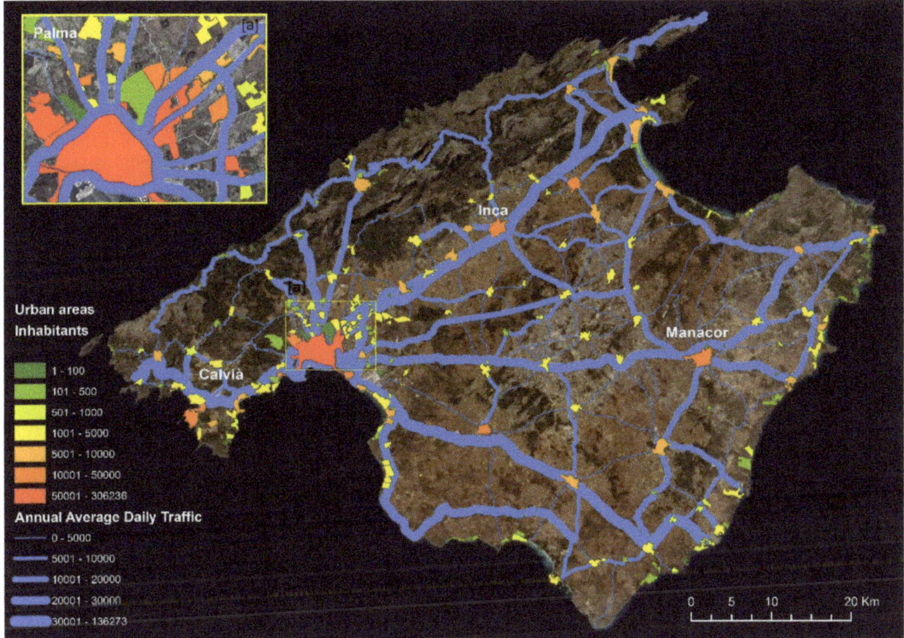

Figure 2. Population in the urban centers of Mallorca and average daily intensity of vehicles on roads. Source: Consell Insular de Mallorca 2020. [75].

The island's population grew by approximately 15% in the period 2005–2019 [72], and the demographic transition model has been established and matured in recent decades. Mallorca's demographic structure shows an intense aging process that is greater in the interior municipalities. The migratory processes are more evident in some centers, especially those that still maintain agricultural activity and receive young immigrants from North Africa and Eastern Europe. However, these movements do not manage to avoid the structural–demographic dynamics of aging.

Most of the island's facilities are located in Palma, which has the most essential transport infrastructure of the island (airport, port), healthcare facilities (hospitals), leading educational facilities (university, schools), commercial and industrial facilities, public administration buildings, etc. This location has a powerful effect on attracting travelers from all over the island to the capital. As a result of this dynamic of mobility, Mallorca's road and train network has always maintained a radial model that links Palma with the rest of the island's municipalities (Figure 3).

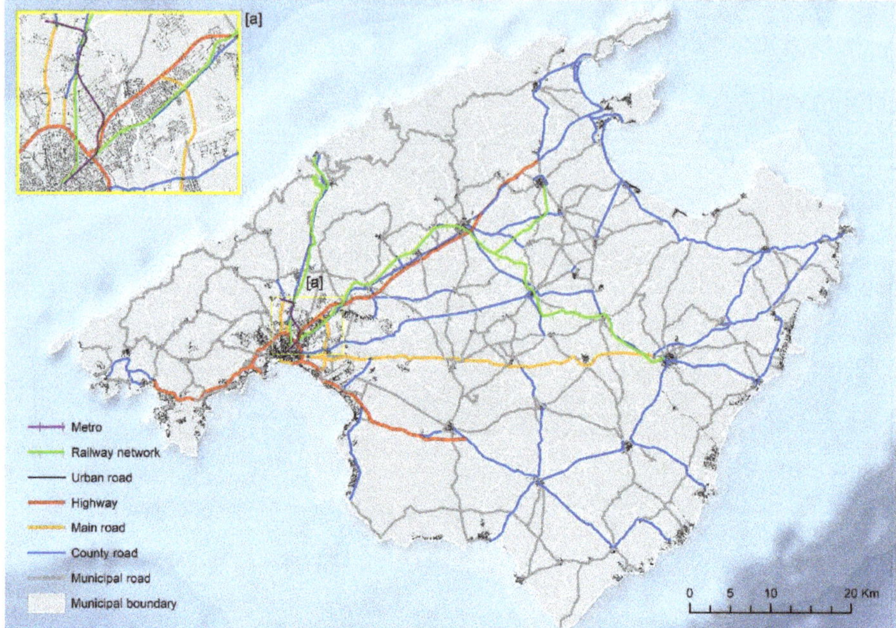

Figure 3. Transport network of Mallorca, [**a**] Detail of Palma city. Source: Consell de Mallorca visor 2020 [76].

The management of island mobility and transport is shared by the Government of the Balearic Islands (GOIB) through the Direcció General de Mobilitat i Transport Terrestre (Conselleria de Mobilitat i Habitatge, GOIB) [77] and the island government of the Consell de Mallorca through the Department of Mobility and Infrastructures [78]. The GOIB manages public transport by road through the Consorci de Transports de Mallorca [79] and the railway and underground through the Serveis Ferroviaris de Mallorca [80]. The Consell de Mallorca is responsible for the regional roads and the traffic of the island. The Palma Town Council manages municipal public transport and infrastructure through its Department of Mobility [81], the Empresa Municipal de Transport (EMT) [82], and the Municipal Parking and Projects Company (SMAP) (http://www.mobipalma.mobi/). Some municipalities have a Sustainable Urban Mobility Plan (SUMP), although at present, they are few (Manacor, Palma, Sóller, Inca, Andratx, Llubí, Capdepera, and Pollença) [83]. The participation of the Palma City Council in the European initiative CIVITAS is noteworthy [84]. It has led to considerable improvement in the public bicycle system, traffic management, a mobility app, the city's parking payment system, the acquisition of a municipal fleet of electric vehicles, and the SUMP of Palma [85].

The regulations on mobility began with the 2004 Law on Land Transport and Sustainable Mobility of the Balearic Islands [86] and culminated in the Balearic Islands Sectorial Mobility Master Plan (2019) [87].

Since the 1960s, Mallorca has based its economy and social development in the tourist industry. In 2019, Mallorca received a total of 11,874,835 tourists (1,597,915 from Spain and 10,276,921 from abroad) [72]. The distribution of the hotels in the island's municipalities is not regular. The coastal municipalities (550 km of coastline) concentrate most of the accommodation, although agrotourism and holiday accommodation are also noteworthy and spread across the interior (Figure 4).

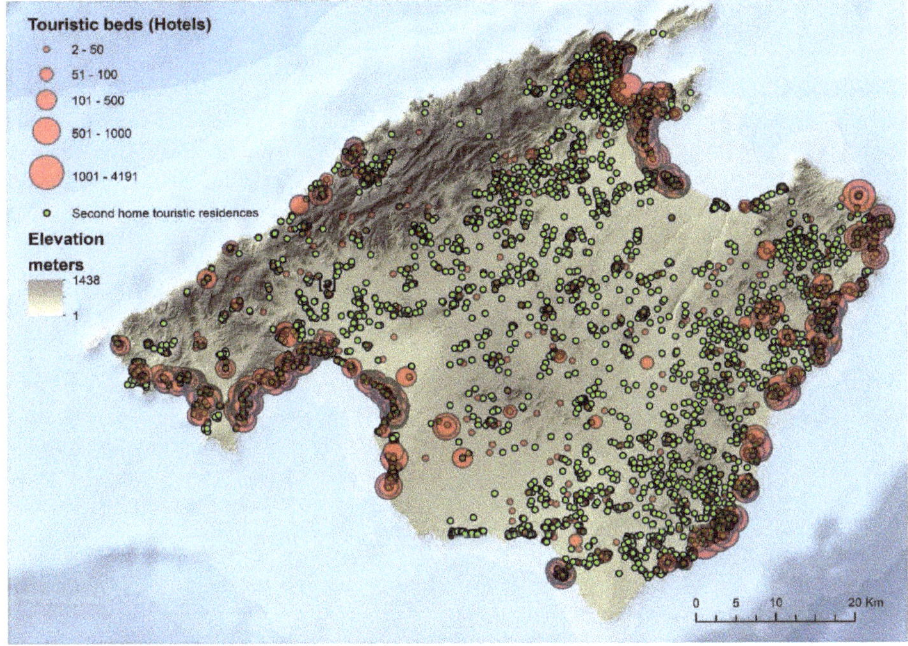

Figure 4. Distribution of hotel equipment in Mallorca on a topographic basis. Source: Govern de les Illes Balears, 2020 [88]; Inside Airbnb (2020) [89].

The topography of the island has a direct influence on the land transport and mobility model. In fact, the mountainous areas, corresponding to the Tramuntana region, are those with the least developed road and rail infrastructure, the smallest population, and the least tourism and economic activity. This means that in this area, the mobility of the resident population is reduced a priori.

At present, the dynamics of population growth, together with the intense development of seasonal tourism, have produced an extraordinary increase in negative externalities derived mainly from private mobility, leading to an increase in traffic and associated problems of congestion and pollution. The average daily intensity of vehicles has experienced continuous growth in recent years. Traffic increased by 42% between 2005 and 2015. The transport model is becoming highly inefficient. Congestion is continuous, especially on roads that serve Palma and the main tourist areas (Figure 2). The large vehicle fleet contributes to this (1136 vehicles/1000 inhabitants in the Balearic Islands and 736 cars/1000 inhabitants in 2020) [90], to which a total of approximately 75,000 rental vehicles must be added that are incorporated into the transport network during the high tourist season in the summer months.

In this geographical framework, it is a priority to deepen our knowledge of the resident population's modal behavior to identify the key factors to guide policies, plans, and projects towards promoting sustainable modes.

3.2. Data Sources

The study used the daily mobility survey corresponding to 2009/2010 from the Government of the Balearic Islands [91]. This survey was undertaken by the Mallorca Transport Consortium under the Balearic Government, and was implemented through telephone interviews (13,905). It was carried out among people over 14 years of age, stratified according to age, gender, and the municipality of residence, in two campaigns between April and June 2009 for Palma and between November 2009 and February 2010 for the rest of the municipalities of the island. The margin of error is ±1.04% for a confidence interval of 95.5%. The survey included questions on the number of trips made the previous day, their origin and destination, the means of transport used, the time of travel, the duration (in minutes) of the trip, the reason, and other data relating to the socio-demographic profiles of the interviewees. The survey results estimate the number of movements made in one working day across the island, broken down by origin and destination, for a total of 53 municipalities. This survey was subsequently expanded with updated demographic information to provide information on the mobility of the resident population at the island level.

For the drafting of the recently approved Sectoral Master Plan for Sustainable Mobility in the Balearic Islands [87], another telephone mobility survey was carried out. This represents an extensive effort recently carried out by the company DOYMO (https://doymo.com/) for the Balearic Islands Government. In this article, it was not possible to obtain the information corresponding to this survey's original data, although its main results were considered.

The data from both sources were compared to obtain more complete information on the mobility situation. Despite the seven-year interval between the two surveys, it can be seen that, a priori, there were no relevant changes in the mobility habits of the population of Mallorca, so the conclusions drawn from the 2009 survey can be reasonably extrapolated to the present day (Table 2). It is essential to point out that 2009 corresponds to a peak moment in the last world economic crisis, which significantly affected the Balearic Islands, and better reflects the island's dynamics as a baseline in terms of mobility.

Table 2. Modal split on Mallorca. Surveys from 2009 and 2017.

			2009	2017
	Mode	**Trips**	%	%
Public	Bus	128,727	5.7%	
	Train	17,401	0.8%	
	Metro	7616	0.3%	
	Cab	6038	0.3%	
	Subtotal	159,788	7.1%	10%
Private	Car	1,214,051	53.4%	
	Motorbike	52,304	2.3%	
	Subtotal	1,266,355	55.7%	55%
Healthy transportation	Bike	30,149	1.3%	2%
	Pedestrian	816,776	35.9%	33%
	Subtotal	846,925	37.2%	35%
Total		2,278,436	100.0%	100%

Source: Conselleria de Territori, Energia i Mobilitat.

Demographic and tourist information was obtained from the National Institute of Statistics (http://ine.es), the Balearic Institute of Statistics [72], and the Government of the Balearic Islands. Finally, information about the traffic flow on the roads of Mallorca was used, which was provided by the Mobility Department of the Consell de Mallorca [78], as well as information about the vehicle fleet published by the General Directorate of Traffic (Ministry of the Interior, Government of Spain) [90].

3.3. Analytic Process

The analysis of the influence of social, economic, and territorial factors on the modal choice and type of travel was made exclusively based on data from the 2009 mobility survey. The phases of analysis are as follows.

3.3.1. Binary Analysis of the Relationship between Modality and Other Variables

Pearson's χ^2 test of independence is used to study whether there is an association between two categorical variables. It is a hypothesis test contrasting qualitative variables. The test supports the null hypothesis, H_o: The variables are independent, so one variable does not vary between the different categories of the other variable; H_a: The variables are dependent, and one variable varies between the different levels of the other variable.

$$\chi^2 = \sum\nolimits_{i,j} \frac{\left(Observed_{ij} - Expected_{ij}\right)^2}{Expected_{ij}} \tag{1}$$

Each group's expected value is obtained by multiplying the marginal frequencies of the row and column in which the cell is located and dividing by the total number of observations. The differences at all levels are added up. The chi-square distribution has only one parameter, the degrees of freedom, which determines its center and dispersion shape.

$$df = (levels\ variable\ A) - 1\ (levels\ variable\ B - 1) = (columns - 1)\ (rows - 1) \tag{2}$$

The chi-square distribution is positive, so the p-value calculation only takes into account the upper tail.

Since the test contrasts whether the variables are related, the effect's size is known as the strength of the association. In our case, we analyze the measures of association of phi or Cramer's V. The limits used for their classification are 0.1 for a small association, 0.3 for a medium association, and 0.5 for a large association. Independence in the sample observations is required.

3.3.2. Integrated Analysis of the Modal Choice

Binary logistic regression was used for the integrated analysis of the modal choice. This statistical technique allows the analysis of the relationship between a dichotomous qualitative dependent variable and one or more independent explanatory variables, or covariates, whether qualitative or quantitative [92]. It is a multivariate predictive regression technique. The basic equation of the binary logistic regression model is exponential, although its logarithmic transformation (logit) allows its use as a linear function. The aim is to model the influence of a set of variables on a dichotomous event's appearance.

In our case, we considered the use of different modes of transport as the dichotomous event. To do so, we transformed a qualitative variable of modes of transportation into a dichotomous variable. The following modes were considered: car, motorbike, bus, metro, pedestrian, bicycle, and sustainable mode, in which all modes were grouped except for car and motorbike.

The logistic regression model makes it possible to quantify the importance of the relationship between the covariates and the dependent variable. First, the interaction between covariates is analyzed concerning the dependent variable (odds ratio). Second, individuals are classified in the dependent variable categories according to the probability they obtain.

It is necessary to ensure that there is no multicollinearity between the predictor variables, so the choice of predictor variables must be made efficiently. The final model should be as small as possible and explain the functional dependencies as much as possible (principle of parsimony).

The covariates' preparation required their corresponding transformation into binary variables (dummy variables), which represent each of the selected categories. Specifically, the following variables are used: sex (a single group is included in which 1 represents men and 0 represents women), age group, occupation, reason for the trip, proximity to Palma of the location of origin, region, and coastal municipality.

The equation of the logistic regression is represented by the natural logarithm of Equation (3):

$$Y = \beta_0 + \beta_1 X_1 + \beta_2 X_2 + \ldots + \beta_k X_k. \quad (3)$$

β_0 and β_k are the coefficients estimated by regression from the data $\beta_0\beta_k{}'' X_{1\ldots k}$.

The probability of using sustainable modes of transport may be presented in an equation, as reflected in Equation (4).

$$p = \frac{1}{1+e^{-Y}} = \frac{1}{1+e^{-(\beta_0 + X_1\beta_1 + \ldots + X_k\beta_k)}} \quad (4)$$

The covariates X_1 and X_k are represented by the set of selected variables.

The binary logistic regression considers two events of the phenomenon and is expressed as Equations (5) and (6):

$$\Pr(y=1) = \frac{1}{1+e^{-(a+bx)}} = P \quad (5)$$

$$\Pr(y=0) = 1 - \frac{1}{1+e^{-(a+bx)}} = 1 - P. \quad (6)$$

The logit transformation arises from considering the probability ratio between two events, called the advantage or ratio (odds). The ratio/odds of an event are the quotient between the probability of it happening and the probability of it not happening (Equation (7)):

$$odds = \frac{P}{1-P} = \frac{Probability\ of\ P}{Probability\ of\ no\ P} \quad (7)$$

Odds are interpreted as ratios, i.e., the number of times something can happen over something that cannot occur. They are a measure of association between two variables that indicate the strength of the relationship. When the odds ratio is close to 1, there is no association. Values less than 1 indicate a negative association, and values greater than 1 indicate a positive association.

The logistic regression function of the statistical package SPSS version 25 was used for the analysis. The method was configured with the "Enter mode" directly incorporating the whole set of variables selected in the calculation process. This allows us to identify the role played by each variable.

The great utility of logistic regression is to interpret the covariates' effects independently on the dependent variable. This is done by reading the ODDS ratio coefficients (Coef. B, Exp (B)). One of the fundamental problems that arises when several variables are involved in a process is to determine the contribution of each one of them, assuming that the rest of the variables do not change.

3.3.3. Analysis of the Modal Choice at the Municipal Level

In addition to the overall analysis of the factors that influence the modal choice for the island of Mallorca, one of the objectives of this article is to analyze spatial patterns in the modal choice. The modal split distribution of each municipality of Mallorca is assessed, and a new index of sustainable modes of transport is generated (Equation (8)):

Sustainable Transport Indicator (ITS) = % healthy modes (% Pedestrian + % Bicycle) + % Collective modes (% Bus + % Metro). (8)

The ITS index is then normalized as follows (Equation (9)):

ITS = (value − min)/(max − min). (9)

This index makes it possible to establish a hierarchy of municipalities on the basis of the weight that sustainable modes have in each case. The overall analysis of the index facilitates the identification of the areas most dependent on private vehicles as places where sustainable modes should be promoted.

Moreover, using a geographical information system, a database was created that includes territorial information at a municipal level expressed as a total percentage of transport demand (population, tourist places) and the negative externalities of transport (average daily intensity, number of vehicles in the vehicle fleet).

ArcMap 10.5 (Esri©) software was used for the cartographic representation and analysis of geographical data.

4. Results and Discussion

The large figures for the modal split of the residents in Mallorca obtained from the 2009 survey, contrasted with the more recent information from the 2017 mobility survey, show a high dependence on motorized modes for all types of trips (Table 2).

A total of 2.2 million journeys are made daily on the island. This figure has remained unchanged in recent decades.

The pattern of modal choice shows a predominance of the use of private vehicles (55.7%), low use of collective public modes (7.1%), and a moderate level (37.2%) of healthy modes (pedestrian and bicycle). A priori, this distribution shows an evident deficiency in all modes of public transport.

4.1. Modal Relationships

The statistical analysis carried out shows that the modal choice is a multifactorial process in which many variables are involved. The chi-square analysis from contingency tables between the transport modes (qualitative variable: ordinal scale) and various socio-demographic and territorial variables (qualitative variable: ordinal scale) shows significance in the multiple relationships (Table 3):

- Gender and modes: The female population is more likely to use both pedestrian modes (56% female/44% male) and bus transport (61.9% female/38.1% male). Bicycles (35.2% women/64.8% men) and motorbikes (16.5% women/83.5% men), on the other hand, are more commonly used by men. These results show a social model of female segregation, in which women are forced to use public transport rather than private vehicles because they do not have the resources to access their own cars or because of a cultural model in which it is difficult for them to have both a driver's license and their own vehicle.
- Age groups and modes: The private vehicle is typically selected as the preferred mode of transport (55.5% for young people aged 14–29, 62.2% for adults aged 30–44, 51.5% for those aged 45–64), except among the population group over 65, where the pedestrian mode accounts for 61.6% of their journeys. The young population group, from 14 to 29 years old, mostly selects the bus (37.2%), train (39.5%), and metro (53%). The young people's modal choices of bus and metro are related to the segment of the youth population traveling to school. A special case is the metro, which is mainly used by students from the University of the Balearic Islands, who travel to the campus (located 7.5 km away) from Palma by the only metro line available on the island. The group of adults aged between 30 and 44 years has the highest use of private vehicles (40.5%), motorbikes (43.3%), and bicycles (37%).
- Travel time and modes: Ninety-one percent of trips in Mallorca are local, not exceeding 30 min by car. This interval of travel time corresponds to 83% of trips by train, 80.8% of trips by metro, 93% of trips by car, 98.2% of trips by motorbike, 84.5% of trips by bicycle, and 92.4% of trips by foot. Trips of 30 to 60 min account for 7.6% of the total. Train use is predominant at this interval. Most of the movements around the island comprise recurrent journeys close to the city of Palma from peripheral municipalities, while there are also journeys between inland municipalities and nearby coastal areas. In the interval of trips lasting over 60 min, movements are mostly made on

foot or by car (with these modes accounting for 62.6% of trips lasting from 60 to 90 min, 59.2% of trips lasting from 90–120 min, and 64.3% of trips lasting over 120 min).

- Proximity to Palma and modes: In total, 54.8% of trips are concentrated in areas less than 15 km from Palma. In addition, 22.53% of train trips are made in areas between 30 and 50 km from Palma, corresponding mainly to movements in Inca and Manacor. These results confirm a radial model of movements to and from Palma as a unique center of attraction and trip generation.
- Regions and modes: The Badia de Palma region concentrates the largest number of journeys (62.9%) in Mallorca. It is worth noting that 95% of all bus journeys and 100% of all trips by metro are made in the Badia de Palma area. The majority of train use is in the Badia de Palma region (56.2%) and in the Raiguer region (23.5%). This distribution shows the great reputation of the city of Palma and its surrounding municipalities (Marratxí, Calvià, Llucmajor) as areas that generate and attract journeys.
- Motive for the journey and modes: Home travel (53.2%), personal arrangements (20.4%), and access to work (16.8%) are the main reasons for travelling. Some 18.1% of bus journeys are made for work purposes. The majority of trips are for traveling home (58.9%) and personal management (24.5%). It should be noted that people preferably start their journeys from their homes in private vehicles and use this mode for almost all their movements.
- Occupation and travel modes: Active people who are employed mainly use cars for their journeys (64.3%), and students also use private vehicles (45.5%). Of the total number of metro users, 49.6% are students. The retired population has reduced use of motorized modes and concentrates 58.9% of their mobility on pedestrian journeys. The bus is used mostly by the working population (37.9%) and students (23.8%).

Table 3. Chi-square values of Cramer's factor V and significance coefficients obtained from the derivation of contingency tables between modes of transport and different variables.

	Pearson Chi-Square	df	Cramer's V
Gender (Categories: men/women)	47,808.306 *	7	0.145 *
Age groups (Categories: 14–29/30/44/45/64, +64)	144,839.779 *	21	0.146 *
Activity (Categories: working, retired, student, unemployed, domestic work, other)	217,340.395 *	35	0.139 *
Trip motivation (Categories: leisure, work, doctor, study, mixed, home)	99,176.263 *	35	0.100 *
Travel time (0–30/30–60/60–90/90–120 min)	126,206.872 *	28	0.118 *
County of origin (Categories: Badia Palma, Llevant, Nord, South, Raiguer, Pla, Tramuntana)	81,004.755 *	42	0.077 *
Distance to Palma	87,084.148 *	21	0.113 *
Coastal (coastal/non-coastal)	31,424.402 *	7	0.118 *

(*) Significant at 0.99 for Sig values < 0.01.

The low degree of randomness of the set of relations analyzed is noted (Table 2). The significance of all the chi-square tests is relevant, so the null hypothesis must be rejected. Despite this, the factors analyzed do not provide high Cramer's values in general, so the modal choice's multifactorial character is evident. The variables with the greatest relation to the selection of modes are gender, age group, and people's activity. Second, the reason for the trip and its duration are highlighted. The region of origin of the journey is less related to the modes.

4.2. Factors in Modal Selection

The travel time variable was eliminated from all the analyses because it has no significance.

4.2.1. Motorized Modes

Table 4 shows the results of the logistic regression using private vehicles as the dependent variable. We can observe that almost all the variables considered play a significant role in predicting the use of private vehicles (significance < 0.00). No variable was extracted from the process.

Table 4. Results of the logistic regression of private vehicles.

	CAR		MOTORBIKE	
	B	**Exp (B)**	**B**	**Exp (B)**
Man	0.352 *	1.422	1.560 *	4.758
14–29 years	0.848 *	2.335	1.153 *	3.167
30–44 years	0.963 *	2.620	1.223 *	3.396
45–64 years	0.600 *	1.822	0.753 *	2.123
Working	0.326 *	1.386	−0.517 *	0.596
Retired	−0.433 *	0.649	−0.926 *	0.396
Student	−1,182 *	0.307	−0.248 *	0.780
Unemployed	−0.215 *	0.807	−0.752 *	0.471
Other	−0.256 *	0.774	−0.138 *	0.871
Domestic work	−0.236 *	0.790	−1.878 *	0.153
Leisure	0.893 *	2.443	0.128 *	1.137
Work	0.960 *	2.611	0.329 *	1.390
Doctor	0.934 *	2.545	−0.642 *	0.526
Study	0.675 *	1.964	−0.419 *	0.658
Mixed activities	0.841 *	2.318	−0.234 *	0.791
Home	0.662 *	1.939	0.051 *	1.052
Badia de Palma	0.625 *	1.869	−1.260 *	0.284
Llevant	0.829 *	2.291	−1.549 *	0.212
Nord	0.778 *	2.178	−1.327 *	0.265
South	0.910 *	2.484	−1.831 *	0.160
Raiguer	0.086 *	1.090	−1.196 *	0.302
Pla	0.206 *	1.228	−0.611 *	0.543
0–30 min	0.934 *	2.546	0.677 *	1.968
30–60 min	0.626 *	1.869	−0.688 *	0.502
60–90 min	0.385 *	1.470	−0.244 *	0.784
90–120 min	0.184 *	1.202	−0.425 *	0.654
Non-coastal	0.345 *	1.413	−1.044 *	0.352
Constant	−2.496 *	0.082	−3.742 *	0.020

(*) Significant at 0.99 for p-value < 0.01.

First, it should be noted that men, in general, have a positive tendency to use private vehicles compared to other modes (Car: B = 0.352, Exp (B) = 1.422; Motorcycle: B = 1.56, Exp (B) = 4.75). This result is relatively common in Spain's mobility and gender studies [93,94]. Women have a type of dependent mobility. This indicator reflects that the society is not very advanced in terms of gender and that there is a social pattern of women's dependency on men.

Concerning the analysis by age group, the majority use of private vehicles by the adult population (30–44 years) is significant (Car: B = 0.352, Exp (B) = 1.422; Motorcycle: B = 1.56, Exp (B) = 4.75), and there is a radical reduction in the use of these vehicles by the elderly. This circumstance confirms the adult population's purchasing power for the acquisition of a private vehicle and its majority use in daily mobility. From this perspective, the demographic dynamics of aging in the coming years could lead to a significant decrease in the young population, especially in the more rural municipalities in the island's interior, which would have a direct effect on reducing the use of private vehicles.

People's work activity also affects their modal choice. People who are working make the most use of private vehicles, while students and retired people seem least inclined to choose cars as their means of transport.

We want to emphasize the positive sign of the B coefficients for all the reasons for the trip. This indicates the high social orientation towards the use of private vehicles. Regarding the reason for the trip, the use of private vehicles for trips to work (Car: B = 0.960; Exp (B) = 2.61) and to health centers (Car: B = 0.934; Exp (B) = 2.54) is prominent. This is because public reference hospitals do not have an efficient transport service in terms of frequency or accessibility to all locations on the island. For this reason, the use of private vehicles is required for these types of trips. This factor shows a high level of inequity in the island's transport service to the facilities. The travel motive for leisure also appears to be differentiated. Therefore, in many cases, these are journeys to urban centers and natural areas.

In terms of the territorial analysis carried out, Mallorca's southern region stands out as being highly dependent on private vehicles (B = 0.91; Exp (B) = 2.48). This tendency is reproduced in the rest of the regions of the island with more or less significant intensity. However, the Raiguer region appears with a very low B coefficient (B = 0.086) and little relevance (Exp (B) = 1.09). It is also important to note that the Serra de Tramuntana region does not appear relevant.

The proximity to the city of Palma is also a factor that increases the use of private vehicles. The residents in the areas closest to the capital are also those who use private vehicles most often. This fact reveals the structural unsustainability of the city of Palma's efforts to encourage the use of alternative modes.

Finally, it can be seen that inland areas of the island (not along the coast) are also more dependent on private vehicles for their journeys.

4.2.2. Collective Modes: Bus, Train, and Metro

The bus and train collective modes are used mostly by women (Bus: B = −0.53 Exp(B) = 0.58). Students use the bus more often (B = 0.318, Exp(B) = 1.374) than other groups (Table 5). Bus trips for medical visits (B = 1383, Exp(B) = 3.98) are particularly noteworthy, followed by work activities (B = 0.826, Exp(B) = 2.28) and access to educational centers (B = 0.816, Exp(B) = 2.26). The coefficients of significance for the use of the train eliminate two age groups (20–44 and 45–64 years) and the variables that refer to the activity carried out by the people. The use of the train by the youngest group (14–29 years old) for any activity in general, but particularly for study, leisure, and work, is remarkable. Access to the railway infrastructure reflects the fact that the Raiguer region is the most privileged (B = 3.653, Exp(B) = 38.59, followed by Pla (B = 2.94, Exp(B) = 19). Journeys by train tend to be between 0 and 15 min from Palma (B = 2.55, Exp(B) = 12.84).

Table 5. Results of logistic regression for collective vehicles.

	BUS		TRAIN		METRO	
	B	**Exp (B)**	**B**	**Exp (B)**	**B**	**Exp (B)**
Man	−0.537 *	0.585	−0.284 *	0.753	0.024	1.024
14–29 years	0.076 *	1.079	0.474 *	1.606	−0.956 *	0.385
30–44 years	−0.207 *	0.813	0.001	1.001	−0.424 *	0.654
45–64 years	0.077 *	1.080	−0.004	0.996	−0.209	0.811
Working	−0.763 *	0.466	15.959	-	1.778	-
Retired	−0.287 *	0.751	16.407	-	14.824	-
Student	0.318 *	1.374	16.667	-	17.858	-
Unemployed	−0.313 *	0.731	16.079	-	14.311	-
Other	−0.115 *	0.892	14.462	-	11.13	3.044
Domestic work	−0.523 *	0.593	16.445	-	13.740	-
Leisure	0.302 *	1.353	1.409 *	4.093	0.449 *	1.566
Work	0.826 *	2.284	1.226 *	3.408	1.544 *	4.685
Doctor	1383 *	3.988	1.088 *	2.968	−13.764	0.000
Study	0.816 *	2.261	1.249 *	3.486	1.879 *	6.545
Mixed activities	0.101 *	1.106	0.307 *	1.359	0.906 *	2.473
Home	0.327 *	1.386	0.894 *	2.446	0.871 *	2.390
Badia de Palma	−0.274 *	0.760	−1.060 *	0.346	−0.949 *	0.387
Llevant	−2.801 *	0.061	1.809 *	6.107	−5.102	0.006
Nord	−2.253 *	0.105	0.693 *	2.001	−16.491	0.000
South	−1.559 *	0.210	−1.752 *	0.173	−14.906	0.000
Raiguer	−0.419 *	0.658	3.653 *	38.591	−5.002 *	0.007
Pla	−0.236 *	0.790	2.945 *	19.009	−3.623 *	0.027
0–30 min	−2.122 *	0.120	2.553 *	12.849	−4.534 *	0.011
30–60 min	−1.750 *	0.174	−2.141 *	0.117	−2.652 *	0.070
60–90 min	−1.867 *	0.155	−2.040 *	0.130	−0.741 *	0.477
90–120 min	−0.479 *	0.620	−2.741 *	0.064	−11.630	0.000
Non-coastal	−0.246 *	0.782	−1.216 *	0.297	1.710 *	5527
Constant	−1.635 *	0.195	−21.013	0.000	−20.732	0.000

(*) Significant at 0.99 for p-value < 0.01.

Concerning the use of the metro, the most important conditions are its preferential use for access to the educational centers (University of the Balearic Islands) (B = 1.87, Exp(B) = 6.54) and the work centers (B = 1.54, Exp(B) = 4.68).

4.2.3. Healthy or Active Modes

The reasons affecting the choice of healthy modes are multiple (Table 6). However, those that have the greatest significance are the following: The pedestrian mode is used preferentially by women, retired persons (B = 0.879, Exp(B) = 2.4), unemployed persons (B = 0.7, Exp(B) = 2.013), and those engaged in domestic work (B = 0.69, Exp(B) = 2). Likewise, the most relevant journeys on foot exceed 90 min. The use of the bicycle is preferable for men (B = 0.6, Exp(B) = 1.83), for trips to work (Exp(B) = 1.4) and to study centers (Exp(B) = 1.25), and for journeys lasting 60 to 90 min.

Table 6. Results of the logistic regression of healthy or active modes.

	PEDESTRIAN		BIKE	
	B	**Exp (B)**	**B**	**Exp (B)**
Man	−0.303 *	0.738	0.606 *	1.833
14–29 years	−0.903 *	0.406	0.108 *	1.114
30–44 years	−0.776 *	0.460	0.041	1.041
45–64 years	−0.408 *	0.665	−0.103 *	0.902
Working	0.167 *	1.181	0.355 *	1.426
Retired	0.879 *	2.408	−0.185 *	0.831
Student	0.598 *	1.818	0.229 *	1.258
Unemployed	0.700 *	2.013	0.041	1.042
Other	0.301 *	1.351	0.496 *	1.641
Domestic work	0.695 *	2.004	0.197 *	1.217
Leisure	−1.113 *	0.329	0.035	1.036
Work	−1.234 *	0.291	−0.287 *	0.750
Doctor	−1.871 *	0.154	−2.097 *	0.123
Study	−1..034 *	0.355	−0.530 *	0.588
Mixed activities	−0.649 *	0.523	−0.625 *	0.535
Home	−0.668 *	0.513	−0.243 *	0.784
Badia de Palma	−0.546 *	0.579	−0.266 *	0.766
Llevant	−0.512 *	0.600	0.468 *	1.597
Nord	−0.455 *	0.634	0.566 *	1.761
South	−0.668 *	0.513	1.082 *	2.950
Raiguer	0.064 *	1,066	−1.394 *	0.248
Pla	−0.109 *	0.897	−0.537 *	0.584
0–30 min	−0.843 *	0.430	−1.680 *	0.186
30–60 min	−0.181 *	0.834	0.090 *	1.094
60–90 min	−0.058 *	0.944	0.629 *	1.876
90–120 min	0.115 *	1.122	−0.629 *	0.533
Non-coastal	−0.349 *	0.706	0.645 *	1.906
Constant	0.921 *	2.512	−4.603 *	0.010

(*) Significant at 0.99 for p-value < 0.01.

4.2.4. Sustainable Modes

An analysis of the factors involved in the modal choice of sustainable modes (bus, train, metro, pedestrian, and bicycle) shows that all the factors considered have some significance, although few stand out in a special way (Table 7). Remarkably, the sustainable modes are chosen by women, especially among retired adults, students, and the unemployed. These modes are rarely selected for travel to work, leisure, or medical facilities. Sustainable methods are used more in the Raiguer region than in other regions. In general, the analysis reflects that compared to private vehicles, the sustainable modes represent a small fraction for all groups and all travel reasons.

Table 7. Results of the logistic regression of sustainable modes.

	B	Exp (B)
Man	−0.374 *	0.688
14–29 years	−0.852 *	0.427
30–44 years	−0.806 *	0.447
45–64 years	−0.427 *	0.653
Working	−0.136 *	0.873
Retired	0.650 *	1.915
Student	0.635 *	1.886
Unemployed	0.428 *	1.534
Other	0.136 *	1.146
Domestic work	0.404 *	1.498
Leisure	−0.978 *	0.376
Work	−0.936 *	0.392
Doctor	−1200 *	0.301
Study	−0.631 *	0.532
Mixed activities	−0.658 *	0.518
Home	−0.597 *	0.550
Badia de Palma	−0.609 *	0.544
Llevant	−0.699 *	0.497
Nord	−0.649 *	0.522
South	−0.781 *	0.458
Raiguer	0.028 *	1.028
Pla	−0.106 *	0.900
0–30 min	−1.066 *	0.345
30–60 min	−0.569 *	0.566
60–90 min	−0.380 *	0.684
90–120 min	−0.209 *	0.811
Non-coastal	−0.388 *	0.678
Constant	1.751 *	5.763

(*) Significant at 0.99 for p-value < 0.01.

4.3. Modal Choice at the Municipal Level

The analysis of the modal choice's geographical distribution allows the identification of significant differences at the municipal and regional levels. Some of the results obtained in this section could be reviewed more generically in the previous section when reference was made to the "Region" variable. Significant patterns could be detected about the regions of origin of the movements concerning the population's use of the different modes.

Figure 5 shows the number of journeys in different modes at the municipal level and the percentage of each mode's use. In the case of the use of private vehicles, a massive generation of journeys can be observed from the populated areas of the municipalities of Badia de Palma (Calvià, Palma, Llucmajor, and Marratxí). Noteworthy cities include Inca in the island's center, Pollença and Alcúdia in the north, and Manacor, Felanitx, and Capdepera in the east. The map shows the island's population distribution model, maintaining the relationship between the number of movements and population density. Certain inland areas of the island (Pla de Mallorca) and the Serra de Tramuntana present the lowest number of journeys. The map of percentages of trips by car (5a′) shows the main dependence on private vehicles for municipalities such as Calvià and Marratxí (Badia de Palma), Capdepera (Llevant), Ariany (Pla), and Escorca (Serra Tramuntana). There are clear patterns of dependence on private vehicles. These are zones in which collective sustainable modes are not accessible, and sustainable mobility must be promoted through investment to improve infrastructure and services.

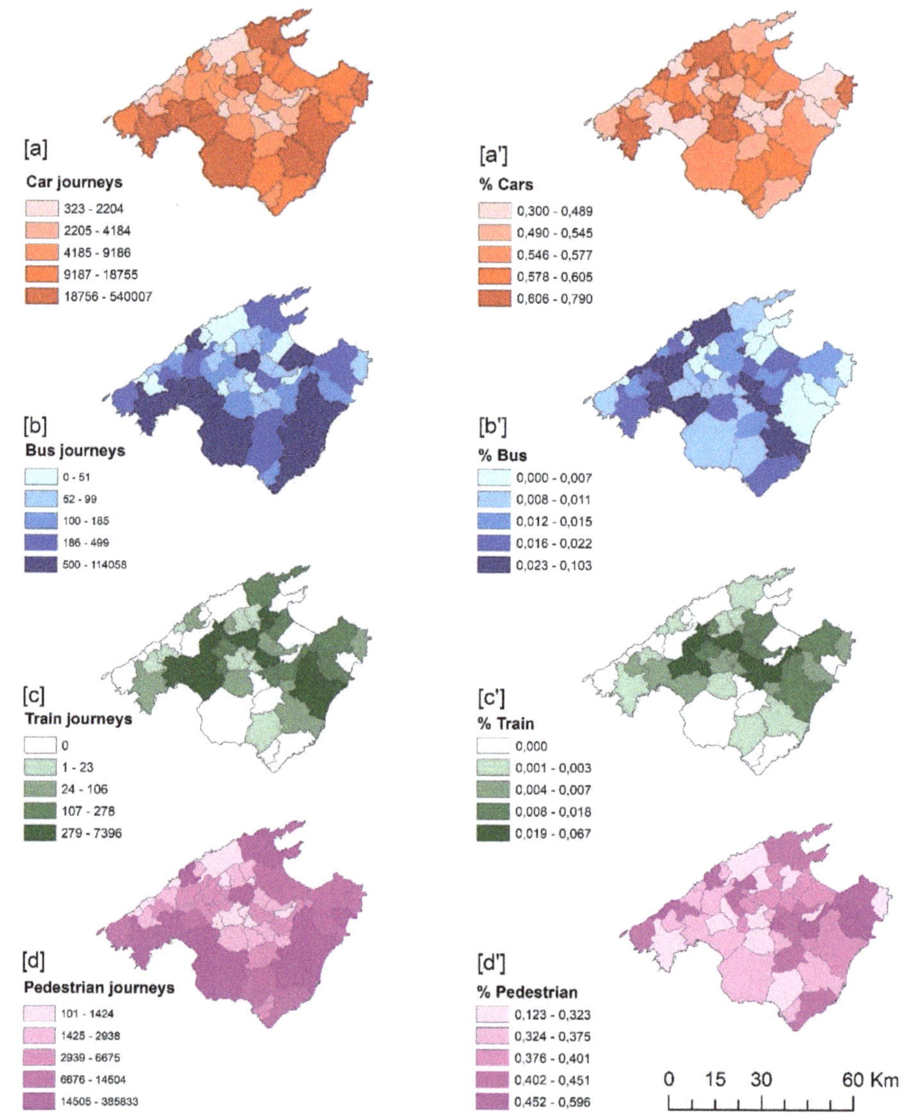

Figure 5. Municipal modal distribution. Number and percentage of movements by mode.

Bus transport at the municipal level, in terms of the number of trips, also reveals a pattern closely related to the distribution of the population (Figure 5b). The percentage distribution of journeys highlights Palma, mainly because it has a municipal bus service (the EMT). The areas without rail infrastructure force people who do not have a private vehicle to use public buses (Figure 5b′). This is a geographical distribution model that highlights the peripheral municipalities of the main centers of travel attraction on the island (Palma, Inca, Manacor).

The distribution of train journeys shows the routes of Mallorca's rail networks (Figure 5c,c'). The municipalities in the Raiguer region obtain the highest values both in absolute figures for the number of journeys and in percentages of preferential use of the railway. The presence of large areas of the Serra de Tramuntana and in the south of Mallorca (Llucmajor, Santanyí, Ses Salines) without railway service is evident.

Pedestrian mobility is more common in municipalities that suffer from certain isolation from collective modes (bus, train). The municipalities of Sóller, Andratx, Estellenç, and Puigpunyent in the Serra de Tramuntana and the municipalities in the eastern (Artà, Son Servera, Sant Llorenç) and the southern parts of Mallorca (Santanyí) are worthy of note. It has been demonstrated that the obligation of having a vehicle for travel motivates pedestrian mobility within the municipalities. Therefore, this represents a factor of isolation, rather than a preference for sustainable modes.

The municipal distribution of sustainable modes (Figure 6) shows a very irregular pattern. On the one hand, there are areas with low values for the use of sustainable modes in municipalities very close to Palma (Calvià, Marratxí, Algaida, Sencelles). They present a model of a dormitory town in which private vehicles are preferred. There are also areas in the Serra de Tramuntana (Escorca, Selva, Deià) whose mobility is highly restricted to cars due to the lack of access to other modes. Capdepera and Ariany also stand out in eastern Mallorca due to their clearly isolated situation. The municipalities that are most highly rated in terms of the use of sustainable modes are also scattered. There are municipalities with a small population, generally older, in which the majority of trips are made on foot within the town itself (Pla), and they have railway stops (Binissalem). The municipalities with an economic sector that encourages travel within the municipality without the need to move for work or professional reasons also obtain high values (Sóller, Artà, Santanyí).

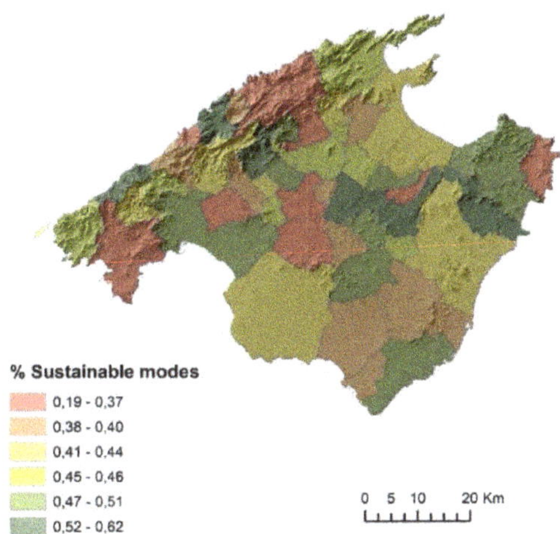

Figure 6. Municipal distribution of sustainable modes (bus, train, metro, pedestrian, and bike modes).

Figure 7 provides an integrated view of the municipal distribution of three key factors of mobility: percentage of transport demand (population + tourist places), rate of use of sustainable modes (bus, train, metro, pedestrian, bicycle), and the percentage of negative externalities derived from the use of private vehicles (average vehicle intensity at the municipal level + mobile vehicle park). It can be shown that demand and impacts, in general, maintain a high correlation (Pearson's correlation coefficient 0.84; p-value < 0.01). In other words, in general, the higher the population and tourist activity of the municipality, the higher its level of traffic and congestion. However, there are municipalities that, although their demand for transport is essential, do not show very significant negative externalities of traffic (Alcúdia, Calvià, Capdepera, Sant Llorenç, Santa Margalida, Santanyí, Son Servera). The opposite is also true, where small towns with low demand suffer from significant negative externalities from the traffic on the roads that pass through them (Marratxí, Algaida, Andratx, Ariany, Búger, Binissalem, Inca, Sencelles, Vilafranca, etc.).

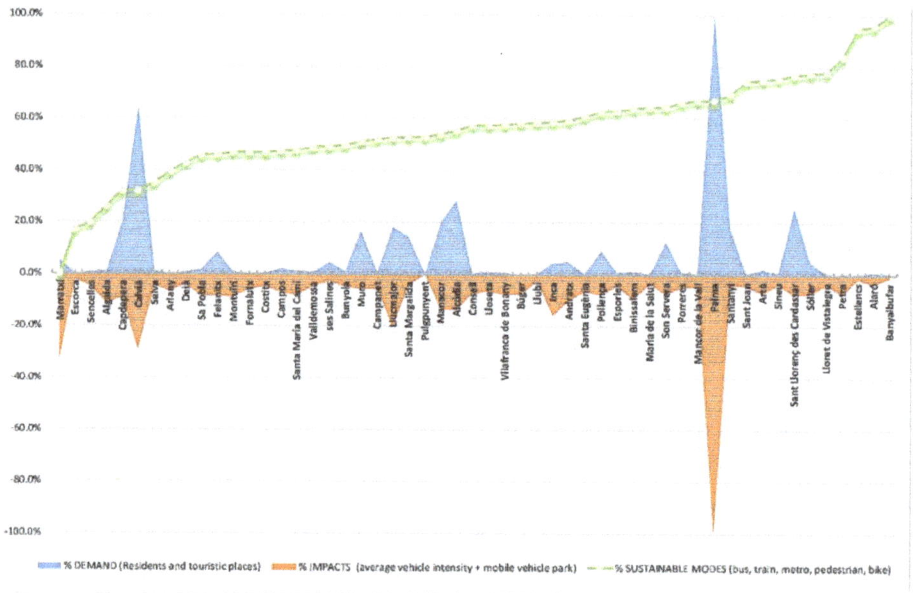

Figure 7. Municipal distribution of % transport demand, % impacts of transport, and % sustainable transport modes.

It is shown that the percentage distribution of the use of sustainable transport at the municipal level has no relationship with the demand for transport or negative externalities. The selection of sustainable modes seems to be influenced by other factors that, as we saw above, are related to variables concerning access to infrastructure and social variables specific to each municipality, the topography, or the municipal location itself. This shows that each municipality has specific dynamics, so municipal classification does not provide a global vision of the island's dynamics.

Figure 8 shows the movements from a municipality of origin to the rest of the municipalities. The maps show the mobility pattern of journeys. From each municipality of origin, an area of concentric influence is reproduced, which makes journeys to the nearest municipalities more likely. Palma's municipality appears on all the maps, which corroborates its role as the capital of the island and as an attraction for trips from the rest of the island. The zoning generated by the tourist destinations of Alcúdia (North), Capdepera (East), and the influential areas of Inca and Manacor are well visualized.

A total of 78.5% of journeys are made within the same municipality (Table 8). Of the total number of journeys in private vehicles on the island, 64% are made in the same municipality. These figures show that actions to improve mobility targeted explicitly at the municipal level could significantly improve the path to sustainable mobility. This circumstance gives great importance to municipal sustainable urban mobility plans as critical instruments for redirecting transport to sustainable modes, and advises that they be developed in all municipalities and tourist areas.

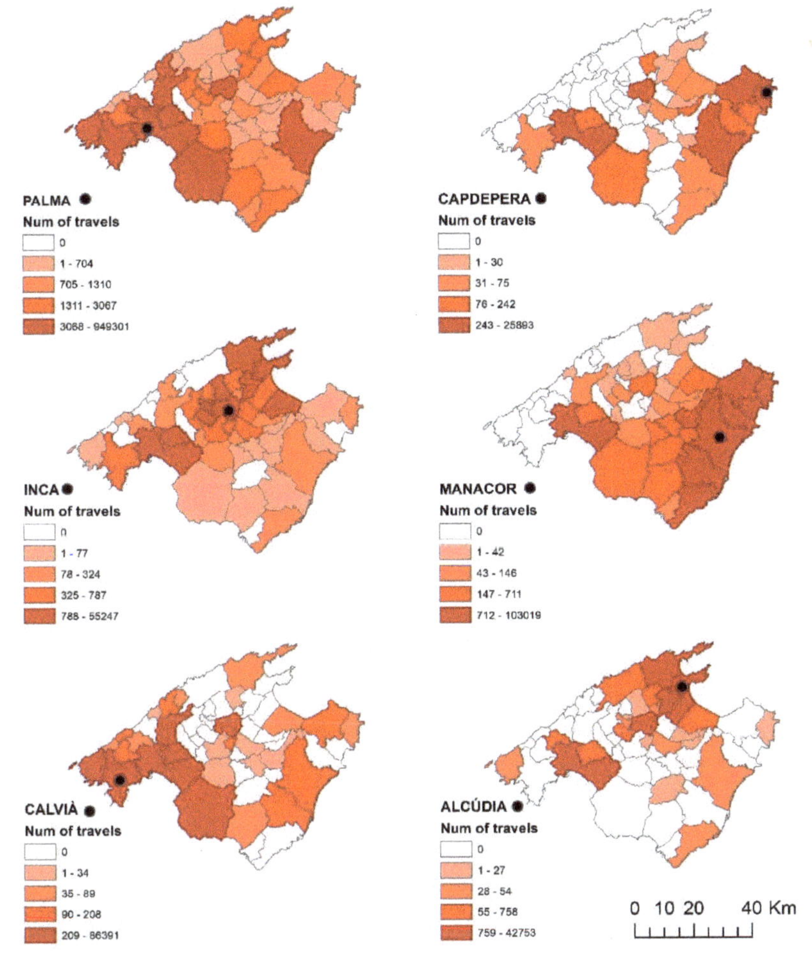

Figure 8. Mobility patterns from the main municipalities. Number of trips from the municipality of origin.

Table 8. Displacements made in the same municipality.

	Mode	% Trips in the Same Municipality
Public	Bus	88.1%
	Train	0.6%
	Metro	82.7%
	Cab	82.0%
Private	Car	64.0%
	Motorbike	87.6%
Healthy transportation	Bike	92.8%
	Pedestrian	99.0%
Total modes		78.5%

The first key issue that was highlighted in this study is the absolute dependence on private vehicles in the daily mobility of residents in Mallorca. Among the leading causes of this are the following:

- Road transport infrastructures are predominant in the island territory, and these roads determine the destinations and their flows. Despite a collective awareness of the promotion of sustainable modes, Mallorca does not have sufficient infrastructure to give proper support to this model. In particular, the rail network is minimal and very poor, as are the metro lines, which are found only in the city of Palma; nor is there any infrastructure for electric trams in urban or peri-urban areas, which could reduce the dependence on cars. The use of railways in certain municipalities in the Raiguer region is significant. Therefore, it is demonstrated that where rail infrastructures exist, their use is also normalized.
- The development of the tourist model in the 1960s was carried out in parallel with the development of motorization, consolidating private transport's predominance and the lack of concern for collective public modes of transport on the part of the authorities.
- The population pattern in Mallorca is seasonal, making the development and maintenance of efficient and sustainable public transport services complex and costly. Most of the connections among population centers are radial through Palma, a historical territorial heritage, so transversal accessibility on the island is not always guaranteed. The areas of new construction are consolidated in territories without transport services, so the use of private vehicles is obligatory for many journeys.
- The dynamics of mobility promoted by seasonal tourism attract workers living in inland areas to coastal areas. This generates a continuous flow of journeys, which increase traffic congestion at the access points to tourist areas. The implementation of holiday homes (Airbnb) in urban and rural areas has added to the pressure on private vehicle transport.
- The size of the island (maximum 100 km) makes it feasible, a priori, to travel by car to all zones in a short time, making the private vehicle the preferred mode of transport for the population. Most journeys (91.7%) do not exceed 30 min. This makes it an aspiration for all inhabitants to have a private vehicle regardless of the externalities it generates.
- The capital, Palma, has the largest population and concentrates the most important infrastructures, equipment, and services of the island. Therefore, it is a unique center of attraction and trip emission. One example is the University of the Balearic Islands, located 7.5 km from Palma, which generates the mobility of a university group that exceeds 15,000 people, and 62.4% of journeys are made by car. This territorial polarization causes imbalances in mobility on an island level that are difficult to resolve.
- The topography of Mallorca makes it difficult to develop sustainable transport infrastructures (railway networks), especially in the Serra de Tramuntana regions. This entails a greater dependence on private vehicles for the residents of this area.

One of the consequences of functional dependence on private vehicles is a social gap of imbalance, equity, and lack of spatial justice, especially in disfavored groups. These groups include the immigrant population without resources, the female population, children, and the elderly. These groups make the most use of (sustainable) collective transport modes. However, we understand that the choice of sustainable collective modes is made in many cases not by choice, but because of the lack of access to the use of private vehicles.

The negative externalities caused by the use of private vehicles, especially congestion and pollution, motivate changes in the mobility habits of residents who are more committed to using sustainable modes, as seen in the evolution of the modal distribution between the surveys in 2009 and 2017. Therefore, we perceive a state of forced dissonance in the selected modes of transport by the resident population [55,95]. The dissonance is between the population's preference for using a mode and its relation to the mode used in practice. The explanation is that there is no possibility for the preferred use of sustainable modes because there are no appropriate infrastructures or services for the citizen, so forced mobility is condemned to be carried out in private vehicles. The dissonance is also of a residential nature because sites are selected for living that do not have efficient public transport, which requires the use of a private vehicle for any type of journey.

The results regarding the modal distribution and inter-municipal mobility show the existence of a generalized pattern of radial mobility towards Palma from all points of the island and the development of sub-graphs of mobility from/to the main population centers (Inca, Manacor, Calvià, Alcudia, etc.). For this reason, island mobility must be understood in a holistic way. This is the vision that the public administration has taken into consideration over the last few years and has been reflected in the Balearic Islands Mobility Sectorial Master Plan 2019–2026 [88]. The following objectives are prioritized in this plan:

- Guaranteeing access to public transport, especially for vulnerable groups;
- Reducing pollution generated by mobility;
- Reducing accidents;
- Minimizing energy consumption;
- Minimizing the minimum distance of journeys;
- Changing the modal distribution in favor of non-motorized collective modes;
- Making public transport more flexible and giving rigidity to the private transport offer;
- Optimizing the connections between islands.

Despite the fact that these objectives' fulfillment will mean a significant advance in mobility, we consider that there is still much work to be done. In particular, sustainable mobility development requires greater coordination between the various sectoral island plans, regarding not only mobility and infrastructure, but also urban and tourism planning. Likewise, for the development of sustainable modes of transport, it will be necessary to focus on the promotion of sustainable built environments [96] that promote sustainable modes of transport. Furthermore, municipal mobility plans are still embryonic and require comprehensive coordination with other plans.

A major risk is promoting sustainable modes based on the use of electric, gas, or other fuel vehicles. This is because one of the main problems generated by mobility in Mallorca is related to congestion (roads, parking). Therefore, the emphasis of the island and municipal plans should be to promote non-motorized modes (train, underground, tram, bicycle, pedestrian, etc.).

The possibility of having efficient sustainable modes of transport is a determining factor in the choice of tourist destination and represents an important environmental incentive. The design of transport investment strategies that maximize the social and economic return on investment should be encouraged, especially in island regions. A commitment to the construction of road transport infrastructure with a high territorial impact can have negative consequences on the tourist destination. It is advisable to commit to the development of sustainable modes [97] and to monitor the sustainability of proposed transport and mobility projects [4,97].

The challenges for the promotion of sustainable mobility in Mallorca must be addressed considering both residents and tourists. It would be important to deploy economic tools, communication, or physical actions regarding the population and tourism [98,99]. The implementation of shared mobility instruments would also be of interest, especially if electric vehicles were promoted [5,100]. It is considered necessary to develop business synergy between the tourism industry and mobility companies and to promote innovative mobility actions for both residents and tourists that would lead Mallorca towards a model of sustainability, as other tourist destinations have developed [101]. On this road to sustainable mobility, initiatives should be developed that take advantage of the widespread use of social networks and apps among both residents and tourists, which could have great applicability [102].

5. Conclusions

The modal choice of transport on the island of Mallorca depends on a wide range of factors, such as gender, age group, the motive for the trip, activity, region of residence, duration of the trip, or proximity to Palma, the capital of the island. The private vehicle is the primary mode. Its use is preferred by working men aged 30–44, and it is used for journeys to the home and to work that do not exceed 30 min, preferably in areas close to Palma. The motorbike is also an essential mode for men of the same age for work purposes. Women's trips, in general, incorporate more collective and healthy modes. Women, young students, and retired people are the main users of buses for access to school, medical visits, or work activities. Trains are used extensively for trips lasting less than 30 min, especially in the municipalities of the Raiguer region. The metro in Palma provides young students access to the university. Sustainable modes are not widely adopted, and women, young people, and retired people are the main groups that use them.

The municipalities included in the Badia de Palma region, the metropolitan area, generate the most significant number of journeys. Additionally, the centers of Inca, Manacor, Calvià, Alcúdia, and Capdepera stand out as areas that generate and attract trips at a regional level.

The scarce deployment of rail and metro infrastructures limits the use of these modes to a minimum. The most isolated areas of Mallorca, with an aging population, are highly dependent on public transport, which can generate imbalances and inequalities in access to health and educational facilities.

The demand for transport generated by the resident population and tourist activity as well as the negative externalities derived from private vehicles' mobility are closely related at a municipal level (Pearson coefficient 0.84, $p = 0.00$). However, the modal distribution does not seem to be directly related to these factors, but is instead more conditioned by access to infrastructures, the location of the municipality, or the topography.

Sustainable modes of mobility in Mallorca are still in an embryonic state. The main dependence on the use of private vehicles is evident for any trip, whether it is for a long time, for a particular reason, or from a specific place. The failure to adopt a sustainable model in time gives rise to negative impacts derived from the use of private vehicles (congestion/pollution), resulting in a deterioration of Mallorca's image as a tourist destination.

The island's diffuse urbanization model, which is deployed radially from Palma towards coastal towns, contributes to the dependence on private vehicles. The transport development model has been characterized by infrastructure plans with enormous environmental and social costs. The new mobility planning developed by the Regional and Island Government's Sectoral Master Plan for Mobility proposes substantial changes towards sustainable modes. Its implementation will require considerable economic investment to deploy the railway network and to take various actions to promote sustainable modes. In any case, it is considered a priority to guarantee the coordination of the plan with other territorial, urban, infrastructure, and tourism plans and to consider mobility as an integral and important issue.

Finally, to ensure the viability of sustainability planning and the joint work of the different administrations, it will be necessary to promote a cultural change in the population towards the acceptance of and preference for sustainable modes. In this sense, initiatives to use shared vehicles and financial support instruments for sustainable modes may be considered alternatives in the coming years.

Author Contributions: Conceptualization, M.R.-P. and J.M.S.-P.; methodology, M.R.-P.; formal analysis, M.R.-P. and J.M.S.-P.; writing—review and editing, M.R.-P. and J.M.S.-P. All authors have read and agreed to the published version of the manuscript.

Funding: The article was developed in the framework of the project "Technologies and Open Data for the Analysis and Management of Tourist Mobility to the Balearic Islands (TecMoTur)" PRD2018 / 52 financed by the General Directorate of University Policy and Research of the Department of the Education, University, and Research of the Government of the Balearic Islands (GOIB) and the Law 2/2016 of March 30 (LIET).

Acknowledgments: We wish to thank Magdalena Guardia and Cristian Esteva, geographers who participated in the collection of some of the information used in the paper at the first stage of the work. We would like to thank the General Directorate of Transport of the Government of the Balearic Islands for access to the information from the mobility survey of 2009 and the Island Council of Mallorca for the information provided of the Annual Average Daily Traffic and thanks to General Directorate of University Policy and Research of the Department of the Education, University, and Research of the Government of the Balearic Islands (GOIB) and the Law 2/2016 of March 30 (LIET) for the support of the research. Finally, our thanks to the reviewers of the article for their corrections and suggestions.

Conflicts of Interest: The authors declare no conflict of interest.

References

1. Banister, D. The sustainable mobility paradigm. *Transp. Policy* **2008**, *15*, 73–80. [CrossRef]
2. Banister, D. Cities, mobility and climate change. *J. Transp. Geogr.* **2011**, *19*, 1538–1546. [CrossRef]
3. Litman, T.; Burwell, D. Issues in sustainable transportation. *Int. J. Glob. Environ. Issues* **2006**, *6*, 331. [CrossRef]
4. Bueno, P.; Vassallo, J.; Cheung, K. Sustainability Assessment of Transport Infrastructure Projects: A Review of Existing Tools and Methods. *Transp. Rev.* **2015**, *35*, 622–649. [CrossRef]
5. Santos, G. Sustainability and Shared Mobility Models. *Sustainability* **2018**, *10*, 3194. [CrossRef]
6. Moeckel, R.; Garcia, C.L.; Chou, A.T.M.; Okrah, M.B. Trends in integrated land use/transport modeling: An evaluation of the state of the art. *J. Transp. Land Use* **2018**, *11*, 463–476. [CrossRef]
7. Bierlaire, P.; De Palma, M.; Hurtubia, A.; Waddell, R. *Integrated Transport and Land Use Modeling for Sustainable Cities*; EPFL Press: Lausanne, Switzerland, 2015.
8. Meilă, A.D. Sustainable Urban Mobility in the Sharing Economy: Digital Platforms, Collaborative Governance, and Innovative Transportation. *Contemp. Read. Law Soc. Justice* **2018**, *10*, 130–136.
9. Ko, J.; Lee, S.; Byun, M. Exploring factors associated with commute mode choice: An application of city-level general social survey data. *Transp. Policy* **2019**, *75*, 36–46. [CrossRef]
10. Buehler, R. Determinants of transport mode choice: A comparison of Germany and the USA. *J. Transp. Geogr.* **2011**, *19*, 644–657. [CrossRef]
11. Van Acker, V.; Van Wee, B.; Witlox, F. When transport geography meets social psychology: Toward a conceptual model of travel behaviour. *Transp. Rev.* **2010**, *30*, 219–240. [CrossRef]
12. Hunecke, M.; Haustein, S.; Grischkat, S.; Böhler, S. Psychological, sociodemographic, and infrastructural factors as determinants of ecological impact caused by mobility behavior11The results are based on research conducted by the junior research group MOBILANZ, which was supported by the German Federal Min. *J. Environ. Psychol.* **2007**, *27*, 277–292. [CrossRef]
13. Barr, S.; Prillwitz, J. Green travellers? Exploring the spatial context of sustainable mobility styles. *Appl. Geogr.* **2012**, *32*, 798–809. [CrossRef]
14. Johansson, L.; Schipper, O. Measuring the long-run fuel demand of cars: Separate estimations of vehicle stock, mean fuel intensity, and mean annual driving distance. *J. Transp. Econ. Policy* **1997**, *31*, 277–292.
15. Keyvanfar, A.; Shafaghat, A.; Muhammad, N.Z.; Ferwati, M.S. Driving behaviour and sustainable mobility—Policies and approaches revisited. *Sustainability* **2018**, *10*, 1152. [CrossRef]
16. Mraihi, R.; Harizi, R.; Mraihi, T.; Bouzidi, M.T. Urban air pollution and urban daily mobility in large Tunisia's cities. *Renew. Sustain. Energy Rev.* **2015**, *43*, 315–320. [CrossRef]
17. Prideaux, B. The role of the transport system in destination development. *Tour. Manag.* **2000**, *21*, 53–63. [CrossRef]
18. Nutsugbodo, R.Y.; Amenumey, E.K.; Mensah, C.A. Public transport mode preferences of international tourists in Ghana: Implications for transport planning. *Travel Behav. Soc.* **2018**, *11*, 1–8. [CrossRef]
19. Gronau, W.; Kagermeier, A. Key factors for successful leisure and tourism public transport provision. *J. Transp. Geogr.* **2007**, *15*, 127–135. [CrossRef]
20. Le-Klähn, D.-T.; Gerike, R.; Hall, C.M. Visitor users vs. non-users of public transport: The case of Munich, Germany. *J. Destin. Mark. Manag.* **2014**, *3*, 152–161. [CrossRef]
21. Dickinson, J.E.; Robbins, D. Representations of tourism transport problems in a rural destination. *Tour. Manag.* **2008**, *29*, 1110–1121. [CrossRef]
22. European Commission Libro Verde. *Hacia una Nueva Cultura de la Movilidad Urbana*; European Commission Libro Verde: Bruselas, Belgium, 2007; p. 25.
23. De Witte, A.; Hollevoet, J.; Dobruszkes, F.; Hubert, M.; Macharis, C. Linking modal choice to motility: A comprehensive review. *Transp. Res. Part A Policy Pract.* **2013**, *49*, 329–341. [CrossRef]

24. Assi, K.J.; Nahiduzzaman, K.M.; Ratrout, N.T.; Aldosary, A.S. Mode choice behavior of high school goers: Evaluating logistic regression and MLP neural networks. *Case Stud. Transp. Policy* **2018**, *6*, 225–230. [CrossRef]
25. Eluru, N.; Chakour, V.; Elgeneidy, A.M. Travel mode choice and transit route choice behavior in Montreal: Insights from McGill University members commute patterns. *Public Transp.* **2012**, *4*, 129–149. [CrossRef]
26. Alfonzo, M.A. To walk or not to walk? The hierarchy of walking needs. *Environ. Behav.* **2005**, *37*, 808–836. [CrossRef]
27. Commins, N.; Nolan, A. The determinants of mode of transport to work in the Greater Dublin Area. *Transp. Policy* **2011**, *18*, 259–268. [CrossRef]
28. Dill, J.; Wardell, E. Factors Affecting Worksite Mode Choice. *Transp. Res. Rec. J. Transp. Res. Board* **2007**, *1994*, 51–57. [CrossRef]
29. Uriel, E. *Regresión lineal múltiple: Estimación y propiedades*; Universidad de Valencia: Valencia, Spain, 2013.
30. Fajardo, H.C.L.C.L.; Gómez, S.A.M.A.M. Análisis de la elección modal de transporte público y privado en la ciudad de Popayán. *Territios* **2015**, *17*, 157–190. [CrossRef]
31. Bergström, A.; Magnusson, R. Potential of transferring car trips to bicycle during winter. *Transp. Res. Part A Policy Pract.* **2003**, *37*, 649–666. [CrossRef]
32. Sabir, M. Weather and Travel Behaviour in Netherlands. Ph.D. Thises, Vrije Universiteit, Amsterdam, The Netherlands, October 2011; pp. 1–152.
33. Sabir, M.; Koetse, M.J.; Rietveld, P. The impact of weather conditions on mode choice decisions: Empirical evidence for the Netherlands. *Proc. BIVEC-GIBET Transp. Res. Day* **2007**, 512–527.
34. Liu, C.; Susilo, Y.O.; Karlström, A. The influence of weather characteristics variability on individual's travel mode choice in different seasons and regions in Sweden. *Transp. Policy* **2015**, *41*, 147–158. [CrossRef]
35. Böcker, L.; Prillwitz, J.; Dijst, M. Climate change impacts on mode choices and travelled distances: A comparison of present with 2050 weather conditions for the Randstad Holland. *J. Transp. Geogr.* **2013**, *28*, 176–185. [CrossRef]
36. Schafer, A.; Victor, D.G. The future mobility of the world population. *Transp. Res. Part A Policy Pract.* **2000**, *34*, 171–205. [CrossRef]
37. Scheiner, J.; Huber, O.; Lohmüller, S. Children's mode choice for trips to primary school: A case study in German suburbia. *Travel Behav. Soc.* **2019**, *15*, 15–27. [CrossRef]
38. Hong, J.; Shen, Q.; Zhang, L. How do built-environment factors affect travel behavior? A spatial analysis at different geographic scales. *Transportation* **2014**, *41*, 419–440. [CrossRef]
39. Ewing, R.; Cervero, R. 'Does Compact Development Make People Drive Less?' The Answer Is Yes. *J. Am. Plan. Assoc.* **2017**, *83*, 19–25. [CrossRef]
40. Braza, M.; Shoemaker, W.; Seeley, A. Neighborhood design and rates of walking and biking to elementary school in 34 California Communities. *Am. J. Health Promot.* **2004**, *19*, 128–136. [CrossRef]
41. Curtis, C.; Babb, C.; Olaru, D. Built environment and children's travel to school. *Transp. Policy* **2015**, *42*, 21–33. [CrossRef]
42. Berrigan, D.; Troiano, R.P. The association between urban form and physical activity in US adults. *Am. J. Prev. Med.* **2002**, *23*, 74–79. [CrossRef]
43. Klinger, T.; Lanzendorf, M. Moving between mobility cultures: What affects the travel behavior of new residents? *Transportation* **2015**, *43*, 243–271. [CrossRef]
44. Zhou, X.; Yu, Z.; Yuan, L.; Wang, L.; Wu, C. Measuring Accessibility of Healthcare Facilities for Populations with Multiple Transportation Modes Considering Residential Transportation Mode Choice. *ISPRS Inter. J. Geo-Inf.* **2020**, *9*, 394. [CrossRef]
45. Tyrinopoulos, Y.; Antoniou, C. Factors affecting modal choice in urban mobility. *Eur. Transp. Res. Rev.* **2012**, *5*, 27–39. [CrossRef]
46. Müller, S.; Tscharaktschiew, S.; Haase, K. Travel-to-school mode choice modelling and patterns of school choice in urban areas. *J. Transp. Geogr.* **2008**, *16*, 342–357. [CrossRef]
47. Zhang, R.; Yao, E.; Liu, Z. School travel mode choice in Beijing, China. *J. Transp. Geogr.* **2017**, *62*, 98–110. [CrossRef]
48. Searcy, S.E.; Findley, D.J.; Huegy, J.B.; Ingram, M.; Mei, B.; Bhadury, J.; Wang, C. Effect of residential proximity on university student trip frequency by mode. *Travel Behav. Soc.* **2018**, *12*, 115–121. [CrossRef]
49. Scheiner, J.; Holz-Rau, C. Gendered travel mode choice: A focus on car deficient households. *J. Transp. Geography* **2012**, *24*, 250–261. [CrossRef]

50. Kemperman, A.A.; Timmermans, H.H. Influences of built environment on walking and cycling by latent segments of aging population. *Transp. Res. Rec. J. Transp. Res. Board* **2009**, *2134*, 1–9. [CrossRef]
51. Van Middelkoop, M.; Borgers, A.; Timmermans, H. Inducing heuristic principles of tourist choice of travel mode: A rule-based approach. *J. Travel Res.* **2003**, *42*, 75–83. [CrossRef]
52. Kuo, A.; Kao, M.-S.; Chang, Y.-C.; Chiu, C.-F. The influence of international experience on entry mode choice: Difference between family and non-family firms. *Eur. Manag. J.* **2012**, *30*, 248–263. [CrossRef]
53. Abdulsalam, M.A.; Miskeen, B.; Alhodairi, A.M.; Atiq, R.; Bin, A.; Rahmat, K. Modeling a Multinomial Logit Model of Intercity Travel Mode Choice Behavior for All Trips in Libya. *Civ. Environ. Struct. Constr. Archit. Eng.* **2013**, *7*, 636–645.
54. Al-Ahmadi, H.M. Development of intercity work mode choice model for Saudi Arabia. *Earthq. Resist. Eng. Struct. VI* **2007**, *96*, 677–685.
55. De Vos, J.; Derudder, B.; Van Acker, V.; Witlox, F. Reducing car use: Changing attitudes or relocating? The influence of residential dissonance on travel behavior. *J. Transp. Geogr.* **2012**, *22*, 1–9. [CrossRef]
56. Lumsdon, L.M. Factors affecting the design of tourism bus services. *Ann. Tour. Res.* **2006**, *33*, 748–766. [CrossRef]
57. Guiver, J.; Lumsdon, L.; Weston, R.; Ferguson, M. Do buses help meet tourism objectives? The contribution and potential of scheduled buses in rural destination areas. *Transp. Policy* **2007**, *14*, 275–282. [CrossRef]
58. Kaufmann, V.; Bergman, M.M.; Joye, D. Motility: Mobility as Capital. *Int. J. Urban Reg. Res.* **2004**, *28*, 745–756. [CrossRef]
59. Albalate, D.; Bel, G. Tourism and urban public transport: Holding demand pressure under supply constraints. *Tour. Manag.* **2010**, *31*, 425–433. [CrossRef]
60. Attard, M. Reforming the urban public transport bus system in Malta: Approach and acceptance. *Transp. Res. Part A Policy Pract.* **2012**, *46*, 981–992. [CrossRef]
61. Enoch, M.P.; Warren, J.P. Automobile use within selected island states. *Transp. Res. Part A Policy Pract.* **2008**, *42*, 1208–1219. [CrossRef]
62. Gil, A.; Calado, H.; Bentz, J. Public participation in municipal transport planning processes—The case of the sustainable mobility plan of Ponta Delgada, Azores, Portugal. *J. Transp. Geogr.* **2011**, *19*, 1309–1319. [CrossRef]
63. Martín-Cejas, R.R.; Sánchez, P.P.R. Ecological footprint analysis of road transport related to tourism activity: The case for Lanzarote Island. *Tour. Manag.* **2010**, *31*, 98–103. [CrossRef]
64. Bojković, N.; Petrovic, M.; Parezanović, T. Towards indicators outlining prospects to reduce car use with an application to European cities. *Ecol. Indic.* **2018**, *84*, 172–182. [CrossRef]
65. Gerlach, J.; Richter, N.; Becker, U.J. Mobility Indicators Put to Test—German Strategy for Sustainable Development Needs to be Revised. *Transp. Res. Procedia* **2016**, *14*, 973–982. [CrossRef]
66. Gillis, D.; Semanjski, I.; Lauwers, D. How to monitor sustainable mobility in cities? Literature review in the frame of creating a set of sustainable mobility indicators. *Sustainability* **2015**, *8*, 29. [CrossRef]
67. Haghshenas, H.; Vaziri, M. Urban sustainable transportation indicators for global comparison. *Ecol. Indic.* **2012**, *15*, 115–121. [CrossRef]
68. Macedo, J.; Rodrigues, F.; Tavares, F. Urban sustainability mobility assessment: Indicators proposal. *Energy Procedia* **2017**, *134*, 731–740. [CrossRef]
69. Mozos-Blanco, M. Ángel; Pozo-Menéndez, E.; Arce, R.; Baucells-Aletà, N. The way to sustainable mobility. A comparative analysis of sustainable mobility plans in Spain. *Transp. Policy* **2018**, *72*, 45–54. [CrossRef]
70. Moeinaddini, M.; Asadi-Shekari, Z.; Shah, M.Z. An urban mobility index for evaluating and reducing private motorized trips. *Measurement* **2015**, *63*, 30–40. [CrossRef]
71. Muñoz, M.D.; Cantergiani, C.; Salado, M.; Rojas, C.; Gutiérrez, S. Propuesta de un Sistema de para Indicadores de Sostenibilidad para la movilidad y el transporte urbanos. Aplicación mediante SIG a la ciudad de Alcalá de Henares. *Cuad. Geogr.* **2007**, *81*, 31–50.
72. IBESTAT. Demografia/Turismo. 2020. Available online: https://ibestat.caib.es/ibestat/ (accessed on 20 September 2020).
73. Sebastián, J.B. Caracterització, Tipificació i Pautes de Localització de les Àrees Rururbanes y L'illa de Mallorca. Ph.D. Thesis, Universitat de les Illes Balears, Illes Balears, Spain, 1996.
74. Mateu, J.; Seguí, J.M.; Ruiz, M. Mallorca and its metropolitan dynamics: Daily proximity and mobility in an island-city. *Eure* **2017**, *43*, 129.

75. Consell Insular de Mallorca: Tráfico. 2020. Available online: https://web.conselldemallorca.cat/documents/774813/882786/20200309-AFOROS_2019.pdf/fc89cd80-22bd-9ae5-47de-eccc49075c48?t=1583748426651 (accessed on 20 September 2020).
76. de Mallorca, C. Carreteres. 2020. Available online: http://www.conselldemallorca.info/sit/incidencies/ (accessed on 20 September 2020).
77. Direcció General de Mobilitat i Transport Terrestre. Conselleria de Mobilitat i Habitatge. 2020. Available online: http://www.caib.es/govern/organigrama/area.do?coduo=200&lang=ca (accessed on 20 September 2020).
78. de Mallorca, C. Departament de Mobilitat i Infraestructures. 2020. Available online: https://web.conselldemallorca.cat/es/web/web1/movilidad-infraestructuras (accessed on 20 September 2020).
79. Govern de les Illes Balears (GOIB). Consorci de Transports de Mallorca. 2020. Available online: https://www.tib.org/es/web/ctm/ (accessed on 20 September 2020).
80. Govern de les Illes Balears (GOIB). Serveis Ferroviaris de Mallorca. 2020. Available online: http://www.trensfm.com/ (accessed on 20 September 2020).
81. de Palma, A. MobiPalma. 2020. Available online: http://www.mobipalma.mobi/es/ (accessed on 20 September 2020).
82. de Palma, A. Empresa Municipal de Transports de Palma. 2020. Available online: http://www.emtpalma.cat/ca/inici (accessed on 20 September 2020).
83. Vega Pindado, P. *Los Planes de Movilidad Urbana Sostenible en España (PMUS): Dos Casos Paradigmáticos: San Sebastián-Donostia y Getafe*; Universidad Complutense de Madrid: Madrid, Spain, 2019.
84. Union, E. CIVITAS Dyn@mo Project. 2020. Available online: https://civitas.eu/content/palma (accessed on 20 September 2020).
85. de Palma, A. Pla de Mobilitat Urbana Sostenible. CIVITAS UE. 2014. Available online: http://www.mobipalma.mobi/es/pla-mobilitat-urbana-sostenible-pmus/ (accessed on 20 September 2020).
86. Govern de les Illes Balears (GOIB). Ley 4/2014, de 20 de junio, de transportes terrestres y movilidad sostenible de las Illes Balears. Boletín Oficial del Estado, 172 de 16 de julio de 2014. *Actual. Jurídica Ambient.* **2014**, *38*, 51–53.
87. Govern de les Illes Balears (GOIB). *Decret 35/2019 de 10 de maig, D'aprovació del Pla Director Sectorial de Mobilitat de les Illes Balears*; Butlletí Of. les Illes Balear: Palma, Spain, 2019; pp. 19412–19418.
88. Govern de les Illes Balears (GOIB). Portal de Dades Obertes. 2020. Available online: http://www.caib.es/sites/opendatacaib/ca/inici_home/?campa=yes (accessed on 20 September 2020).
89. Airbnb. Inside Airbnb. 2020. Available online: http://insideairbnb.com/ (accessed on 20 September 2020).
90. Dirección General de Tráfico. Ministerio del Interior. Gobierno de España. Parque de Vehículos. Available online: http://www.dgt.es/es/explora/en-cifras/parque-de-vehiculos.shtml (accessed on 20 August 2020).
91. Govern de les Illes Balears (GOIB). *Enquestes de Mobilitat 2009/2010*; Internal Report; Govern de les Illes Balears (GOIB): Illes Balears, Spain.
92. Peng, C.-Y.J.; Lee, K.L.; Ingersoll, G.M. An introduction to logistic regression analysis and reporting. *J. Educ. Res.* **2002**, *96*, 3–14. [CrossRef]
93. Camarero, L.; Oliva, J. Exploring the social face of urban mobility: Daily mobility as part of the social structure in Spain. *Int. J. Urban Reg. Res.* **2008**, *32*, 344–362. [CrossRef]
94. Miralles-Guasch, C.; Melo, M.M.; Marquet, O. A gender analysis of everyday mobility in urban and rural territories: From challenges to sustainability. *Gender Place Cult.* **2016**, *23*, 398–417. [CrossRef]
95. Schwanen, T.; Mokhtarian, P.L. What if you live in the wrong neighborhood? The impact of residential neighborhood type dissonance on distance traveled. *Transp. Res. Part D Transp. Environ.* **2005**, *10*, 127–151. [CrossRef]
96. Castanheira, G.; Bragança, L. The Evolution of the Sustainability Assessment Tool SBToolPT: From Buildings to the Built Environment. *Sci. World J.* **2014**, *2014*, 491791. [CrossRef]
97. Pan, S.-Y.; Gao, M.; Kim, H.; Shah, K.J.; Pei, S.-L.; Chiang, P.-C. Advances and challenges in sustainable tourism toward a green economy. *Sci. Total. Environ.* **2018**, *635*, 452–469. [CrossRef]
98. Scheepers, C.; Wendel-Vos, G.; Broeder, J.D.; Van Kempen, E.; Van Wesemael, P.; Schuit, A. Shifting from car to active transport: A systematic review of the effectiveness of interventions. *Transp. Res. Part A Policy Pract.* **2014**, *70*, 264–280. [CrossRef]

99. Schneider, R.J. Theory of routine mode choice decisions: An operational framework to increase sustainable transportation. *Transp. Policy* **2013**, *25*, 128–137. [CrossRef]
100. Firnkorn, M. Selling Mobility instead of Cars: New Business Strategies of Automakers and the Impact on Private Vehicle Holding. *Strat. Dir.* **2012**, *28*, 264–280. [CrossRef]
101. Ma, Y.; Rong, K.; Mangalagiu, D.; Thornton, T.F.; Zhu, D. Co-evolution between urban sustainability and business ecosystem innovation: Evidence from the sharing mobility sector in Shanghai. *J. Clean. Prod.* **2018**, *188*, 942–953. [CrossRef]
102. Gabrielli, S.; Forbes, P.; Jylhä, A.; Wells, S.; Sirén, M.; Hemminki, S.; Nurmi, P.; Maimone, R.; Masthoff, J.; Jacucci, G. Design challenges in motivating change for sustainable urban mobility. *Comput. Hum. Behav.* **2014**, *41*, 416–423. [CrossRef]

Publisher's Note: MDPI stays neutral with regard to jurisdictional claims in published maps and institutional affiliations.

Article

Does High-Speed Railway Influence Convergence of Urban-Rural Income Gap in China?

Weidong Li [1], Xuefang Wang [1] and Olli-Pekka Hilmola [2,*]

[1] School of Economics and Management, Beijing Jiaotong University, Beijing 100044, China; wdli@bjtu.edu.cn (W.L.); 18120729@bjtu.edu.cn (X.W.)

[2] Industrial Engineering and Management, LUT Kouvola, LUT University, Prikaatintie 9, FIN-45100 Kouvola, Finland

* Correspondence: olli-pekka.hilmola@lut.fi; Tel.: +358-40-761-4307

Received: 27 March 2020; Accepted: 18 May 2020; Published: 21 May 2020

Abstract: Transportation is an important factor affecting the balance of regional economic pattern. The construction of high-speed railway enhances the mobility of population, capital, technology and information resources between urban and rural areas. Will it further affect the income gap between urban and rural areas? Based on the nonlinear time-varying factor model, this paper analyzes the convergence of urban-rural income gap with the angle of high-speed railway. After rejecting the assumption of overall convergence in the traditional four economic regions, three convergence clubs of urban-rural income gap were found. For these ordered logit regression model is used to explore the initial factors that may affect the formation of "convergence club". Empirical results show that the construction of High-speed railway has effectively narrows the urban-rural income gap in China, but it is not the cause of the formation of the three convergence clubs. The convergence effect of High-speed railway on the urban-rural income gap in China is still relatively weak.

Keywords: high-speed railway; income gap; club convergence; nonlinear time-varying factor model

1. Introduction

Since the reform and opening up, China's economy has maintained rapid growth. The income level of residents in urban and rural areas also has increased significantly. In recent years, China's urban-rural income gap has shown a long-term gradual decline, but in absolute terms, it is still substantial. According to the National Bureau of Statistics, the disposable income ratio of urban and rural residents in China during 2017 was 2.71:1 [1]. Although it has declined compared with previous years, it is still one of the important factors that constraints China's economic development. In 2014, China's economic development has entered a new normal era. Under the background of the new economic development in the future, China's economic development will pay more attention to structural adjustment, regional development and income distribution. Under the grim situation of China's urban-rural income gap, it is debated how to find a solution. Approaching this gap is an important issue of concern to academics and government.

Due to the complexity of economic development and the variability of social conditions, there are many factors affecting the income gap between urban and rural areas. In the past decade, China's high-speed railway has developed rapidly. Chen et al. (2018) used high-speed rail parking frequency data in 274 cities in China from 2007 to 2014 to examine the impact of high-speed rail development on the income gap between urban and rural residents. The research results show that the development of high-speed rail is generally conducive to the narrowing of the income gap between urban and rural residents in China. Moreover, population flow and capital flow are important mechanisms for the development of high-speed rail to narrow the income gap between urban and rural residents in China [2].

Does high-speed railway affect the income gap between urban and rural areas to form a convergence club? The research goal of this paper is to evaluate the impact of the opening of high-speed railway on urban-rural income gap based on club convergence and deeply analyze the mechanism of high-speed railway's economic convergence effect on urban-rural income gap. The high-speed railway is both capital-intensive and labor-intensive large-scale transportation infrastructure. Its investment in construction and operation can have a profound impact on social and economic activities. The purpose of this paper is to find the convergence characteristics of urban-rural income gap under high-speed rail network situation. The contributions of the paper are as follows. At first, the paper firstly discusses the impact of the opening of high-speed railway on club convergence of urban-rural income gap in China with the nonlinear time-varying factor model. Secondly, three convergence clubs of urban-rural income gap in China are found. Thirdly, it is found that the construction of High-speed railway has effectively narrows the urban-rural income gap in China, but it is not the cause of the formation of the three convergence clubs. The convergence effect of High-speed railway on the urban-rural income gap in China is still relatively weak. It will enrich the impact of traffic on regional balance development. At the same time, it provides reference for alleviating the income gap between urban and rural areas in China. It also provides a policy basis for the implementation of regional balanced development strategy.

The rest of this research is organized as follows: In Section 2 we provide literature review follows, and Section 3 describes heterogeneity of urban-rural income gap within China's cities. The convergence club identification is being introduced in Section 4. The causes of convergence club are discussed in Section 5. Finally, conclusions are drawn in Section 6.

2. Literature Review

High-speed rail is a type of rail transport that operates significantly faster than traditional rail traffic. New lines in excess of 250 km/hr and existing lines in excess of 200 km/hr are considered to be high-speed rail [3]. High-speed railway has brought great impact on many aspects of social & economic life. Many scholars have conducted numerous research works on the influence of HSR. Zhang, et al., and Chai, et al. focus on the impact of HSR on aviation [4–6]. Li, et al., and Vickerman, et al. Discuss on the impact of HSR on economic growth [7–16]. Besides, some scholars pay attention to the social impact of HSR [17,18]. Chen et al., and Jiang, et al. take explorative analysis on the impact of HSR on regional income disparities and found that HSR can accelerate regional economic convergence and reduce the regional income disparities [19,20].

With regard to the urban-rural income gap, Chinese scholars have carried out many constructive studies on this theme. The researches mainly focus on the status quo of urban-rural income gap, analysis of its causes, measurement and convergence issues In the related research on the convergence of urban-rural income gap, Wang et al. (2016) used the panel data of 71 counties in Wuling Mountain District from 2000 to 2012 to empirically examine the urbanization process, financial development and financial expenditure. Other factors promote the convergence of urban-rural income gap [21]. Deng (2016) used Hansen's panel threshold model to construct a nonlinear panel threshold model using China's 1997-2014 provincial panel data to study the impact of urbanization and industrialization on urban-rural income gap [22]. Song and Wang (2018) used data in 262 prefecture-level cities from 2000 to 2016 to explore the dynamic convergence of urban-rural income gap from the perspective of urban scale and analyze the influences of urban-rural income gap between population migration and household registration [23]. Wu (2019) used panel data from 1999 to 2017 in 30 provinces in China to construct a convergence model between government investment and urban-rural income gap. Research shows that the convergence of government investment structure promotes the convergence of urban and rural income gaps. Therefore, preventing government investment from being too fast can effectively reduce the urban-rural income gap [24]. Xu and Zhang (2019) used panel data from 30 provinces other than Tibet in China from 2000 to 2016 to analyze the dynamic evolution of China's urbanization and urban-rural income gap. The spatial and temporal effects of new urbanization on China's urban-rural income gap are also discussed. The study clarifies whether the transformation of urbanization can promote the convergence of the urban-rural income gap in China [25]. Besides China, some scholars conduct research on club convergence about income in other regions. Bartkowska, M., & Riedl, A. (2012) used two-step method and ordered logit model to analyze convergence clubs in per capita incomes of European regions. It is found European regions form six separate groups converging to their own steady state paths. The level of initial conditions such as human capital and per capita income plays a crucial role in determining the formation of convergence clubs among European regions [26]. Canova, F. (2004) developed a predictive density approach to identify convergence clubs in Organization for Economic Co-operation and Development (OECD) countries [27]. Galor, O. (2007) used Unified Growth Theory to explain the emergence of convergence clubs [28].

In the study of the convergence of urban-rural income gap, most scholars focus on the impact of urbanization, population migration, regional economic growth, human capital and other factors on the convergence of urban-rural income gap [29–38]. However, the impact of high-speed rail on the convergence of urban-rural income gap has not yet been discussed. The research will shed some light on the understanding of how HSR influence the convergence of urban-rural income gap. The research results can provide reference for investment construction of high-speed rail and narrowing the income gap between urban and rural areas in China.

3. Heterogeneity of Urban-Rural Income Gap in China's Cities

The factors affecting the income gap between urban and rural areas arise from many aspects such as nature, economy, society and culture [39–41]. Consumption expenditure ratio of urban and rural residents, urbanization level, population migration, regional economic growth and human capital are all important factors affecting the income gap between urban and rural areas. Based on the feasibility of data collection, we first calculate the mean and coefficient of variation of the urban-rural income gap of 274 prefecture-level cities in China (see Figure 1). The ratio of the per capita disposable income of urban residents to the per capita disposable income of rural residents is used as a measure of the income gap between urban and rural areas. The selected research period is from 2005 to 2017. The data are collected from the China Regional Statistical Yearbook and China's City Statistical Yearbook.

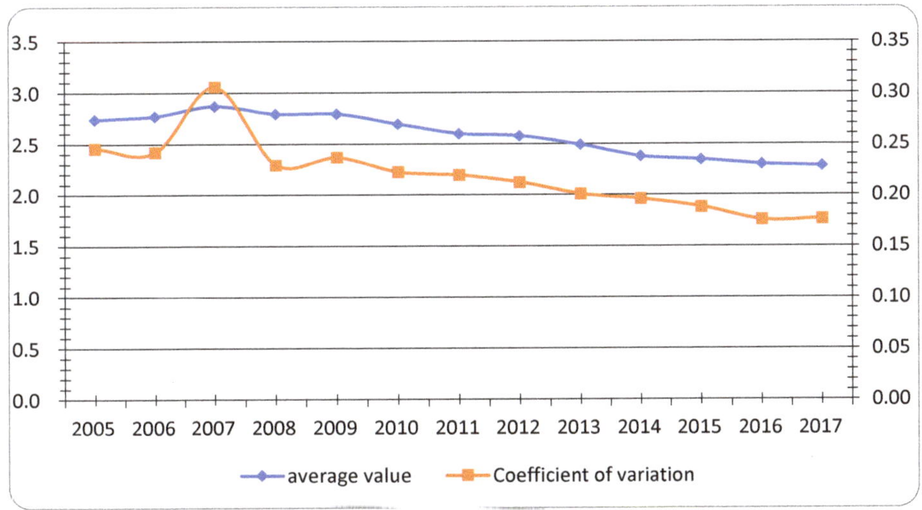

Figure 1. Mean and coefficient of variation of urban-rural income gap in 274 administrative regions in China.

From Figure 1 it can be identified the average income gap between urban and rural areas that the urban-rural income gap in China continued to grow before 2007. However, it has showed clear downward trend thereafter. In terms of the difference in urban-rural income gap, the coefficient of variation is an indicator reflecting σ convergence. The larger the value, the higher the dispersion of urban-rural income gap among cities. The trend of coefficient of variation of urban-rural income gap in prefecture-level cities in China has experienced a process of increasing first and then decreasing. The coefficient of variation reached its highest value in 2007. It shows that the difference in urban-rural income gap between cities shows a process of σ convergence.

4. Convergence Club Identification

The basic concept of convergence came first from Solow's neoclassical growth model (Solow, 1956), starting with Baumol (1986) [42,43]. It has been used in different economic processes involving groups of regions or units. Generally there are three types of convergence: β-convergence, σ convergence and club convergence. β-convergence implies that the units analyzed converge to one another in the long run. σ convergence implies convergence that is conditional on the units having similar characteristics. Club convergence means that a set of economies with similar conditions and structural characteristics will tend to converge to the same steady state [44]. Phillips and Sul (2007) proposed the concept of relative convergence, which considers the transition path of each country together with its growth performance to find convergence [44].This concept can be easily extended to the field of convergence in income gap.

4.1. Club Convergence Test Method Based on Nonlinear Time-Varying Factor Model

This paper first explains the logt test method based on nonlinear time-varying factor model [45]. Then the paper gives the steps to identify the convergence club according to the logt test method. The identification steps of the convergence club was obtained by simplifying on the basis of Phillips and Sul (2007) [44].

4.1.1. Logt Test

At first a simple factor model is built as follows:

$$X_{it} = \delta_i \mu_t + \epsilon_{it} \tag{1}$$

The model is designed to capture the evolution of individual X_{it} with respect to μ_t under the influence of two heterogeneity factor system terms δ_i and error term εit. Among them, δ_i measures the heterogeneity distance between the public factorμ_t and the system part X_{it}, μ_t represents the common behavior of X_{it} aggregation, which reflects the influence of individual behavior plus common factors. It should be pointed out that although the system termδ_i expresses the heterogeneity of individuals, this heterogeneity does not change with time. Therefore, when considering the heterogeneity of individuals with time-varying characteristics, the following improvements are needed:

$$X_{it} = \delta_{it} \mu_t \tag{2}$$

Among them, δ_{it} expresses the change of individual heterogeneity with time, including the random component ε_{it}. Therefore, equation (2) contains the time-varying characteristics of individual heterogeneity and the common trend term of the system, which is the nonlinear time-varying factor model.

To model the time-varying parameter δ_{it}, we define a relative transfer coefficient:

$$h_{it} = \frac{x_{it}}{N^{-1}\sum_{i=1}^{N} x_{it}} = \frac{\delta_{it}}{N^{-1}\sum_{i=1}^{N} \delta_{it}} \tag{3}$$

$$H_t^2 = N^{-1}\sum_{i=1}^{N}(h_{it}-1)^2 \tag{4}$$

Specifically, the h_{it} relative transfer coefficient, H_t^2 is its variance. Since the common growth path, that is, the common factor portion is eliminated, when the convergence occurs, $h_{it}\to 1$, $H_t^2\to 0$.

Further, in order to verify the null hypothesis of convergence, it is also necessary to construct a semi-parametric model:

$$\delta_{it} = \delta_i + \delta_i \zeta_{it}/L(t)t^{-\alpha} \tag{5}$$

where δ_i is fixed and does not change with time t; σ_i is a heterogeneous scale parameter; ξ_{it} is subject to iid(0,1); L(t) is a slowly changing function and when $t\to\infty$, $L(t)\to\infty$; α is the decay rate. The semi-parametric model shows that as long as $\alpha\geq 0$ is satisfied, $\delta_{it}\to\delta_i$, that is, the convergence is established.

Based on the above derivation process, the null hypothesis of the test convergence can be written as follows:

$$H_0{:}\delta_i = \delta \text{ and } \alpha \geq 0 \tag{6}$$

$$H_A{:}\delta_i \neq \delta \text{ or } \alpha < 0 \tag{7}$$

In order to verify the original hypothesis of convergence, regression is performed using the following formula:

$$\log\left(\frac{H_1}{H_t}\right) - 2\log L(t) = a + b\log t + \mu_t \tag{8}$$

where $L(t) = \log(t+1)$; $t = [\gamma T], [\gamma T]+1,\ldots,T$; γ is the parameter determining the starting time t, according to Monte Carlo simulation experiment of Phillips and Sul. As a result, $\gamma = 0.3$, $a = 0.5b$, and then heteroscedasticity and autocorrelation (HAC) one-sided T test. If $t_b \geq -1.65$, the null hypothesis cannot be rejected. Otherwise the null hypothesis is rejected. Since the above method utilizes a cross-sectional variance ratio to perform a linear regression of the logt time series, this convergent regression test is called a logt test.

4.1.2. Convergence Club Identification Step Based on Logt Test

When the null hypothesis of the convergence of all cities is rejected, it can be further identified according to the principle of logt test whether there is a convergence club. The identification method is an endogenous convergence club identification method. It is based entirely on urban-rural income disparity data for each city and is not based on any external criteria such as geographic zoning. The specific identification steps are as follows:

Step 1: Sort the individual segments. The basis is:

$$(T-[Ta])^{-1}\sum_{t=[Ta]+1}^{T} X_{it}, a = 1-f \tag{9}$$

The observations are sorted in descending order, and the time span f of the observations is selected. The value of f is chosen as 1/2.

Step 2: Identify the initial core group. Select the first two cities in the results of step 1 and do the logt test. If the original hypothesis is rejected, the first city is removed, the second and third cities are grouped into new groups, and the logt test is performed. Repeat above processes until you find two cities that cannot reject the null hypothesis. These two cities are the initial core groups and jump to step 3. If you traverse all two cities, you can't find the initial core group, indicating that there are no converging clubs and all cities are not aligned with each other.

Step 3: Identify the new core group members. According to the ranks arranged in the first step, the cities that have not been inspected together with the initial core group members are added to the initial core group one by one for logt test. In order to cluster members of similar nature as much as possible, the standard c of the selected club is set. c will be significantly higher than the threshold of the convergence club - 1.65, generally recommended c = 0. Here c is chosen as 0. According to this cycle, after traversing all cities, all members of the core group are identified, and one convergence club is obtained. Phillips recommends numerical simulations, selecting a 50% significance level for a sample of 20 or 50; a 40% significance level for a sample of 100; and a 20% significant level for a sample of 200. The significant level chosen in this paper is 20%.

Step 4: Do a logt test on all cities that do not belong to the Convergence Club in Step 3. If you cannot reject the null hypothesis, then these cities are considered as a convergence club. If you reject the null hypothesis, repeat steps 1 through 3. Circulate this until you can identify all existing convergence clubs.

Step 5: Convergence clubs formed according to the clustering process are carried out with the members entering the core group criteria. Therefore, at a significant level of 5%, there may still be convergence between different clubs, which need to be combined to merge the clubs with $t_k > -1.65$ into one group. The specific integration methods are as follows:

The first is the column value. In the club group, you get M Convergence Clubs: Club1, ..., ClubM. A logt test is performed for every two adjacent clubs. Let m = 1, 2, ..., M−1, and obtain the tm value of the (M−1) group. The second is the combination: starting from the first group, if $t_m > -1.65$ is satisfied and $t_m > t_{m+1}$, the two groups are merged and the first step is repeated; if not, the remaining group remains as a single group.

4.2. Identification of the Convergence Club

4.2.1. Judgment of Overall Convergence

Before identifying the Convergence Club, the overall urban-to-urban income gap convergence was judged by performing logt tests on 274 prefecture-level cities in China. Using the methods and steps in Section 4.1, 274 cities were fitted to obtain the following estimation results.

$$\log\left(\frac{H_1}{H_t}\right) - 2\log t = 0.4934 - 0.3163\log t \quad (5.34)\ (-3.20) \tag{10}$$

It can be seen that b = −0.3163, $t_b = -3.2$, far less than the convergence threshold of −1.65 at the significant level of 5%. Therefore, there is no overall convergence in the urban-rural income gap in China. It is necessary to explore whether there is a region where economic growth converges to the club. According to the division standard of the National Bureau of Statistics for economic regions in 2011, the convergence test of the clubs in the four major economic zones of northeast, eastern, central and western regions was obtained with the regression results shown in Table 1. It can be found that there is a convergence of the urban-rural income gap in the northeastern region and the western region, while the eastern and central regions do not have the convergence of the urban-rural income gap. This shows that although the division of the traditional economic zone reflects the facts of urban and rural areas to some extent. The spatial correlations of the development of income gap can be identified, but it cannot accurately reflect the internal relationship and similarity of regional income gap between urban and rural areas. Only some economic regions can meet the criteria for judging the convergence of urban-rural income gap. The study of Xu and Zhang analyzed the dynamic evolution of urbanization and urban-rural income gap in China, as well as the spatiotemporal effects and differences of new urbanization on urban-rural income gap in China. It was found in this study that there is an obvious regional heterogeneity in the impact of urbanization transformation on urban-rural income gap. With the in-depth transformation of urbanization, the urban income gap in the eastern region is still significantly expanding, while the expansion trend in the central region is not obvious, and the expansion trend in the western region is restrained [25].The convergence of the income gap between urban and rural areas in the western region further verifies the conclusion of Xu and Zhang that the widening trend of the income gap between urban and rural areas in the western region has been restrained.

Table 1. Convergence test of the four major economic zones.

Region	Provinces, Municipalities, Autonomous Regions	t Value	Convergence
North-east area	Heilongjiang, Jilin, Liaoning	−0.73	Yes
East area	Beijing, Tianjin, Hebei, Shanghai, Jiangsu, Zhejiang, Fujian, Shandong, Guangdong, Hainan	−3.40	No
Central Region	Shanxi, Anhui, Jiangxi, Henan, Hubei, Hunan	−6.30	No
Western Region	Inner Mongolia, Guangxi, Chongqing, Sichuan, Guizhou, Yunnan, Shaanxi, Gansu, Qinghai, Ningxia, Xinjiang	−1.02	Yes

4.2.2. Convergence Club Judgment

(1) Divide 274 cities into 4 groups. If $t > -1.65$, it means that the individual in the club has a convergence trend.

(2) Add the data of the remaining cities to the core group in turn. At this time, the t value is increased from −1.65 to 0, and there is still a possibility of convergence at the 20% significance level between the groups.

(3) Integrate the four groups according to the club integration method. The third largest convergence club for urban and rural income gaps in China was obtained (see Table 2 for detailed club convergence). On the surface of the inspection, the three clubs all showed significant convergence, and most of the cities were located in the first two clubs. According to the club rankings, the average urban-rural income gap of the club members showed a downward trend.

Table 2. China's urban and rural income gap convergence trend club group integration results.

Club	Group	Number of Cities	Coefficient b	t Value	S.E.	Mean of Urban-Rural Income Gap
Club1	1	167	0.2017	1.41	0.1427	2.8050
Club2	2, 3	85	0.3741	1.44	0.2606	2.2011
Club3	4	5	0.3972	1.68	0.3972	2.1033
Divergence	5	17	−1.3583	−14.31	0.9494	2.5420

4.2.3. Analysis of Empirical Results

Table 3 shows the distribution of convergence clubs in China's four major economic zones. In general, most cities in the western region are located in the first club. The cities in northeast region are mainly distributed in the First Club and the Second Club. The central and eastern regions are distributed in all three major clubs.

Table 3. Distribution of clubs in the four economic zones.

Club	Number of Cities	Distribution Area
Club1	167	western region (67), central region (51), eastern region (34), northeastern region (15)
Club2	85	eastern region (39), central region (27), northeastern region (11), western region (8)
Club3	5	eastern region (3), central region (2)

It could be observed the variation coefficient of the urban-rural income gap among different clubs. As shown in Figure 2 below, we can clearly see that the coefficient of variation of urban-rural income gap among different clubs fluctuates constantly. Overall, we have experienced the process of expanding and then shrinking. Since 2009, the three clubs have shown a process of continuous convergence. It shows that the income gap between urban and rural areas within the club is constantly narrowing.

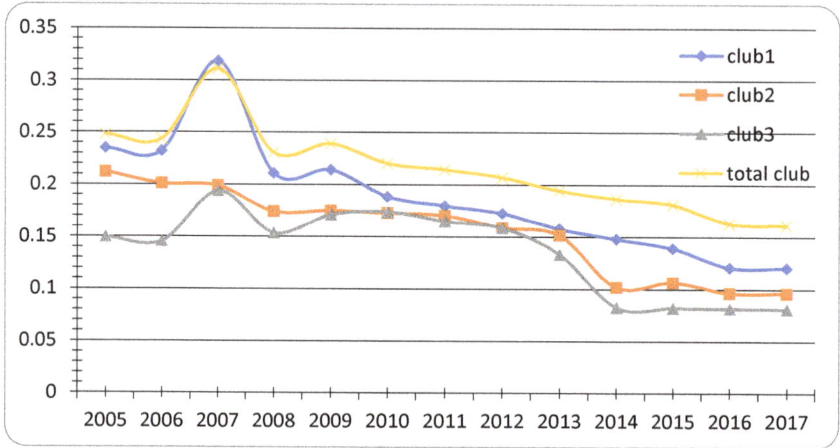

Figure 2. Coefficient of variation of urban-rural income gap among clubs.

5. Causes of Club Convergence

For studying further the factors affecting the urban and rural income gap club grouping, this paper will use the ordered logit model to conduct regression analysis on the club grouping and explore the effects of internal determinants of the urban-rural income gap convergence club formation.

5.1. Variable Selection and Model Construction

According to the needs of the empirical research and considering the availability of data, this paper selects: (1) The level of local economic development (devel), expressed by the per capita GDP of each city. The level of local economic development is an important influencing factor of social development, such as the urban–rural income gap. (2) The degree of local intervention (gbeha), measured by the ratio of local fiscal expenditure to local GDP. The degree of local intervention denotes the government management capacity, which is crucial to social development.(3)Foreign direct investment (fdi), measured by the proportion of actual use of foreign capital in GDP in the year of Chinese Yuan (CNY) conversion. The fdi reflects the openness of a city. It is the representative of the government performance. The fdi brings the introduction of foreign enterprises and causes the hiring of more employees. It determines the income levels of different residents. It is also an important factor of the urban–rural income gap. (4) High-speed rail effect (gt), measured by the number of high-speed railway stations opened. (5)The proportion of urban population to total population, measured by the level of urbanization(urb). Urbanization is an important factor that will influence the income gap between urban and rural residents.(6)The industrial structure, measured by the sum of the added value of the secondary industry and the added value of the tertiary industry as a percentage of GDP (ind). Industrial structure determines the income level of residents. It is an influencing factor for the income gap. The sample interval is from 2005 to 2016. The data mainly come from the Economy Prediction System (EPS) China database, the statistical yearbook of each city, the website of the China Railway Corporation, the news reports or announcements of the National Railway Administration, and the website of the China Railway Corporation 12306. All the data except devel and gtare relative indexes. To avoid the difference of units, the variable devel is calculated with 10,000 RMB.Here we focus on the High-speed railway effect on the urban-rural income gap. Based on the literature review and above analysis, the hypotheses are constructed as follows:

Hypothesis 1. *High-speed railway effect of urban-rural income gap is significant.*

Hypothesis 2. *High-speed railway effect of convergence of urban-rural income gap is significant.*

According to the regression results estimated by the panel fixed effect model, as shown in Table 4, we have determined the influencing factors of the urban-rural income gap. It can be seen that the regression coefficient is negative and significant at 1% level. The results shows High-speed railway effect (gt) have significant negative effects on the urban-rural income gap. It shows the hypothesis 1 is correct. High-speed railway effect of urban-rural income gap is significant. Besides, foreign direct investment (fdi), local economic development level (devel), and urbanization rate (urb) also have significant negative effects on the urban-rural income gap. In other words, the construction of High-speed railway, foreign direct investment, urbanization, and the improvement of regional development levels are conducive to narrowing the urban-rural income gap in China. The degree of local intervention (gbeha) and industrial structure changes (ind) have significant positive effects on the urban-rural income gap. It can be found that the intervention of local governments is an important factor in expanding the urban-rural income gap in China. Moreover, the industrial structure upgrade has widened the urban-rural income gap.

Table 4. Fixed effect model regression results.

Model	1	2	3	4
devel	−0.0762359*** (0.0046241)	−0.067033*** (0.0045977)	−0.0620697*** (0.0046505)	−0.0631602*** (0.0050142)
urb	−0.2297582*** (0.0484925)	−0.1480799*** (0.0489607)	−0.1391756*** (0.0487126)	−0.2171518*** (0.0499533)
gt	−0.0634502*** (0.0196414)	−0.0520909*** (0.0192339)	−0.0686494*** (0.0193361)	−0.0683473*** (0.0191887)
fdi		−4.840153*** (0.5251888)	−4.290662*** (0.5306807)	−4.351761*** (0.5274679)
gebha			0.8348368*** (0.1429672)	0.9678393*** (0.1452194)
ind				0.3343478*** (0.1278829)
_cons	2.984747*** (0.0206154)	3.009603 *** (0.0211706)	2.851538 *** (0.0342922)	2.569233*** (0.1102339)
Prob > F	0.0000	0.0000	0.0000	0.0000

*** $p < 0.01$, ** $p < 0.05$, * $p < 0.1$.

We use the indicators that significantly affect the urban-rural income gap in the above regression analysis as the independent variables of the ordered logit model. Then the ranked logit regression method is used to test the factors that affect the three major convergence clubs. The club categories (club1 ~ club3) as explained variables are ordered discrete variables. In this paper, ordered logit model is used to explain the mechanism of club classification. Through the model analysis of various factors on the probability of the formation of convergence club. In this paper, the club's category value is 1, 2, 3, the model is set as follows:

$$Y^* = x'\beta \ + \varepsilon \tag{11}$$

Among them, y^* is the latent variable corresponding to the explained variable "club category (y)". x' is the set of explained variables, including devel, gbeha, fci, gt, urb and ind. β is the parameter to be estimated. ε is the random disturbance term. Its selection rules meet the following conditions:

$$y = \begin{cases} 1 & y^* \geq a_0 \\ 2 & a_0 \leq y^* \leq a_1 \\ 3 & y^* \geq a_1 \end{cases} \tag{12}$$

Among them, $\alpha_0 < \alpha_1 < \alpha_2$ is the parameter to be estimated, also known as threshold or cut. y is the rating score, which is 1, 2,3. y represents the category of the club. For example, y = 1 represents club1. If $\beta > 0$, with the increase of x ′, the probability of y * in the higher category will increase, which means that the category of the club will increase. Based on this, the probability effect of each independent variable on the observation value y of the dependent variable can be calculated.

5.2. Empirical Results and Discussion

Table 5 shows the orderd logit model parameter estimation results with y* as the dependent variable. It can be seen that fdi, indand gebha have a significant effect on the latent variable y*. The estimated β of fdi is 11.6, which is significantly positively correlated at the 1% level. The estimated value of β for gebha is −3.957, which is significantly negatively correlated at the 1% level. The estimated value of β for ind is −1.275, which is significantly negatively correlated at the 5% level.The independent variables gt, devel, and urb have no significant effect on the latent variable y*. In other words, although the construction of High-speed railway, the improvement of regional development level and urbanization can effectively reduce the urban-rural income gap between cities, it is not an important factor in forming the three convergence clubs. It shows hypothesis 2 is not correct. High-speed railway effect of convergence of urban-rural income gap is not significant. This indicates that the current impact of China's High-speed railway construction on the urban-rural income gap is relatively weak compared to the impact of local government intervention and foreign direct investment.

Table 5. Estimation results of the ordered logit model.

Club	Coef.	St.Err.	t-Value	p-Value	[95% Conf	Interval]	Sig
gt	0.035	0.073	0.48	0.634	−0.109	0.179	
fdi	11.738	1.963	5.98	0.000	7.892	15.585	***
gdp	0.021	0.019	1.11	0.268	−0.016	0.058	
gebha	−4.500	0.685	−6.57	0.000	−5.842	−3.158	***
czl	0.173	0.195	0.89	0.374	−0.209	0.556	
ind	−1.275	0.502	−2.54	0.011	−2.259	−0.292	**
cut1	−0.793	0.440	.b	.b	−1.655	0.068	
cut2	2.539	0.456	.b	.b	1.646	3.433	

*** $p < 0.01$, ** $p < 0.05$, * $p < 0.1$.

It should be pointed out that since the ordered logit regression uses the maximum likelihood method to estimate the regression coefficient, the size of the above regression coefficient does not have the meaning of the marginal effect, and the direction of the influence of the variable on the interpreted variable can only be explained according to its symbol analysis. Further exploration of the marginal effects of each influencing factor on the probability of each club is shown in Table 6.

According to the estimation results in Table 6, it can be seen that with the increase of foreign investment, the possibility of the city belonging to the club1 will decrease, and the possibility of belonging to the club2 and club3 will increase. As the level of local government intervention increases and the industrial structure upgrades, the possibility of the city belonging to the club1 will increase, and the possibility of belonging to the club2 and club3 will increase. In other words, foreign investment will help cities converge to clubs with lower average urban-rural income gaps. Local government intervention and industrial structure upgrade will promote cities to converge to clubs with high average urban-rural income gap.

Table 6. Average marginal effect results of the ordered logit model.

	dy/dx	Std.Err.	z	P > z	[95%Conf.	Interval]
fdi _predict						
1	−2.582	0.422	−6.12	0	−3.409	−1.755
2	2.348	0.383	6.13	0	1.597	3.098
3	0.234	0.049	4.76	0	0.138	0.331
gebha _predict						
1	0.99	0.147	6.73	0	0.702	1.278
2	−0.9	0.134	−6.74	0	−1.162	−0.638
3	−0.09	0.018	−5.07	0	−0.125	−0.055
ind _predict						
1	0.281	0.11	2.55	0.011	0.065	0.496
2	−0.255	0.1	−2.55	0.011	−0.451	−0.059
3	−0.025	0.011	−2.41	0.016	−0.046	−0.005

6. Conclusions

Since the reform and opening up, China's economy has shifted from high-speed development to a new normal of medium-to-high-speed growth in the latter period [46]. Whether China can cross the current middle-income trap after the economic slowdown and achieve regional coordinated development is still a puzzle. Exploring the convergence of urban-rural income gaps can provide a basis for the formulation of national and regional industrial policies.

This paper examines the urban-rural income gap of prefecture-level cities from 2005 to 2016 through the nonlinear time-varying factor model for automatic identification and grouping, then tests the conditions and internal mechanisms that determine the convergence of urban-rural income gaps. This study found the following results: In the traditional four economic regions of China, the convergence of urban-rural income gap exists in the northeast and the west, while the convergence of urban-rural income gap does not exist in the east and the middle. Although the division of traditional economic zones reflects the spatial correlation of the development of urban-rural income gap to some extent, and it cannot accurately reflect the internal relationship and similarity of the change of regional urban-rural income gap. Only some economic zones meet the criteria of convergence of urban-rural income gap. However, cities are divided into three major convergence clubs by logt test. The eastern and central regions are scattered in three clubs. The western region and the northeast region are concentrated in the largest club.

Secondly, the construction of High-speed railway has effectively narrows the urban-rural income gap in China, but it is not the cause of the formation of the three convergence clubs. The convergence effect of High-speed railway on the urban-rural income gap in China is still relatively weak. Related research shows that the impact of High-speed railway on urban economic growth is relatively lagging behind. Moreover, in some cities, the opening time of the High-speed railway is relatively late, and the role has not yet appeared. Therefore, the sample data still has certain limitations.

Thirdly, foreign direct investment promotes cities to converge to high-class clubs with lower average urban-rural income gap. Local government intervention and industrial structure upgrades will promote cities to converge to low-class clubs with higher average urban-rural income gap.

Based on the above research conclusions, we propose the following policy recommendations. In the construction of High-speed railway, it is necessary to carry out overall planning considering the changes of the industrial structure and the equalization of welfare. Providing farmers with supporting policies and other opportunities to enable the High-speed railway to bring better urban back-feeding effects and narrow the urban-rural income gap. The government needs to reverse the urban bias policy in terms of fiscal expenditure and investment. To achieve coordinate development in urban and rural areas, China's government should provide more infrastructure construction and financial investment in education, healthcare, social security or other social undertakings in rural areas.

Author Contributions: Conceptualization, W.L.; Methodology, W.L. and X.W.; Software, X.W.; Validation, W.L., O.-P.H. and X.W.; Formal Analysis, W.L., X.W. and O.-P.H.; Investigation, W.L., O.-P.H. and X.W.; Resources, W.L., O.-P.H.; Data Curation, W.L. and X.W.; Writing-Original Draft Preparation, W.L.,X.W.; Writing-review & Editing, W.L., O.-P.H. and X.W.; Visualization, X.W.; Supervision, W.L., O.-P.H..; Funding acquisition, W.L.; Project Administration, W.L.. All authors have read and agreed to the published version of the manuscript.

Funding: This research was funded by the China National Key Research and Development Plan, grant number 2018YFB1201400.

Conflicts of Interest: The authors declare no conflict of interest.

References

1. National Statistical Bureau. *China's Statistical Yearbook 2018;* China Statistics Press: Beijing, China, 2018.
2. Chen, F.; Xu, K.; Wang, M. The development of high-speed rail and the income gap between urban and rural residents: Evidence from Chinese cities. *Econ. Rev.* **2018**, *2*, 59–73.
3. National Railway Bureau. *Introduction of HSR Economics;* China Railway Publishing House: Beijing, China, 2018.
4. Zhang, A.; Wan, Y.; Yang, H. Impacts of high-speed rail on airlines, airports and regional economies: A survey of recent research. *Transp. Policy* **2019**, *81*, A1–A19. [CrossRef]
5. Chai, J.; Zhou, Y.; Zhou, X.; Wang, S.; Zhang, Z.G.; Liu, Z. Analysis on shock effect of China's high-speed railway on aviation transport. *Transp. Res. Part. A: Policy Pract.* **2018**, *108*, 35–44. [CrossRef]
6. Liu, L.; Zhang, M. High-speed rail impacts on travel times, accessibility, and economic productivity: A benchmarking analysis in city-cluster regions of China. *J. Transp. Geogr.* **2018**, *73*, 25–40. [CrossRef]
7. Li, H.; Strauss, J.; Shunxiang, H.; Lui, L. Do high-speed railways lead to urban economic growth in China? A panel data study of China's cities. *Q. Rev. Econ. Financ.* **2018**, *69*, 70–89. [CrossRef]
8. Vickerman, R. Can high-speed rail have a transformative effect on the economy? *Transp. Policy* **2017**, *62*, 31–37. [CrossRef]
9. Li, X.; Huang, B.; Li, R.; Zhang, Y. Exploring the impact of high speed railways on the spatial redistribution of economic activities—Yangtze River Delta urban agglomeration as a case study. *J. Transp. Geogr.* **2016**, *57*, 194–206. [CrossRef]
10. Jia, S.; Zhou, C.; Qin, C. No difference in effect of high-speed rail on regional economic growth based on match effect perspective? *Transp. Res. Part. A: Policy Pract.* **2017**, *106*, 144–157. [CrossRef]
11. Ke, X.; Chen, H.; Hong, Y.; Hsiao, C. Do China's high-speed-rail projects promote local economy? New evidence from a panel data approach. *China Econ. Rev.* **2017**, *44*, 203–226. [CrossRef]
12. Lynch, T.A.; Sipe, N.; Polzin, S.E.; Chu, S.E. An analysis of the economic impacts of Florida high speed rail. *Employment* **1997**, *8*, 1–16.
13. Masson, S.; Petiot, R. Can the high speed rail reinforce tourism attractiveness? The case of the high speed rail between Perpignan (France) and Barcelona (Spain). *Technovation* **2009**, *29*, 611–617. [CrossRef]
14. Diao, M. Does growth follow the rail? The potential impact of high-speed rail on the economic geography of China. *Transp. Res. Part. A: Policy Pract.* **2018**, *113*, 279–290. [CrossRef]
15. Chen, C.L.; Hall, P. The impacts of high-speed trains on British economic geography: A study of the UK's InterCity 126/226 and its effects. *J. Transp. Geogr.* **2011**, *19*, 689–706. [CrossRef]
16. Sasaki, K.; Ohashi, T.; Ando, A. High-speed rail transit impact on regional systems: Does the Shinkansen contribute to dispersion? *Ann. Reg. Sci.* **1997**, *31*, 77–98. [CrossRef]
17. Chen, Z.; Xue, J.; Rose, A.Z.; Haynes, K.E. The impact of high-speed rail investment on economic and environmental change in China: A dynamic CGE analysis. *Transp. Res. Part. A Policy Pract.* **2016**, *92*, 232–245. [CrossRef]
18. Preston, J.; Wall, G. The Ex-ante and Ex-post Economic and Social Impacts of the Introduction of High-speed Trains in South East England. *Plan. Pract. Res.* **2008**, *23*, 403–422. [CrossRef]
19. Chen, Z.; Haynes, K.E. Impact of high-speed rail on regional economic disparity in China. *J. Transp. Geogr.* **2017**, *65*, 80–91.

20. Jiang, M.; Kim, E. Impact of high-speed railroad on regional income inequalities in China and Korea. *J. Int. J. Urban. Sci.* **2016**, *20*, 393–406. [CrossRef]
21. Wang, Y.; Tang, S.; Huang, Y.; Bai, S. Analysis of the effect of urbanization promoting the convergence of urban and rural income gap—Based on the empirical test of 71 counties in Wuling mountain area. *Mod. Manag. Sci.* **2016**, *6*, 42–44.
22. Deng, J. Does urbanization and industrialization promote the convergence of urban and rural income gaps? An Empirical Test Based on Panel Threshold Model. *Hanjiang Acad.* **2016**, *35*, 78–84.
23. Song, J.; Wang, J. Dynamic convergence analysis of population migration, household registration urbanization and urban-rural income gap—Evidence from 262 prefecture-level cities. *Popul. J.* **2018**, *40*, 86–99.
24. Wu, H. Research on the convergence of urban-rural income gap in China—Based on the perspective of government investment. *East China Econ. Manag.* **2019**, *33*, 115–121.
25. Xu, J.; Zhang, D. The Transformation of urbanization and the convergence of income gap between urban and rural areas in China. *Reg. Res. Dev.* **2019**, *38*, 17–21.
26. Bartkowska, M.; Riedl, A. Regional convergence clubs in Europe: Identification and conditioning factors. *Econ. Model.* **2012**, *29*, 22–31. [CrossRef]
27. Canova, F. Testing for convergence clubs in income per capita: A predictive density approach. *Int. Econ. Rev.* **2004**, *45*, 49–77. [CrossRef]
28. Galor, O. Multiple growth regimes—Insights from unified growth theory. *J. Macroecon.* **2007**, *29*, 470–475. [CrossRef]
29. Tian, X.; Zhang, X.; Zhou, Y.; Yu, X. Regional income inequality in China revisited: A perspective from club convergence. *Econ. Model.* **2016**, *56*, 50–58. [CrossRef]
30. Wang, Y.F.; Yang, H.B.; Tang, S. The function mechanism and dynamic analysis of urbanization and industrial structure adjustment to urban-rural income gap. *Mod. Econ. Manag.* **2015**, *3*, 56–63.
31. Sun, F.; Mansury, Y.S. Economic Impact of High-Speed Rail on Household Income in China. *Transp. Res. Rec. J. Transp. Res. Board* **2016**, *2581*, 71–78. [CrossRef]
32. Vickerman, R. High-speed rail and regional development: The case of intermediate stations. *J. Transp. Geogr.* **2015**, *42*, 157–165. [CrossRef]
33. Wang, F.; Wei, X.; Liu, J.; He, L.; Gao, M. Impact of high-speed rail on population mobility and urbanisation: A case study on Yangtze River Delta urban agglomeration, China. *Transp. Res. Part. A: Policy Pract.* **2019**, *127*, 99–114. [CrossRef]
34. Pan, X.; Liu, Q.; Peng, X. Spatial club convergence of regional energy efficiency in China. *Ecol. Indic.* **2015**, *51*, 25–30. [CrossRef]
35. Ivanovski, K.; Churchill, S.A.; Smyth, R. A club convergence analysis of per capita energy consumption across Australian regions and sectors. *Energy Econ.* **2018**, *76*, 519–531. [CrossRef]
36. Khan, H.U.; Siddique, M.; Zaman, K.; Yousaf, S.U.; Shoukry, A.M.; Gani, S.; Sasmoko; Khanj, A.; Hishan, S.S.; Saleem, H. The impact of air transportation, railways transportation, and port container traffic on energy demand, customs duty, and economic growth: Evidence from a panel of low-, middle-, and high-income countries. *J. Air Transp. Manag.* **2018**, *70*, 18–35. [CrossRef]
37. Wang, S.; Tan, S.; Yang, S.; Lin, Q.; Zhang, L. Urban-biased land development policy and the urban-rural income gap: Evidence from Hubei Province, China. *Land Use Policy* **2019**, *87*, 91–103. [CrossRef]
38. Su, C.W.; Liu, T.Y.; Chang, H.L.; Jiang, X.Z. Is urbanization narrowing the urban-rural income gap? A cross-regional study of China. *Habitat Int.* **2015**, *48*, 79–86. [CrossRef]
39. Li, Y.; Wang, X.; Zhu, Q.; Zhao, H. Assessing the spatial and temporal differences in the impacts of factor allocation and urbanization on urban–rural income disparity in China, 2004–2010. *Habitat Int.* **2014**, *42*, 76–82. [CrossRef]
40. Mora, T.; Vayá, E.; Suriñach, J. Specialisation and growth: The detection of European regional convergence clubs. *Econ. Lett.* **2005**, *86*, 181–185. [CrossRef]
41. Li, R.; Pantelous, A.A.; Yang, L. Robust analysis for premium-reserve models in a stochastic nonlinear discrete-time varying framework. *J. Comput. Appl. Math.* **2020**, *368*, 112592. [CrossRef]
42. Solow, R.M. A contribution to the theory of economic growth. *Q. J. Econ.* **1956**, *70*, 65–94. [CrossRef]
43. Baumol, W.J. Productivity growth, convergence, and welfare: What the long-run data show. *Am. Econ. Rev.* **1986**, *76*, 1072–1085.

44. Peter, C.B.P.; Sul, D. Transition Modeling and Econometric Convergence Tests. *Econometrica* **2007**, *75*, 1771–1855.
45. Yu, Y.; Pan, Y. Has the opening of high-speed rail narrowed the income gap between urban and rural areas? An explanation from the perspective of Heterogeneous Labor Transfer. *China's Rural Econ.* **2019**, *1*, 79–95.
46. Li, W.; Hilmola, O.-P.; Wu, J. Chinese High-Speed Railway: Efficiency Comparison and the Future. *PROMET Traffic Transp. J.* **2019**, *31*, 693–702. [CrossRef]

Article

Business Models Amid Changes in Regulation and Environment: The Case of Finland–Russia

Oskari Lähdeaho [1] and Olli-Pekka Hilmola [1,2,*]

[1] Industrial Engineering and Management, LUT Kouvola, LUT University, Prikaatintie 9, FIN-45100 Kouvola, Finland; oskari.lahdeaho@lut.fi

[2] Estonian Maritime Academy, Tallinn University of Technology (Taltech), Kopli 101, 11712 Tallinn, Estonia

* Correspondence: olli-pekka.hilmola@taltech.ee

Received: 13 March 2020; Accepted: 17 April 2020; Published: 21 April 2020

Abstract: Changes in regulation are affecting the international business environment. In this study the impact of regulation changes and ways to benefit from those in Finland and Russia are examined. Logistics and manufacturing companies are studied using the case study approach including ten semi-structured interviews (Finland and Russia) and a survey (Southeast Finland), further supported by an additional survey for logistics sector companies (Southeast Finland). The changes in the business environment have created a fragmented market with a growing number of actors. Three business models (blockchain-based, platform-based and innovative subcontracting-based), capitalizing on the growing number of actors, were incepted in the interview phase and evaluated in the survey phase with companies. These models are integrable with the circular economy, a relevant practice according to the studied companies. Blockchain was perceived as a still immature technology. Further study revealed that the companies are not well prepared for environmental demands in logistics, and the overall volumes and business climate between the analyzed countries have not improved. Additionally, those companies do not actively pursue the possibilities of new technologies. The impact of regulatory changes in this region has not been examined closely with a case study approach. This study helps to explain the current trends in an established market.

Keywords: business models; regulation; logistics; supply chains; Finland; Russia

---

1. Introduction

The international railway connection between Finland and Russia (then Union of Soviet Socialist Republics) was formalized in 1948 when these countries signed a contract on interconnection via railways [1]. Since then, policymaking between these countries has progressively been facilitating international trade, and furthermore connecting eastern and western markets. More recent changes in legislation and regulation have been gone along the lines of European Union's (EU) railway legislation renewal. These so-called railway packages seek to open railway transports for competition and to enable fluent international operations [2]. However, some restrictive legislative changes have also emerged between Finland and Russia, driving the affected industries to adapt to the risen challenges. Thereupon, the changing business environment requires involved companies to revise their respective business models.

Recent turbulence in the political environment between the EU and Russia has led to sanctions by the EU and respectively counter sanctions by Russia. The sanctions affect the exports to and imports from Russia of military equipment, dual-use goods, energy-related equipment (mainly concerning oil exploration and production) and supporting services for the mentioned equipment and goods [3], whereas the counter sanctions are mainly imposed against the import of food products (such as milk, dairy products and meat), but also various other industries and sectors of the economy [4], e.g., agricultural equipment and pharmaceutical goods [5].

Recently, the EU has introduced strategies and regulations to lower emissions produced by transport, which is globally responsible for a major share of all produced emissions, i.e., approximately a quarter of all carbon dioxide emissions [6]. The general goal for carbon dioxide emissions from transport is to reduce them by 2050 to 60% lower in comparison to the base year of 1990 [7]. Furthermore, the EU has committed to reducing other emissions that also occur from transport by 2020 and beyond in comparison to 2005 levels, namely sulfur (by 59%), nitrogen oxides (by 42%), ammonia (by 6%), volatile organic compounds (by 28%) and atmospheric particulate matter (diameter less than 2.5 micrometers; by 22%) are targeted [8]. In addition, the International Maritime Organization (IMO) has set the cap of sulfur content in fuels used by ships to 0.5% starting from January 2020 [9]. A more demanding sulfur regulation of 0.1% was implemented in the Baltic and North Sea already in 2015 [10]. Moreover, nitrogen oxide emissions are strictly controlled by the IMO in the Baltic and North Sea emission control area with a reduction target of 80% in comparison to 2016 by requiring all the new ships built after January 2021 to have either installed catalyst converters for emissions or use Liquefied Natural Gas (LNG) as their fuel [11]. IMO also prohibits the discharge of waste from maritime vessels into the sea and requires monitoring of the waste while on-board until proper disposal [12]. These type of regulations incentivizes companies to optimize their supply chains and transport mode selection correspondingly [13]. Moreover, while the presented regulations cover specific regions, the environmental policies can be expected to be implemented in other areas depending on the policies' effectiveness and the given area's economic and political situation [14]. Therefore, adapting to emission regulations is vital for companies operating internationally.

While the presented legislation changes tighten the business environment for the involved companies, the transformed environment could facilitate or even necessitate the deployment of new business models. Thus, this article seeks to provide insight into the following questions:

- RQ1: What effects do legislative changes have on the business environment between Finland and Russia?
- RQ2: What types of new business models are enabled by the changing environment?
- RQ2.1: How are innovation and new technologies enabling these business models?

The rest of the article follows a structure of a literature review on sustainability in supply chains and logistics as well as sustainable business model innovation in Section 2, succeeded by explaining the used methodology to study the impact of the changing business environment in Southeast Finland and its implications for manufacturing and transportation companies in Section 3. Thereafter, the results of the empirical study are presented in Section 4. Lastly, this article is concluded in Section 5, where the results are discussed and reflected upon the theoretical baseline, and possible directions for future research are drawn.

2. Sustainable Business Models in Logistics

While sustainability has been established as an essential topic in supply chain research [15], generally it remains inadequately regarded in practical supply chain operations [16]. A modest shift of the aspects creating value in logistics has been observed, from a rigid cost orientation to other factors, environmental sustainability being one [17,18]. In maritime traffic, environmental sustainability is increasingly regarded due to tightening emission regulations as well as demand from stakeholders, customers and business partners [19]. Furthermore, strong orientation toward environmental sustainability within a company could improve that company's overall competitiveness [20]. Additionally, disregard or inaction concerning environmental sustainability could impose unexpected costs on a company [21], and it could be in companies' best interest to implement proactive measures to mitigate these possible costs [22].

Environmental sustainability and its possible competitiveness benefits in transportation can be enhanced by transport mode selection (e.g., utilizing multimodal transport chains with a larger share of less-emitting transport modes) and emphasizing collaboration within supply chains [23]. However, even though railways can be utilized to transport large amounts of freight conveniently and ecologically, road transports are often preferred due to higher mobility and flexibility [24]. Moreover, road transports are often used to support other modes of transport, such as in the pre- and post-haulage of railway transports [25]. To benefit from multimodal transports and the involvement of multiple actors within the chain, adequate intermodality and information sharing are required [26]. Additionally, an integrated supply chain requires a certain degree of trust between the involved actors [27]. Moreover, research by Ayoub and Abdallah [28] suggested that the benefits from flexibility are reached through innovativeness and responsiveness within a supply chain. It should also be kept in mind that multimodal transport chains are imposed to transaction cost every time a transport mode is changed [23,29,30]. Technologies such as Radio Frequency Identification (RFID) [31,32] and Wireless Sensor Networks (WSN) [32] could be utilized to support supply chains involving multiple separate actors through efficient information exchange. The tracking of goods offered by such technologies also enables monitoring of the reverse logistics, e.g., recycling of the packages [31].

To maintain the profitability of a business in an ever-changing world, deployed business models should evolve accordingly [33]. Furthermore, introduction of innovations to business should be conducted so that said innovations are woven into the company's business model [34]. In the case of incumbent companies, the existing business models and assets must be addressed in the business model renewal to avoid conflicts between new and established practices [35]. In addition to business model design, the implementation of the model is a significant factor in its profitability [36]. Boons and Lüdeke-Freund [37] claim a strong relationship between a company's business model and its innovation activities, where those activities enable not only innovative outputs but also business model renewal for competitive advantage. As established, the role of business model innovation is of high importance in economically sustainable business; however, often this process is not successful [38]. Due to this relationship of need and associated uncertainty, companies' dynamic capabilities towards renewing business models and their specific industry remain important factors in the business model innovation [39]. Especially in business model innovation aiming towards higher environmental sustainability, the surroundings of a company (e.g., industry, other actors and society) should be regarded in their business model through network orientation [40].

Hence, logistics service provision interconnects with environmental sustainability and business model innovation, mainly due to emerging trends in legislation and regulation as well as stakeholder demands. As stated, environmental sustainability is a growing issue in logistics [17,19] and in order for a company to reap the benefits from innovations, their business model should be designed in a manner allowing that [34,37,40]. This trajectory suggests that the logistics service providers should revise their business model design and include environmental sustainability as a factor of value for their offering. Furthermore, the increased competition and new entrants in railway traffic due to the changing business environment between Finland and Russia suggest that the companies related to this field should find ways to cope with the introduction of numerous new actors. Furthermore, environmentally sustainable business models, such as those based on circular economy practices, require extensive cooperation between separate actors [41].

3. Materials and Methods

This article combines a research from the latter part of 2018 (see [42]) with newer empirical study from 2019. The former research investigated the possible new trends in business models associated with manufacturing and transportation companies situated in Southeast Finland. Furthermore, the interplay of relevant innovations and new business models in the given context were studied. The previous study was conducted as a case study (e.g., [43,44]) with 10 semi-structured interviews of Finnish and Russian transportation professionals and experts, and a survey for manufacturing and transportation companies in Southeast Finland. A qualitative approach to studying emerging markets was used, as proposed by Guillotin [45]. The surveyed companies were mostly small- or medium-sized companies (SMEs) handling raw materials or low manufacturing value products. Findings from the previous study were used as a baseline to investigate the diffusion of the most promising business models and innovations since the last period of investigation. From the point of view of the studied companies, contemporary relevant and feasible innovations lie in the sphere of environmental sustainability. In transportation, these include technologies such as alternative fuels (e.g., LNG and electricity), and operations improvement (e.g., transport mode selection and multimodal transport chains).

In order to examine the validity of previous findings and to further study the development of the transportation industry in Southeast Finland, a newer survey study conducted in the autumn of 2019 concerning road transportation between Finland and Russia was conducted. This study of international road transportation was executed in the form of a web-based survey, which was distributed to 919 companies operating in the field of transportation, logistics and forwarding situated in South Finland alongside the highway E18 (European Road, a main serving road between Finnish and Russian trade). This research scope covers regions from Finland's west coast to its capital area and furthermore to its eastern border with Russia (starting in the west from the greater Turku region, and continuing to the capital region of Helsinki and from there onwards to the eastern border, ending at the Vaalimaa border-crossing point). A sample size of 56 recipients responded to the survey, setting the response rate for this survey at approximately 6%. This survey was more successful in terms of response rate, possibly due to its more specific scope and the shorter time required to fill the survey. The conducted survey focused on the past performance and future projections of local, national, international and transit road traffic in Finland, as well as perception towards alternative fuels in road transports and the usage of multimodal transport chains.

In addition, secondary data from an open-access database on road traffic near border crossing points between these countries, provided by the Finnish Transport Infrastructure Agency [46], was used to examine road freight traffic between Finland and Russia on a macroscale. A map with these border crossing points can be found in Appendix A. The described approaches were used jointly to gain a deeper understanding through triangulation on the complex problem setting established by the research questions [47].

4. Case Study Findings

The findings are presented in such a way that the results from the previous study are presented briefly first in order to prime the reader to the context of this case study. Thereafter, the results from the succeeding study are examined against the backdrop of the information discovered in the preceding research. While the first study had a broader scope in terms of the studied industries and factors in their respective environment, the results from this research indicate the need to focus on the transportation operations. The evidence pointed out that the companies in Southeast Finland focus their innovation activities to promote environmental sustainability on the operations, and transportation practices were the most obvious target for optimization.

4.1. Previous Findings from Manufacturing and Transportation Companies

4.1.1. Semi-Structured Interview Results

The research began with semi-structured interviews to map the present situation within the studied region and to let the involved experts and professionals share their vision about the current situation as well as the direction of future development. The central topics that emerged during the interviews are summarized in Table 1. The trends among the interviewees seem to focus on the growing demand of subcontractors and competition, as well as the fragmentation of the market into a multitude of separate actors and a higher number of smaller customers than before. The innovation activities seem to emphasize solutions to environmental challenges and ways to improve collaboration between different actors via communication channels and information flows. As this research focuses on the Finnish–Russian international business environment, the Eurasian Land Bridge, and more specifically the railway connection from China through Kazakhstan and Russia to Finland, plays an important role (other routes to China also exist through Russia and Mongolia, but the Kazakh route is currently used). Regarding this, the interviewees had recognized growing volumes of international freight traffic on railways between Finland and Russia. However, Finnish logistics operators have experienced a decrease in their internal operations on the Russian side, but they remained optimistic regarding future investments toward operations in Commonwealth of Independent States (CIS) member countries. Likewise, interviewees from Russia saw potential in the railway connection of Europe and the Far East through the Eurasian Land Bridge. In addition, the importance of the Northern Sea Route seems to hold more significance amongst Russian interviewees. While the overall trajectory of affairs seems similar on the Finnish and Russian sides, the development in Russia aims to capitalize on the growth potential of the logistics industry by shifting more of the handling of the freight flows to local actors instead of foreign actors, i.e., seeking lower use of transit countries in imports and exports. This development has played a role in diminishing freight traffic on roads between Finland and Russia.

Table 1. Overview of the emerged topics during the interviews (modified from [42]).

	Finnish Interviewees	Russian Interviewees
General remarks on the international logistics industry	Share of railway freight is growing between Finland and Russia. Railway connection from Finland to China has challenges in the intermediary border crossings. International operations target CIS countries, Mongolia and China. The Russian market has fragmented from a few large customers to numerous smaller ones. The Imatra–Svetogorsk border crossing point could be used to relieve pressure from other points.	The Northern Sea Route alongside supporting infrastructure is being developed. High importance of a Russian railway corridor between West Russia and the Far East as an alternative to the conventional sea routes. Containerization rate is still low in comparison to Europe. Balance of imports and exports is offset by decreasing imports and stagnant exports. Local ports are increasingly favored over transit countries.
National logistics infrastructure and competition	Ongoing and planned development of infrastructure. Some disagreements on the emphasis of development. There is demand for new entrants in the railway industry to create more flexible supply networks. Competition on railways is fierce; few actors handling bulk material are realistically competing. Role of the state in stimulating competition on railways. Subcontracting and other supporting services.	Infrastructure in Central and East Russia is not optimal, but it is being developed. Intense competition. Russian railways (RZD) remains as a focal actor in the industry. Political and economic uncertainty is a challenge, but there is development and growth potential.
Innovation in the logistics industry	Research and development activities emphasize environmental sustainability. Blockchain technology could improve communication between separate actors, information exchange and tracking of shipments, and cut costs by reducing unnecessary slack within the logistics operations.	Common platform to unify separate actors is being developed. Academy and businesses show interest toward Blockchain technology. The environmental sustainability of logistics industry is not being actively developed.

Moreover, the railway infrastructure in Southeast Finland is seeing investments to its development, and at the same time the competition of railway traffic has become liberated, allowing other actors into the market in addition to the state-owned operator. However, some of the interviewees saw that the entry barriers to this field are too high for new entrants, expect for few operators specializing in certain types of bulk freight. At the same time, the incumbent actors on railways signaled their demand for new potential partners in the field to develop their network to fit customer demands more flexibly. Therefore, the Finnish interviewees called for state-level initiatives to stimulate the entry of new actors into the market. The vision of interviewees from both sides of the border on relevant innovation in transportation related to the solutions seeking to manage and interact with a multitude of different actors in transportation industry. Additionally, on the Finnish side, interviewees saw the environmental sustainability of operations as a pressing issue due to the tightening regulations and rising demand for sustainability from customers as well as business partners. On the other hand, the industry on the Russian side seems to be in more dire need for solutions to communicate efficiently with the growing number of actors in the transportation field.

4.1.2. Survey Results

In the 2018 survey, approximately half of the respondent companies had engaged in international business. Approximately a quarter (23.1%) of the surveyed companies focused on exporting, whereas approximately 11.5% were focused on importing. A group of 7.7% were both exporting and importing and the rest (7.7%) reported doing other international business. As can be seen in Figure 1, a share of 42.9% of the companies with international operations had those within the EU. The popularity of the EU is most probably due to low barriers for the movement of goods, people and capital within the region. Since the geographical location of the studied region lies in the border area of Finland and Russia, the next largest target for foreign operations was Russia and other CIS countries. The markets were targeted by a group of companies with the shares of 23.8% and 4.8% for Russia and other CIS countries, respectively. Interestingly, despite the historical, cultural and geographical proximity of other Nordic countries, fewer companies indicated them as their target market. Lastly, some of the studied companies also had operations towards China and other Far East regions, but other locations were not mentioned by the respondents. The remaining half of the respondents without foreign operations signaled no interest to establish them.

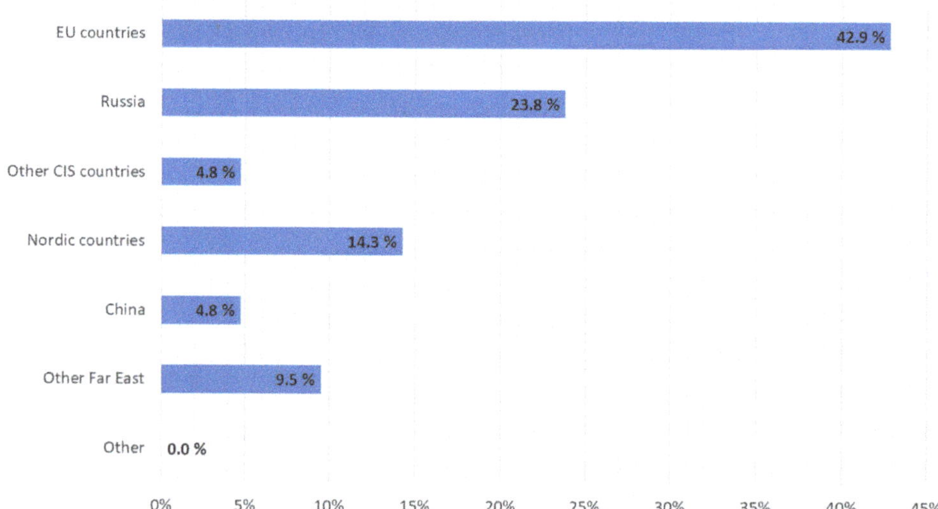

Figure 1. Target countries for international operations of the surveyed companies.

The previous survey was set to investigate the effect of changes in legislation and regulation to the business of the surveyed companies. While many of the presented changes were not regarded as impactful, the carbon dioxide (CO_2) mitigation directive was seen to have a relatively distinct effect by the companies, as illustrated in Figure 2. This is not very surprising, as the transportation sector has a significant stake in generated CO_2 emissions globally. While 42.3% did not grade the effect, and 15.4% saw no impact by the directive to their business, the rest of the group valuated some effect for the

EU's CO_2 mitigation strategy. Since the trajectory of CO_2 emissions originated from transportation, the reduction of those emissions requires most probably some radical changes in the used vehicles or how they are operated (e.g., alternative fuels with lower CO_2 emissions). The impact is difficult to estimate, but transportation companies will have to renew their fleet from its current composition.

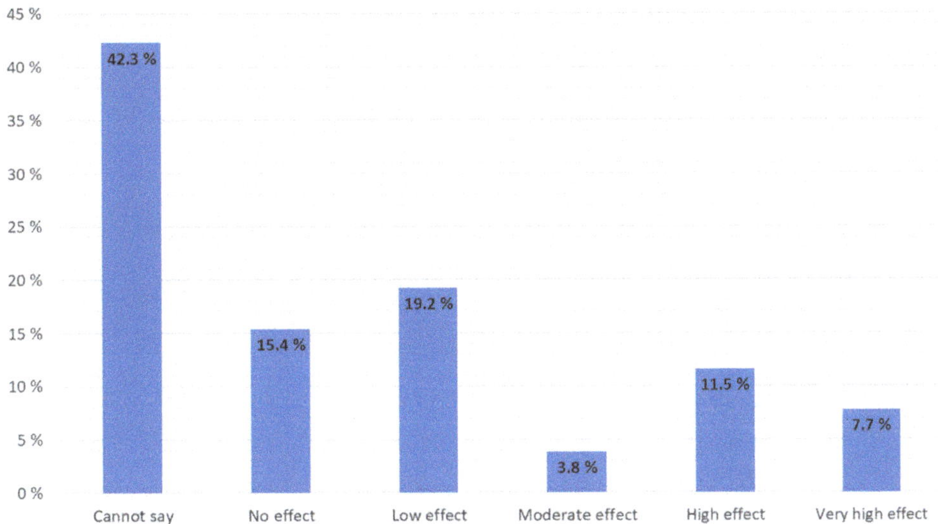

Figure 2. Effect of the EU's carbon dioxide emission mitigation directive on the surveyed companies' activities (Likert scale; 0 = cannot say; 1 = no effect; 2 = low effect; 3 = moderate effect; 4 = high effect; 5 = very high effect).

The transportation and manufacturing industries follow the same development as other sectors of the economy, where service provision has taken a significant share of all transactions in the market. This phenomenon is portrayed in the survey, since approximately half (46.2%) of the surveyed companies indicated that they require subcontracting services to support their business. At the same time, a large share of the respondents (57.7%) was offering subcontracting services to other companies. This overlapping of service provision and consumption can be seen in Figure 3. The survey allowed the respondents to write free-form comments about subcontracting, and some of the responses pinpointed that companies offer subcontracting back and forth to each other, whenever the need arises. This type of behavior has been observed by Hedenstierna et al. [48] in 3-D printing operations in Europe. Moreover, some of the interviewees indicated their need for more subcontractors to help them serve their customers in a more flexible manner. Additionally, the Russian interviewees indicated the growing importance of service provision in the logistics sector in Russia, likewise pointed out in the research by Yakunina [49].

Based on the interviews conducted before the distribution of the survey, the most relevant innovations according to the interviewees were studied with the help of the surveyed companies. These were blockchain, the Internet of Things (IoT), Artificial Intelligence (AI), LNG, catalyst converters for emissions in sea vessels, bio-economy or the utilization of renewable energy sources and circular economy. The surveyed companies were asked to rate their interest towards applying the mentioned innovations in their respective business practices. In addition, the respondents were asked to indicate if they have already implemented some of the mentioned innovations, or if they plan to do so.

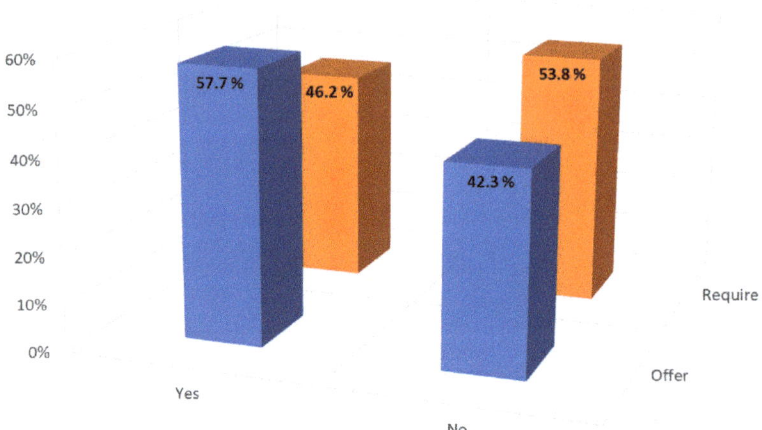

Figure 3. Supply and demand for subcontracting by the surveyed companies.

As illustrated in Figure 4, the innovations related to higher environmental sustainability scored higher grades from the respondents. Exceptions were LNG as fuel and catalyst converters, which are more relevant to maritime transports (although in fact LNG is being experimented with in road transports [50–52]). While shipment monitoring was revealed as a topic of high interest for transportation companies during the interviews, technologies such as blockchain and IoT received a low score on interest during the survey. Blockchain was deemed not interesting for the surveyed companies, possibly due to the debatable maturity of the solutions based on this technology, despite it being already implemented in transportation activities by the Danish container logistics company Maersk [53]. Circular economy and bio-economy were perceived as the most interesting innovations from the ones presented to the surveyed companies, and respectively 42.3% and 23.1% of these companies were planning or had already implemented these in their business activities. From these results it is evident that the manufacturing and transportation companies see environmental sustainability as a main target for their innovation activities.

To conclude the survey, the respondents were asked to grade three distinct business models by their feasibility in the companies' respective business environment. These models are generalized examples of the visions for emerging business models by the interviewees. Moreover, the models strive to capture the benefits from legislation and regulations, the changing business environment of Southeast Finland, and new technology and innovations, which were studied during this research. Thus, the proposed business models for the respondents were innovative subcontracting-based, platform-based and blockchain-based models.

Firstly, the innovative subcontracting-based model in the context of this research refers to a model where the focal company offers subcontracting services on business sections that have not been externalized by the principal companies before. Furthermore, this type of model would enable lateral collaboration between companies to offer subcontracting services to each other, e.g., by order book smoothing as described by Hedenstierna et al. [48]. For example, the gradually liberated competition on railway traffic could offer opportunities for this type of business activities. Secondly, the platform-based model involves a situation where the existing markets operate within a digital platform. The initiative for companies to join this community would be the convenience through a streamlined process of sourcing providers and identifying customers. In addition, a platform could stimulate competition through a less formal contract structure and more transparent tendering. Additionally, as the studied market is more fragmented in terms of number of separate actors; according to the interviewees, a platform could act as a tool to navigate this increasingly complex network. Lastly, the blockchain-based model in this research can be understood similarly to the platform-based one, but the point here is not focusing on digitalizing the marketplace. This model seeks to reinforce the existing networks by allowing a more efficient exchange of information between partners, by making the transactions more transparent by involving the whole supply chain in the information exchange and by verifying the transactions within a supply chain by the participants of said chain.

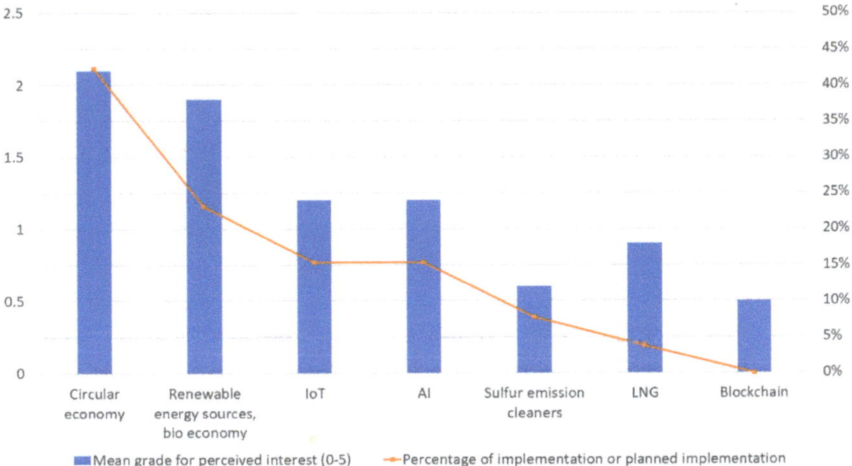

Figure 4. Perceived interest and implementation rate of the studied innovations (modified from [42]; grading for interest was done on a Likert scale; 0 = cannot say; 1 = no interest; 2 = low interest; 3 = moderate interest; 4 = high interest; 5 = very high interest).

The different proposed business models and their perceived feasibility are combined in Figure 5. As was established in the earlier results, the need for subcontracting services in the studied region is significant. This can be seen manifested in the perceived feasibility grade for the innovative subcontracting-based model, which had a feasibility rating of "very high" for 15.4%, "high" for 7.7%, and "moderate" for 19.2% of the respondents. The other two proposed models received a lower feasibility rating, and the platform-based model was seen as slightly more feasible. The modest success of those models can possibly be explained by the low maturity of the required technology in the given context of manufacturing and transportation SMEs in Southeast Finland. A more significant factor in the low perceived feasibility for these models probably is that they require extended trust between the separate actors introduced into the network [54], which is not typical in industries with fierce competition as the ones studied. The interviewees also recognized this challenge in deploying the mentioned business models.

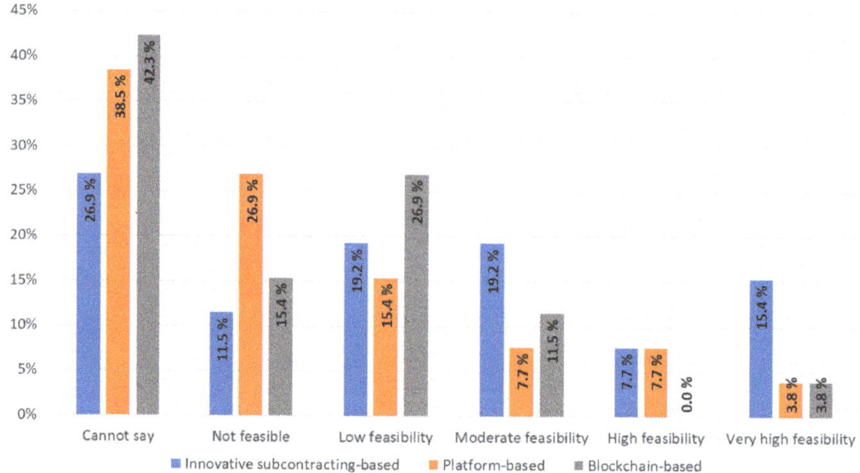

Figure 5. Feasibility of the proposed business models (Likert scale; 0 = cannot say; 1 = not feasible; 2 = low feasibility; 3 = moderate feasibility; 4 = high feasibility; 5 = very high feasibility).

4.2. Findings from Transportation Companies of South Finland through A Second Survey

As established during the previous study, environmental sustainability emerged as the driving factor for business model renewal as well as research and development in the manufacturing and transportation industries of Southeast Finland. The described legislation and regulation changes concerning emissions are currently gradually coming into effect, and their impact on energy intensive industries such as transportation are yet to be observed. Therefore, the succeeding study is scoped to emphasize environmental sustainability and transportation companies. A survey on road transports and the usage of highway E18 in Finland, targeting transportation, logistics and forwarding companies, was conducted in the autumn of 2019. The results of this survey interconnect with the topic and results of the previous survey, although it is focused on the road transportation mode. Based on this survey, the barriers of internationalization still exist and are perceived as strong for Finnish transportation companies considering Russia as a target market. A summary of the topics that emerged from the free-form comments about road freight traffic between these countries by the survey respondents can be found in Table 2.

Table 2. Topics that emerged from the free-form comments on the road freight traffic between Finland and Russia by the surveyed companies.

International road transport has lower volumes of cargo than before.	The renewed highway E18 on the Finnish side is a safe and working road.
The border formalities have become stricter and therefore take more time, which disrupts the traffic flows.	Transport business between Finland and Russia is volatile, thus not very appealing for Finnish companies.
Road use taxation sets challenges for international operations.	Demand for services targeted to the professional users of road E18.
The Russian side of E18 (Scandinavia road) only has one lane near the border; 2–3 lanes would allow for a smoother flow of road traffic and enhance safety.	Russian companies handle most of the international road transports between Finland and Russia.

The operators based in Finland see the Russian market as uncertain and volatile, i.e., the perceived risks are higher than the perceived benefits. The same observation was made in the interview phase, and interviewees from the Russian side share the view of market volatility to a certain degree. One of the more glaring barriers to international road transportation are the strict border formalities, which disrupt the fluency of the traffic flows. Reportedly, another factor in the undesirability of international road transport activities is the inadequateness of the road infrastructure; onwards from the Finnish border station Vaalimaa, the E18 European Road has only one lane, which not only undermines the traffic flow, but also decreases the safety on a busy road. From the Finnish perspective, it must also be acknowledged that the pricing for transports originating from Russia are lower than those from Finland. The majority of the road transports originating from (or transiting through) Finland headed to Russia are undertaken by Russian operators. This shift of emphasis in the responsible companies from Finnish actors to Russian ones could arguably be one of the factors explaining the relatively unenthusiastic view on the prospects of international road transports from the Finnish side. This, connected to the lower import activity of Russia (also concluded from the interviews), could be used to explain why business opportunities in international road transports between these two countries are not exceptionally flourishing.

Furthermore, the development of road transportation in Southeast Finland is seen by the respondents as stagnating. As shown in Figures 6 and 7, the road transports between Finland and Russia through main border crossing points between these countries have been constantly decreasing from 2010. A main contributor towards this development is the disappearance of transit freight traffic via road between Finland and Russia [55]. While in recent years the road transport volumes are nowhere near the amounts of 2010, traffic seems to be returning slowly to Vaalimaa and Nuijamaa border stations, whereas activity at the Imatra border crossing point continues to decrease in both directions. As has come up in the interview phase of this research, material flows originating from and arriving to Russia are increasingly shipped from and to local seaports. Therefore, once-active transit traffic through Finland (ships arriving to Finland and the goods being transported to Russia via road) can be seen decreasing drastically.

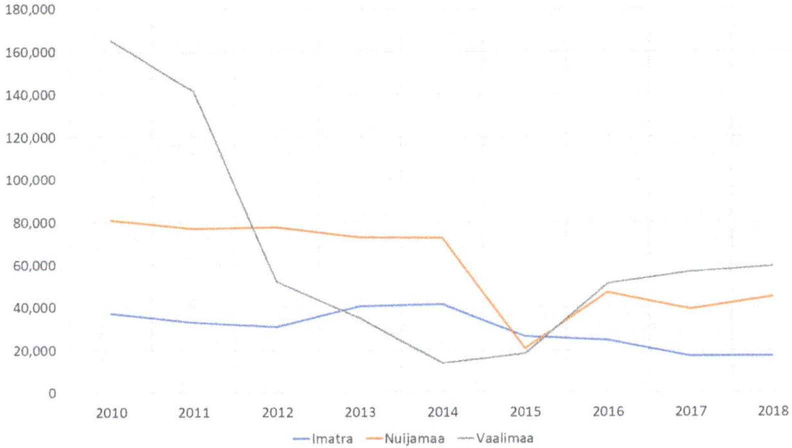

Figure 6. Road freight traffic (number of vehicles) from Finland to Russia [46].

As can be seen in Figure 8, road transport practitioners' perceived feasibility of alternative fuels remains on the low end—similarly to the situation in the earlier survey. While it could be spotted that some companies are actively getting ready to implement a higher biocomponent share in diesel and even LNG in their road transport activities, most of the companies are hesitant to adapt these alternative fuels. Moreover, it seems that electric vehicles are not seen as relevant for transportation activities, i.e., companies do not see electric heavy-duty vehicles penetrating the market just yet. It is a peculiar situation, since alternative fuels require upgraded infrastructure to offer operation range for vehicles running on such fuels, such as service stations with natural gas pumps or electric vehicle charging spots, but on the other hand road traffic support service providers do not experience enough demand to invest in these upgrades. However, it should be noted that when addressing total carbon dioxide emissions originating from transportation, thorough consideration of the large picture regarding environmental sustainability is required. For example, large-scale investments in the electric vehicle fleet could in fact increase total emissions from transportation, since a majority of electricity is still produced with fossil fuels [6]. This challenge acts as a proof for the complexity of sustainability challenges.

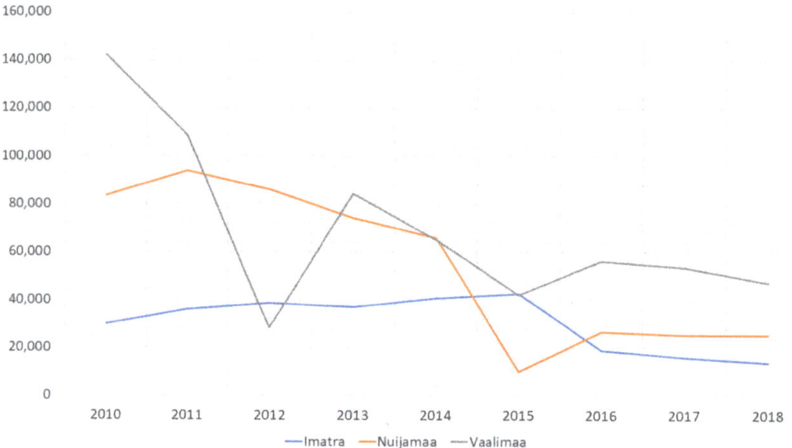

Figure 7. Road freight traffic (number of vehicles) from Russia to Finland [46].

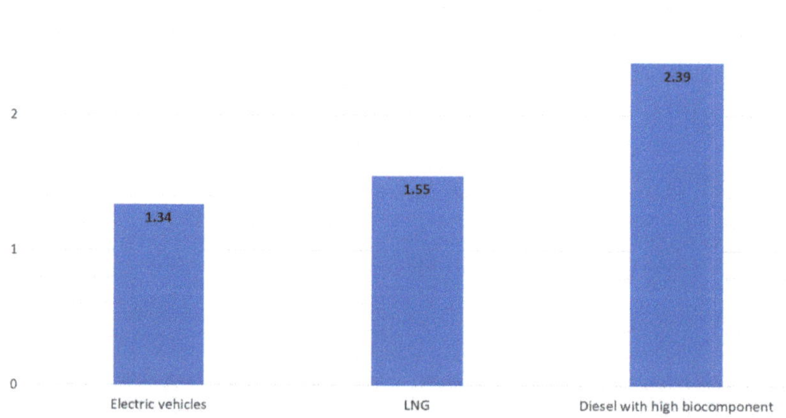

Figure 8. Average grades for feasibility of alternative fuels in road transportation (Likert scale; 0 = cannot say; 1 = not feasible; 2 = low feasibility; 3 = moderate feasibility; 4 = high feasibility; 5 = very high feasibility).

The existence of internal warehousing activities seems to drive the importance of road transports and related infrastructure for a company. A share of 36% respondents from the surveyed companies had internal warehousing activities, and the locations of these warehouses were mostly on the eastern sections of highway E18 (78% of warehouses were either in Kotka, Vantaa or Kouvola). Respondents from this region indicated that highway E18 is important for their operations, more so than those from the western regions (82.4% of companies with warehouses in the eastern parts signaled the importance of E18, whereas the share was 40% for those with warehouses more to the west). This correlation could arise from the fact that most of the warehouses are only accessible by road. Only a certain number of companies overall benefit from the strategic positioning of warehouses, where different transport modes can be used efficiently, i.e., warehouses that are in the vicinity to railway tracks or marine ports. As emphasis on environmental sustainability is growing in transportation, environmental logistics service providers can capitalize on this apparent need of multimodal transport chains. Based on the premise and results of this research, sustainability, especially in terms of the environment, could emerge as an important factor in service provider tendering. Moreover, companies that are future-oriented could capitalize on this need with a state-of-the-art vehicle fleet, i.e., vehicles running on alternative fuels and high capacity transports. High capacity transports in this context refer to road trains, which are heavier and longer in comparison to conventional ones. In Finland these high capacity transports mean road trains that range from 76 to 100 tons in total weight, which have special permissions to operate on certain roads in the Finnish road network.

5. Discussion

One of the most influential legislation changes affecting the studied region is the liberation of competition on railway freight traffic, allowing new entrants to the market. The introduction of new entrants to railways could create demand for new road transport services to support the railway operations in the pre- and post-haulage phases of the transport. In addition, maritime traffic faces tightening global and local (Baltic Sea Emission Control Area) emission regulations, which chip away at this mode's competitiveness against other modes [9–12]. This development creates a need for road transport service providers, but in order to enable prolonged success these actors must meet adequate degree of environmental sustainability [20] to comply with the regulations of the examined region [7,8]. Furthermore, if more actors keep entering the market, a demand for managing the increasingly complex logistics network could arise. As discussed earlier in this article, state-of-the-art solutions exist to satisfy the described need for connecting a multitude of separate actors. However, the presented models seem to require extended trust between the involved actors, which seems to be a considerable barrier for their diffusion.

The international road freight traffic between Finland and Russia has been declining consistently during the last decade. While there are various policy and national strategy-level reasons (e.g., those leading to diminished transit traffic [55]), one reason is the increasing popularity of the railway connection between these countries. Wood, pulp and metal industry products are transported utilizing the railway infrastructure of the Eurasian Landbridge [56]. The railway mode also has the potential for increasing volumes for food (especially meat) transports from Northern Europe to China through Finland and CIS countries. Increased utilization of this route has also spawned businesses in Finland, who specialize in railway freight traffic between Finland and Russia.

The rapidly changing business environment implies the need for equally rapid business model renewal with corresponding managerial decision-making—a challenge for companies who wish to remain competitive [33–40]. Cued by the changing business environment and refined by the business model theory evolution, three general-level business models (innovative subcontracting-based, platform-based and blockchain-based) were proposed and their feasibility was graded by manufacturing and transportation companies of Southeast Finland. The models are designed to exploit the changes in the studied business environment as well as relevant emerging innovations (to acquire competitive advantage [37,39]) as indicated by the interviewees. The trust between the involved actors remains an inherent requirement with varying intensity for these models, which is also seen as a factor for success of supply chains [25,27]. While the interviewed experts voiced the need for efficient shipment tracking, the new technologies and business models enabling that need were not seen as feasible by the surveyed companies, possibly due to a lack of trust between actors in the examined region. The main emphasis of the studied companies' innovation activity lies in promoting their environmental sustainability. As proposed by Jasmi and Fernando [19], the studied companies' informed focus on environmental sustainability is due to the stakeholder demands and tightening legislation and regulation towards emissions. Once again, the companies' business model should accompany this vision in order to create benefit from it [40]. However, the practice in the studied region seems to differentiate from the business model and innovation theory.

6. Conclusions

The changing regulations and environment regarding international business between Finland and Russia are moving the manufacturing and logistics industries towards a freer market with lower barriers for new entrants. At the same time, the requirements concerning the environmental impact of these business activities are becoming stricter. This development drives incumbent companies to re-design their respective business models, and acts as a cue for new actors to enter the market. In addition, the presented changes open avenues for achieving a competitive advantage with adapting the new set of rules over those who do not. Business models that are designed to capture value from growing number of actors in the market and environmentally sustainable business are likely to prosper amid the discussed changes.

New technology and innovation regarding information and communication technology (e.g., blockchain) enable more precise tracking of shipments [31,32] and more efficient communication between actors in a supply chain. If the required degree of trust can be established between the actors, these innovations can reinforce collaboration between them. Moreover, enhanced tracking of shipments and their origin enables the verification of sustainably sourced goods. Some of the studied companies seem to be ready for this kind of commitment, but a majority remains skeptic. Alternative fuels (e.g., LNG and electricity) for transportation could significantly lower the environmental impact of logistics operations. Furthermore, environmental sustainability can be improved via innovative business practices (e.g., circular economy). Means to reduce the negative impact to the environment were seen as more relevant among the studied companies.

The presented case of international transportation between Southeast Finland and Northwest Russia could be used as a reference point for studies concerning other countries that rely on railway transportation in import and export activities. Furthermore, the insights on the railway connection through the Eurasian Landbridge between the Far East and Northern Europe could benefit other regions, such as Central Europe. Especially when companies are looking to ease the environmental impact of their supply chain activities, the railway transport mode could offer means to reduce produced emissions. The studied technologies and innovations should also be considered in other regions. While the reception for the studied digital technologies was not overwhelmingly enthusiastic in the studied region, the case might be different in the context of other regions. The technologies and innovations focusing on reducing negative environmental impacts should be considered by companies looking to achieve a competitive advantage through environmentally sustainable business practices.

While this research is limited to a specific region, it is also limited by the general scope of manufacturing and logistics industries in that region. It would be recommendable to study business model and innovation theory closer to the practice in the future, e.g., through piloting in experimental environments with companies. Furthermore, it is important to study those theories in the context of high physical asset intensive (and low intellectual capital intensive) industries. Pieroni et al. [40] also call for more experimentation and learning from practice, and Poponi et al. [41] call for more generalizable models in environmentally sustainable business model design and innovation. The new situation, where bigger countries aim to produce more locally and import less, should deserve further research from the angle of international logistics companies. Lower demand for logistics services is not the only implication; services are also experiencing higher competition, and structures as well as transportation modes do seem to be changing.

Author Contributions: Conceptualization, O.L. and O.-P.H.; methodology, O.L. and O.-P.H.; validation, O.L. and O.-P.H.; formal analysis, O.L.; investigation, O.L.; resources, O.-P.H.; data curation, O.L.; writing—original draft preparation, O.L.; writing—review and editing, O.-P.H.; visualization, O.L.; supervision, O.-P.H.; project administration, O.-P.H.; funding acquisition, O.L. and O.-P.H. All authors have read and agreed to the published version of the manuscript.

Funding: The interviews and first survey were done as consulting work and the second survey was funded by the Southeast Finland–Russia Cross-Border Cooperation Program.

Acknowledgments: This article is based on a conference proceedings manuscript presented at the 23rd Cambridge International Manufacturing Symposium in September 2019 (Cambridge, UK). The previous research has been extended with new empiric and secondary data. We would like to thank the participants of this conference for their valuable comments and insights.

Conflicts of Interest: The authors declare no conflict of interest.

Appendix A

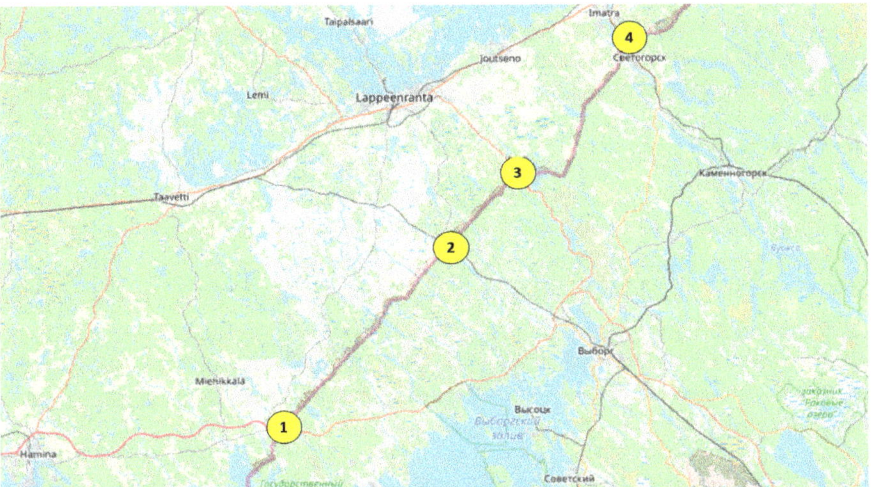

Figure A1. Border crossing points between Southeast Finland (left side) and Northwest Russia (right side; Leningrad region). Numbers in the figure correspond the name of the respective crossing point, as follows: 1 = Vaalimaa–Torfjanovka; 2 = Vainikkala–Buslovskaja (railway border-crossing point); 3 = Nuijamaa–Brusnitsnoje; 4 = Imatra–Svetogorsk [57].

References

1. SopS 1. *Asetus Sosialististen Neuvostotasavaltojen Liiton Kanssa Rautateitse Tapahtuvasta Yhdysliikenteestä Tehdyn Sopimuksen Voimaansaattamisesta (Free Translation by the Authors: Act of Enactment for Railway Interconnection Traffic with Union of Soviet Socialist Republics)*; The Finnish Ministry of Justice: Helsinki, Finland, 1948.
2. European Commission. Railway Packages. Available online: https://ec.europa.eu/transport/modes/rail/packages_en (accessed on 24 February 2020).
3. European Union Newsroom. EU Sanctions against Russia over Ukraine Crisis. Available online: https://europa.eu/newsroom/highlights/special-coverage/eu-sanctions-against-russia-over-ukraine-crisis_en (accessed on 18 November 2019).
4. Korhonen, I.; Simola, H.; Solanko, L. Sanctions, Counter-Sanctions and Russia—Effects on Economy, Trade and Finance. Available online: http://urn.fi/URN:NBN:fi:bof-201805301586 (accessed on 24 February 2020).
5. Romanova, T. Sanctions and the Future of EU–Russian Economic Relations. *Eur.-Asia Stud.* **2016**, *68*, 774–796. [CrossRef]
6. Solaymani, S. CO2 emissions patterns in 7 top carbon emitter economies: The case of transport sector. *Energy* **2019**, *168*, 989–1001. [CrossRef]

7. European Commission. A European Strategy for Low-Emission Mobility. Available online: https://eur-lex.europa.eu/legal-content/en/TXT/?uri=CELEX:52016DC0501 (accessed on 24 February 2020).
8. United Nations Economic Commission for Europe. Protocol to Abate Acidification, Eutrophication and Ground-Level Ozone. Available online: https://www.unece.org/env/lrtap/multi_h1.html (accessed on 24 February 2020).
9. International Maritime Organization. List of Amendments Expected to Enter into Force this Year and in the Coming Years. Available online: http://www.imo.org/en/About/Conventions/Pages/Action-Dates.aspx (accessed on 24 February 2020).
10. Hilmola, O.-P. *The Sulphur Cap in Maritime Supply Chains: Environmental Regulations in European Logistics*; Palgrave Macmillan, Pivot Series: London, UK, 2019.
11. Ministry of Transport and Communications. Nitrogen Emissions from Ships Restricted in the Baltic Sea and North Sea. Available online: https://www.lvm.fi/-/nitrogen-emissions-from-ships-restricted-in-the-baltic-sea-and-north-sea (accessed on 24 February 2020).
12. Svaetichin, I.; Inkinen, T. Port waste management in the Baltic Sea area: A four port study on the legal requirements, processes and collaboration. *Sustainability* **2017**, *9*, 699. [CrossRef]
13. Wang, Q.; Sun, H. Traffic Structure Optimization in Historic Districts Based on Green Transportation and Sustainable Development Concept. *Adv. Civ. Eng.* **2019**. [CrossRef]
14. Arbolino, R.; Carlucci, F.; De Simone, L.; Ioppolo, G.; Yigitcanlar, T. The policy diffusion of environmental performance in the European countries. *Ecol. Indic.* **2018**, *89*, 130–138. [CrossRef]
15. Fahimnia, B.; Sarkis, J.; Davarzani, H. Green supply chain management: A review and bibliometric analysis. *Int. J. Prod. Econ.* **2015**, *162*, 101–114. [CrossRef]
16. Dey, A.; LaGuardia, P.; Srinivasan, M. Building sustainability in logistics operations: A research agenda. *Manag. Res. Rev.* **2011**, *34*, 1237–1259. [CrossRef]
17. Arvis, J.; Ojala, L.; Wiederer, C.; Shepherd, B.; Raj, A.; Dairabayeva, K.; Kiiski, T. *Connecting to Compete 2018 Trade Logistics in the Global Economy: The Logistics Performance Index and Its Indicators*; The International Bank for Reconstruction and Development; The World Bank: Washington, DC, USA, 2018.
18. Macharis, C.; Melo, S.; Woxenius, J.; Van Lier, T. *Sustainable Logistics, Transport and Sustainability*; Emerald Group Publishing: Bingley, UK, 2014; Volume 6.
19. Jasmi, M.F.A.; Fernando, Y. Drivers of maritime green supply chain management. *Sustain. Cities Soc.* **2018**, *43*, 366–383. [CrossRef]
20. Wang, C.H. How organizational green culture influences green performance and competitive advantage: The mediating role of green innovation. *J. Manuf. Technol. Manag.* **2019**, *30*, 666–683. [CrossRef]
21. Oldenkamp, R.; Van Zelm, R.; Huijbregts, M.A.J. Valuing the human health damage caused by the fraud of Volkswagen. *Environ. Pollut.* **2016**, *212*, 121–127. [CrossRef]
22. Nocera, S.; Cavallaro, F. Economic evaluation of future carbon dioxide impacts from Italian highways. *Procedia Soc. Behav. Sci.* **2012**, *54*, 1360–1369. [CrossRef]
23. Das, C.; Jharkharia, S. Low carbon supply chain: A state-of-the-art literature review. *J. Manuf. Technol. Manag.* **2018**, *29*, 398–428. [CrossRef]
24. Catrina, M. Rail Logistics for an Efficient Supply Chain Management. *Valahian J. Econ. Stud.* **2012**, *3*, 99–104.
25. Bergqvist, R.; Monios, J. The last mile, inbound logistics and intermodal high capacity transport—The case of Jula in Sweden. *World Rev. Intermodal Transp. Res.* **2016**, *6*, 74–92. [CrossRef]
26. Bretzke, W.R. Supply chain management: Notes on the capability and the limitations of a modern logistic paradigm. *Logist. Res.* **2009**, *1*, 71–82. [CrossRef]
27. Thunberg, M.; Fredriksson, A. Bringing planning back into the picture—How can supply chain planning aid in dealing with supply chain-related problems in construction? *Constr. Manag. Econ.* **2018**, *36*, 425–442. [CrossRef]
28. Ayoub, H.F.; Abdallah, A.B. The effect of supply chain agility on export performance: The mediating roles of supply chain responsiveness and innovativeness. *J. Manuf. Technol. Manag.* **2019**, *30*, 821–839. [CrossRef]
29. Panova, Y.; Korovyakovsky, E.; Semerkin, A.; Henttu, V.; Li, W.; Hilmola, O.-P. Russian Railways on the Eurasian Market: Issue of Sustainability. *Eur. Bus. Rev.* **2017**, *29*, 664–679. [CrossRef]
30. Wang, M.; Liu, K.; Choi, T.M.; Yue, X. Effects of Carbon Emission Taxes on Transportation Mode Selections and Social Welfare. *IEEE Trans. Syst. Man Cybern. Syst.* **2015**, *45*, 1413–1423. [CrossRef]

31. Cui, L.; Deng, J.; Liu, F.; Zhang, Y.; Xu, M. Investigation of RFID investment in a single retailer two-supplier supply chain with random demand to decrease inventory inaccuracy. *J. Clean. Prod.* **2017**, *142*, 2028–2044. [CrossRef]
32. Badia-Melis, R.; Mc Carthy, U.; Ruiz-Garcia, L.; Garcia-Hierro, J.; Robla Villalba, J.I. New trends in cold chain monitoring applications—A review. *Food Control* **2018**, *86*, 170–182. [CrossRef]
33. Zott, C.; Amit, R. Business model design: An activity system perspective. *Long Range Plan.* **2010**, *43*, 216–226. [CrossRef]
34. Chesbrough, H.; Rosenbloom, R.S. The role of the business model in capturing value from innovation: Evidence from Xerox Corporation's technology spinoff companies. *Ind. Corp. Chang.* **2002**, *11*, 529–555. [CrossRef]
35. Kim, S.K.; Min, S. Business model innovation performance: When does adding a new business model benefit an incumbent? *Strateg. Entrep. J.* **2015**, *9*, 34–57. [CrossRef]
36. Brea-Solís, H.; Casadesus-Masanell, R.; Grifell-Tatjé, E. Business model evaluation: Quantifying walmart's sources of advantage. *Strateg. Entrep. J.* **2015**, *9*, 12–33. [CrossRef]
37. Boons, F.; Lüdeke-Freund, F. Business models for sustainable innovation: State-of-the-art and steps towards a research agenda. *J. Clean. Prod.* **2013**, *45*, 9–19. [CrossRef]
38. Geissdoerfer, M.; Vladimirova, D.; Evans, S. Sustainable business model innovation: A review. *J. Clean. Prod.* **2018**, *198*, 401–416. [CrossRef]
39. Rachinger, M.; Rauter, R.; Müller, C.; Vorraber, W.; Schirgi, E. Digitalization and its influence on business model innovation. *J. Manuf. Technol. Manag.* **2018**, *30*, 1143–1160. [CrossRef]
40. Pieroni, M.P.; McAloone, T.; Pigosso, D.A. Business model innovation for circular economy and sustainability: A review of approaches. *J. Clean. Prod.* **2019**, *215*, 198–216. [CrossRef]
41. Poponi, S.; Arcese, G.; Mosconi, E.M.; Di Trifiletti, M.A. Entrepreneurial drivers for the development of the circular business model: The role of academic spin-Off. *Sustainability* **2020**, *12*, 423. [CrossRef]
42. Lähdeaho, O.; Hilmola, O.; Niiranen, J. New Business Models as Regulation Changes: Case of Finland and Russia. In Proceedings of the 23rd Cambridge International Manufacturing Symposium, University of Cambridge, Cambridge, UK, 26–27 September 2019. [CrossRef]
43. Eisenhardt, K.M.; Graebner, M.E. Theory building from cases: Opportunities and challenges. *Acad. Manag. J.* **2007**, *50*, 25–32. [CrossRef]
44. Ghauri, P.; Grønhaug, K. *Research Methods in Business Studies*, 4th ed.; Pearson Education Limited: Essex, UK, 2010.
45. Guillotin, B. Using Unconventional Wisdom to Re-Assess and Rebuild the BRICS. *J. Risk Financ. Manag.* **2019**, *12*, 8. [CrossRef]
46. Finnish Transport Infrastructure Agency. Traffic Meausrement System data. Available online: https://aineistot.vayla.fi/lam/reports/ (accessed on 24 February 2020).
47. Burton, R.M.; Obel, B. Computational Modeling for What-Is, What-Might-Be, and What-Should-Be Studies—And Triangulation. *Organ. Sci.* **2011**, *22*, 1195–1202. [CrossRef]
48. Hedenstierna, C.P.T.; Disney, S.M.; Eyers, D.R.; Holmström, J.; Syntetos, A.A.; Wang, X. Economies of collaboration in build-to-model operations. *J. Oper. Manag.* **2019**, *65*, 753–773. [CrossRef]
49. Yakunina, Y.S. o : oo a o ooo a (Logistic services: Characteristics and Specificity in the Russian Market). *B oo a* **2014**, *24*, 107–112.
50. Arteconi, A.; Brandoni, C.; Evangelista, D.; Polonara, F. Life-cycle greenhouse gas analysis of LNG as a heavy vehicle fuel in Europe. *Appl. Energy* **2010**, *87*, 2005–2013. [CrossRef]
51. Osorio-Tejada, J.L.; Llera-Sastresa, E.; Scarpellini, S. Liquefied natural gas: Could it be a reliable option for road freight transport in the EU? *Renew. Sustain. Energy Rev.* **2017**, *71*, 785–795. [CrossRef]
52. Cision. Vähälä Yhtiöiden Käyttöön Ensimmäinen Ivecon Kolmiakselinen LNG-Rekkaveturi Suomessa (Free translation by the Author: "Vähälä Company Takes in the Use First LNG Fueled Combined Iveco truck"). Available online: News.cision.com/fi/gasum/r/vahala-yhtioiden-kayttoon-ensimmainen-ivecon-kolmiakselinen-lng-rekkaveturi-suomessa,c2646789 (accessed on 24 February 2020).
53. Maersk. Maersk and IBM Introduce TradeLens Blockchain Shipping Solution. Available online: www.maersk.com/en/news/2018/06/29/maersk-and-ibm-introduce-tradelens-blockchain-shipping-solution (accessed on 24 February 2020).

54. Harvard Business Review. A Study of More Than 250 Platforms Reveals Why Most Fail. Available online: https://hbr.org/2019/05/a-study-of-more-than-250-platforms-reveals-why-most-fail (accessed on 7 April 2020).
55. Hilmola, O.-P.; Hämäläinen, E. Faltering demand and performance of the logistics service sector in two cities of Southeast Finland. *Fenn. Int. J. Geogr.* **2016**, *194*, 135–151.
56. Hilmola, O.-P.; Henttu, V.; Panova, Y. Development of Asian Landbridge from Finland: Current state and future prospects. In Proceedings of the 22nd Cambridge International Manufacturing Symposium, University of Cambridge, Cambridge, UK, 27–28 September 2018.
57. OpenStreetMap Contributors. OpenStreetMap. Available online: https://www.openstreetmap.org/ (accessed on 8 April 2020).

Article

Development of Micro-Mobility Based on Piezoelectric Energy Harvesting for Smart City Applications

Chaiyan Jettanasen, Panapong Songsukthawan and Atthapol Ngaopitakkul *

Faculty of Engineering, King Mongkut's Institute of Technology Ladkrabang, Bangkok 10520, Thailand; chaiyan.je@kmitl.ac.th (C.J.); plugov82@gmail.com (P.S.)

* Correspondence: knatthap@kmitl.ac.th; Tel.: +66-23298330

Received: 21 February 2020; Accepted: 1 April 2020; Published: 7 April 2020

Abstract: This study investigates the use of an alternative energy source in the production of electric energy to meet the increasing energy requirements, encourage the use of clean energy, and thus reduce the effects of global warming. The alternative energy source used is a mechanical energy by piezoelectric material, which can convert mechanical energy into electrical energy, that can convert mechanical energy from pressure forces and vibrations during activities such as walking and traveling into electrical energy. Herein, a pilot device is designed, involving the modification of a bicycle into a stationary exercise bike with a piezoelectric generator, to study energy conversion and storage generated from using the bike. Secondly, the piezoelectric energy harvesting system is used on bicycles as a micro-mobility, light electric utility vehicle with smart operation, providing a novel approach to smart city design. The results show that the energy harvested from the piezoelectric devices can be stored in a 3200 mAh, 5 V battery and power sensors on the bicycle. Moreover, 13.6 mW power can be generated at regular cycling speed, outputting 11.5 V and 1.2 mA. Therefore, the piezoelectric energy harvesting system has sufficient potential for application as a renewable energy source that can be used with low power equipment.

Keywords: shared mobility; piezoelectric; energy harvesting; two-wheelers; smart city

---

1. Introduction

Energy demands at a global level are increasing, due to advances in technology and consumer habits that have resulted in higher energy demands both now and in the future [1]. Alternative energy technology—the use of renewable fuels—to produce electricity is a global solution because it is clean, renewable, and less polluting, thus potentially improving quality of life in the process of producing and recycling fuel. Clean energy sources are commonly used today—solar energy, waterpower, wind energy, and biomass energy [2]. Based on global issues, high energy costs are the result of not being able to meet energy demand with local supply, resulting in importing energy from neighboring countries at increased cost, raising the cost of utilities in the country, and the cost of living. Solving the energy problem is an essential part of developing a cleaner, brighter future.

The government has set energy policy to improve the energy situation in the country by developing power sources and power systems domestically and abroad. The distribution of energy sources and types is being diversified; the Thailand Power Development Plan (2015–2036) [3], aims to replace fossil fuels by at least 25% within 10 years by promoting clean energy development to reduce greenhouse gases and reduce global warming.

The cities of Thailand are trending towards the integration of technology and living in society; as a Smart City of the future, it aims to bring information technology to everyday life for convenience, security, and to raise the quality of life. Key to achieving this is an energy mix that can support various technology systems; renewable energy is essential in order to develop into a Smart City and help the country reduce its demand for imported energy.

While Thailand is advancing towards an intelligent city society, there are many innovations that can be adopted to improve living standards and realise energy goals. Bicycle travel has diminished because of the long travel time and discomfort associated with cycling when compared to an electric vehicle. Therefore, the idea to promote bicycle travel, along with the potential to generate alternative energy, has become an idea of interest in realising meaningful energy solutions and a healthy society. At present, Thailand is an aging society, with a population aged 60 years or over making up more than 10% of the population, and people aged 65 years or over making up more than 7% of the population. The public health plan regarding exercise for good health aims to achieve a healthy population that can effectively develop the country; in this light, micro-mobility has received attention for its health and sustainability benefits.

The vibration-based energy conversion termed piezoelectric energy harvesting has significant advantages when compared to other forms of renewable energy, including low start-up investment and less complex wiring. Piezoelectric energy harvesting has been identified as a candidate for low-power devices such as portable rechargeable devices [4], wireless electronic devices [5], and sensors [6]. However, this research aims to study the feasibility of electric power generation using piezoelectric materials applied to a broader scale, for example, creating a bikeway and pedestrian walkway with a power-generating floor that can be used to generate power through piezoelectric energy conversion. The relationship of piezoelectric materials, conversion of material methodology, matching the resonance frequency and physical parameters are considered to achieve the maximum harvested power [7]. There are two ways to generate electrical power from piezoelectric modules: hitting and vibrating. Generating power through vibration involves the vibrating frequency of the power source being matched with the resonance frequency of the piezoelectric module, but additional instrumentation is required to operate and control the vibration characteristics to achieve the resonance vibration. Hence, this paper focuses on using a hitting method, from an exercise bike, which involves the direct energy transfer by mechanical stress on piezoelectric modules to generate more electric energy than using vibration, owing to the fact that cycling is a low-frequency operation, it is difficult to reduce the piezoelectric natural frequency approaching the cycling frequency [8].

Based on the literature, multi-perspective analysis of the data to evaluate Smart City applications was introduced by Lau et al. [9]., it performed excellently with several smart city applications for different domains that leverage the data fusion techniques. Anjomshoaa et al. [10] presented a model of mobile sensing, that was mounted on top of garbage trucks and collected drive-by data for eight months in Cambridge, thermal irregularities and air pollutant hot spots identified. This technique presents more advantages than the conventional methods of collecting environmental data in urban areas to optimise data collection. Nizetic et al. [11] showed that smart technologies that improve energy efficiency and waste management strategies focus on the utilisation of insufficient resources. The usage of intelligent technologies supported by the IoT concept showed that potentials for improvements are essential, particularly in the smart city framework. Ismagilova et al. [12] focussed on the arrangement of smart cities with the United Nations sustainable development goals, focusing on the limitations of current developments and future trends for the Smart City concept. The technological aspects of Smart cities have been widely investigated. Some recent studies have taken a holistic Information System perspective focusing on issues such as citizens, quality of living, and sustainability. In the research of Korczak and Kijewska [13] Industrie 4.0 and Smart Logistics are studied, it can be seen that Smart Logistics provides one to distinguish the mainstream and the regulations of the further improvement of the perception of intelligent logistics, and Smart City concept. H. Haarstad and M. Wathna [14] examine the connections between Smart Cities and urban energy sustainability, and the feasibility of realising a Smart City. They noticed that city growth might obscure sustainability problems and disguise smart measures to encourage economic growth and innovation as energy sustainability measures. The micro-mobility framework in Smart City strategies are also presented in this paper. Cellina et al. [15] develop a living means in Switzerland within the framework of the Smart City, examining the effectiveness of new intelligent devices such as smartphone apps that influence low power consumption devices and have low carbon mobility to anticipate air pollution. Battarra et al. [16], implement the smart mobility model to encourage more efficient transport, convenience, sustainability, and information and communication technologies (ICT), and show that the efficiency of the transportation system has improved from this concept. The review paper from Peprah et al. [17], reports that Ghana is not a mobility-smart city; information logistic mobility, and information mobility are needed to further improve the people's mobility. This requires a technologically literate and receptive population, and investments in transport infrastructure. Lopez-Carreiro and Monzon study the development of sustainable transport networks in a Smart City [18], based on the identification of quantitative indicators that assess urban mobility through a parameter known as the Smart Mobility Index that can be used to outline appropriate transport arrangements. In research conducted by Din et al. [19], a mobility management protocol is proposed; the proposed scheme gives the most reliable performance in terms of total signalling and handover suspension from currently known protocols. In conclusion, the Smart City concept is one of the new strategies to improve the competitiveness of the country, raise social quality, increase modernity, and become a developed country.

The use of alternative energy or clean energy to improve electricity generation is a widely researched topic, as it can be used as energy for domestic use and may generate revenue for the country if the energy can be exported; furthermore, it is environmentally friendly and benefits wellness and vitality, which may considerably affect tourism businesses and health tourism. In the research of Sun et al. [20], the consequences of power electronics on grid stability, aiming to support a national energy plan emphasising the use of renewable energy, are examined. Rahbar et al. [21], studied the integration and optimisation of renewable energy sources in a microgrid and presented online algorithms for use with multiple microgrids; the simulation results confirm that algorithms perform similarly to the optimal off-line solution under several practical settings. Du et al. [22] purposed an algorithm and model focused on developing solar power technology and delivering it to the power system; the proposed technique provided flexibility to integrate compared to variable renewable energy, which leads to the installation having less renewable capacity to achieve the same level of renewable energy penetration. Zhou et al. [23], proposed the optimal scheduling of various renewable energy sources, analysing multicarrier energy supplies and applying the method to a stand-alone microgrid; the proposed scheme can reduce the battery charging/discharging operations and also confirm its capability to provide a high penetration of variable renewables. Qazi et al. [24], presented a review of renewable energy sources management in the power generation system, and highlighted the influence of public opinion. Huang et al. [25], investigated the optimal configuration for multi-renewable energy systems and considered load profiles, energy prices, and equipment parameters to create optimal consideration systems; the proposed approach can optimize both the equipment selection and the configuration of the multi-energy system. Cao et al. [26] studied polypropylene ferroelectric (PPFE) as nanogenerators that can convert mechanical energy to electrical energy for self-powering devices and energy storage systems; the maximum power output obtained around 0.902 μW. This technique can be used for powering electronic devices and a solid-state battery chip. Li et al. [27] introduced polypropylene ferroelectric (PPFE), which is a lightweight, flexible, foldable, and biocompatible device. The scavenging energy from the touching LCD screen was demonstrated, and the finite element method determining mechanical–electrical energy conversion was presented. Wan et al. [28] demonstrated a porous-structured, PDMS film-based triboelectric nanogenerator device using intrinsically stretchable materials. The output response, different loading force and frequency are determined; this material is suitable for applying self-powered wearable electronic devices. Wang et al. [29] presents a non-resonant with ferrofluid as the liquid used for vibration energy harvesting; the response shows that it is able to harvest vibration energy over a broad frequency band energy harvesting. Wang et al. [30] presents multi-layer coils for vibration-energy harvesters with a magnet array and plastic spring based on electromagnetic energy conversion; the results show that a low resistance coil with a large number of spiral coil turns contributed to increasing power generation. It can be seen that there is a research focus on clean energy technology that supports the integration of renewable energy sources to help generate electricity for energy sustainability.

The piezoelectric effect is a material property that transduces mechanical action, whether compressive or tensile force, into electricity; internal stress causes ions to re-arrange, forming an electric displacement, resulting in an electric field being created as spontaneous polarization in ferroelectrics material. Ali et al. [31] investigates piezoelectric energy harvesting in biomedical applications, i.e., energies created through muscle relaxation, body movement, blood circulation, lung and cardiac motion, discussing the limitations and prospects of the technology. Kulkarni et al. [32] introduced the application of piezoelectric energy harvesting to automotive systems and showed that fuel injectors using piezoelectric technology are more accurate than conventional fuel injectors. Chen et al. [33] investigated the application of piezoelectric energy harvesting in building structures, and review piezoelectric materials in energy harvesters, sensors, and actuators for several construction systems. Tang et al. [34], evaluated piezoelectric technology in machining processes, and proposed a method that improved the stability and performance of the system using piezoelectric patches. Xu et al. [35], studied the electromechanical conversion characteristics of piezoelectric materials applied to the pavement, showing that road energy harvesting is feasible. Yang et al. [36], reviewed piezoelectric applications with a focus on methodologies that deliver a high-power output and wide operational bandwidth. Elhalwagy et al. [37], present a feasibility study applying piezoelectric energy harvesting to building floors, and propose a method to maximise the energy generated from the energy harvesting floor. The approach to generating electrical power from vibration has been discussed by Garimella et al. [38], with the proposed method being able to generate energy from unwanted vibrations. Laumann et al. [39], reviewed piezoelectric energy harvesting, analysing the projection of the technology into the future. Yang et al. [40], did theoretical calculations and experiments on piezoelectric energy conversion from linear and non-linear vibration, with the output energy related to the phase difference between excitations and responses. Wei and Jing [41] reviewed the modelling of piezoelectric vibration energy harvesting, using polymer and ceramic compounds as piezoelectric materials. Zhang et al. [42] focused on bandwidth and non-linear resonance in the vibration of piezoelectric energy harvesting and discussed non-linear effects. Reviews on non-linear vibration control are determined by Tran et al. [43]; moreover, the enhancement technique needed to convert the ambient energy is presented by Yildirim et al. [44], and the resonance tuning method introduced, matching frequency, is essential to enhance the performance of the electromechanical conversion. Cao et al. [45] presents the transverse piezoelectric effects of polypropylene ferroelectric to perform a self-powered and water-resistant bending sensor by analyzing electrical characteristics under the various bending conditions, where linear dependency on voltage and inverse square root dependency on power are determined. Cao et al. [46] performed the fundamental working principles of dipole moments in ferroelectric polymers to represent the energy conversion mechanism. The energy transmission was significantly affected by the internal impedance of instruments and instrument sampling rate.

From the literature review, it was found that there are currently alternative energy sources that are interesting and can be used to harvest ambient energy to be transformed into usable electrical energy. Therefore, this paper presents the use of piezoelectric material to accumulate energy from exercise in order to support energy conservation in a healthy society. The experiment was carried out by installing energy harvesting systems using piezoelectric materials on a stationary exercise bike system to study the feasibility of harvesting electric energy from exercise equipment, and further development of the system for installation on a moving bicycle. The study of electrical characteristics and the design of energy management circuits that are sufficient for the required load is presented in this article for the benefit of further development of the system to be more efficient and to be applied in several applications in the future.

2. Energy Harvesting from a Commuter Bicycle

In this section, the piezoelectric energy harvested from using an exercise bike is examined. The piezoelectric material used in the study is PZT-5H, PZT one of the most popular and commercial synthetic piezoelectric-ceramics owing to its atomic structure resulting in high relative electric displacement and permittivity, capable of working at a high Curie temperature. Moreover, PZT-5H has a better performance over other piezoelectric-ceramics at a comparable price. During its operation the piezoelectric apparatus converts mechanical energy into electrical energy, through an electromotive force created by a magnet mounted on the wheel, which causes a re-orientation of the corresponding magnets due to the polarity alignment causing repulsion between the magnets that are positioned on the tip of the piezoelectric device, as shown in Figure 1. The electric power is generated from both the thickness and shear modes of operation. However, here the emphasis is on the thickness mode because the installation of the piezoelectric devices and magnets, causing the tip displacement along the thickness axis.

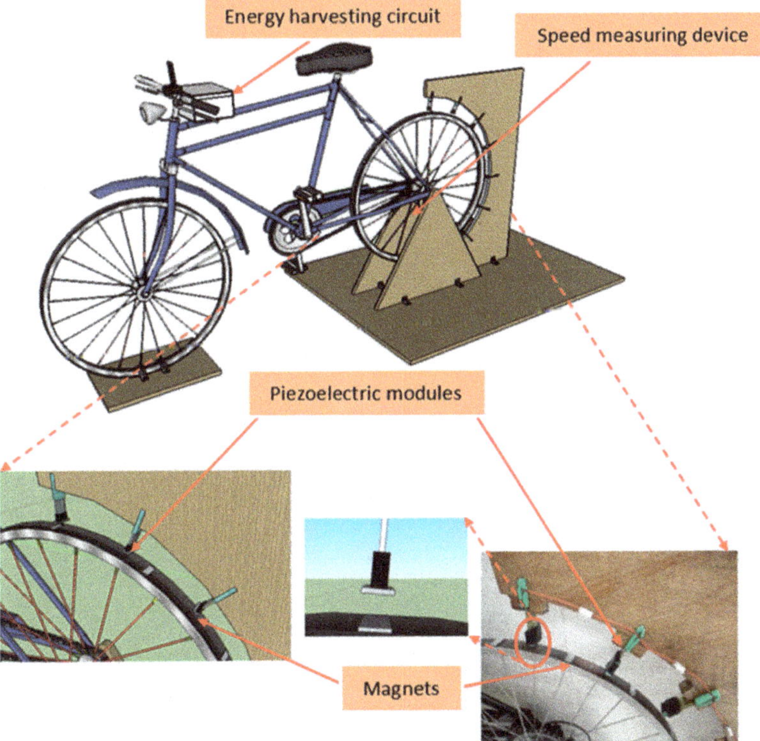

Figure 1. Schematic of exercise bike implemented with energy harvesting system using a piezoelectric material.

The tip displacement of the piezoelectric modules causes a re-arrangement of the internal dipoles and the formation of an electric potential difference between the two sides, resulting in a current flow through the material, forming electrical energy. This electrical energy can be harvested through an energy harvesting circuit and stored in capacitors and batteries for later use. Allowing the energy to be used in low power devices such as sensors, Figure 1 presents the concept of energy harvesting from an exercise bike.

The overall energy harvesting flow chart is shown in Figure 2. A rectifier circuit converts the electrical signal from AC to DC, with a capacitor arrangement employed to reduce the ripple voltage. A voltage regulator circuit is used to reduce and maintain a constant voltage that connects to two batteries; the two batteries work as energy collectors and supply power to the load. Finally, a voltage converter increases or decreases the voltage level to meet load requirements.

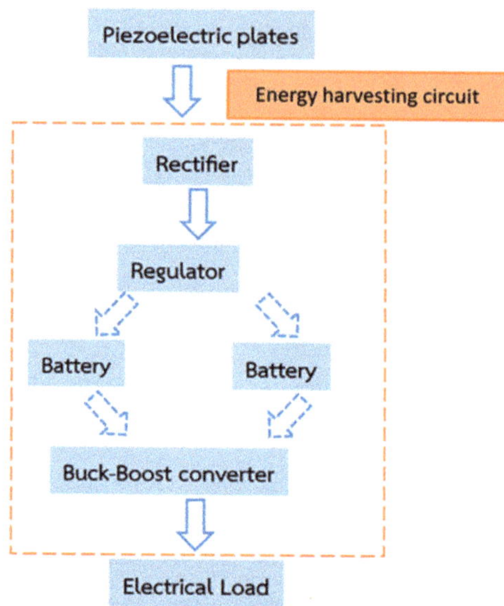

Figure 2. The Piezoelectric energy harvesting circuit flow chart.

Voltage signals generated from the piezoelectric module were investigated by installing piezoelectric devices in varied configurations, from one to four modules with series and parallel circuitry configurations. Experiments were conducted at a cycling speed of 75 and 120 revolutions per minute (rpm). The open circuit voltage signals collected from the oscilloscope are seen in Figure 3, with measured electrical quantities presented in Table 1.

Table 1. Summary of experimental results of electrical quantities from the piezoelectric devices.

Configurations		Average Speed					
		75 rpm			120 rpm		
		Open Circuit Voltage (V)	Current (mA)	Power (μW)	Open Circuit Voltage (V)	Current (mA)	Power (μW)
Base case	One module	14.4	0.09	8.10	22.5	0.10	10.10
Series	Two modules	24.1	0.10	10.10	30.3	0.11	12.10
	Three modules	38.9	0.12	14.40	44.1	0.14	19.60
	Four modules	45.6	0.13	16.90	48.7	0.15	22.50
Parallel	Two modules	17.1	0.19	36.10	23.6	0.23	52.90
	Three modules	17.3	0.28	78.40	24.2	0.33	108.90
	Four modules	17.6	0.37	136.90	26.4	0.42	176.40

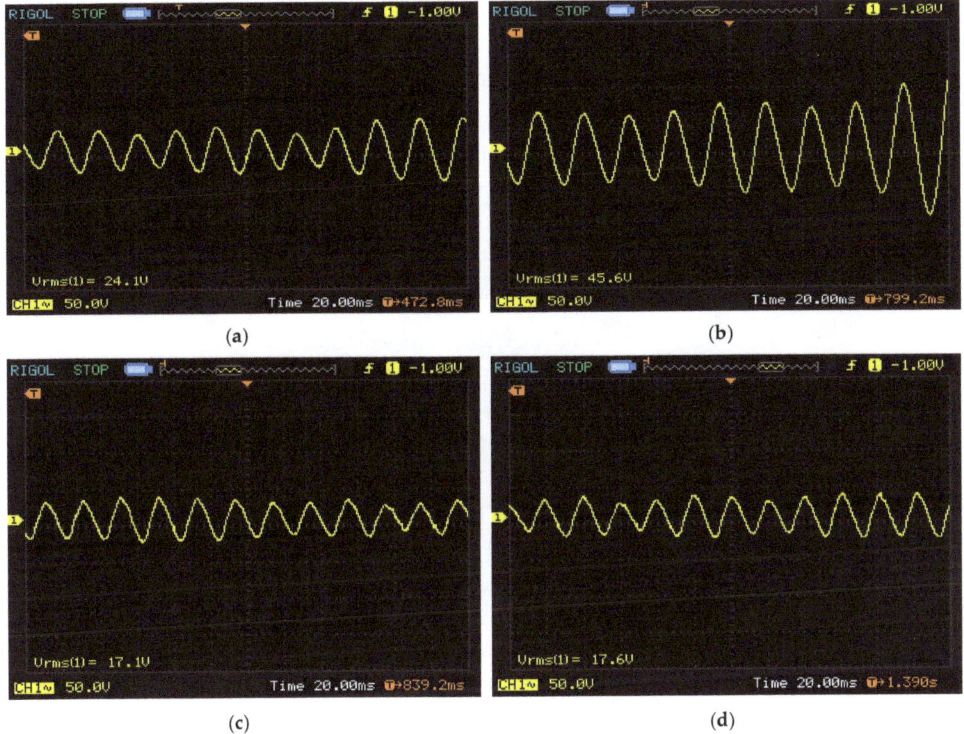

Figure 3. The voltage signal from the various configurations of piezoelectric modules at average speeds 75 rpm. (**a**) Two piezoelectric modules connected in series; (**b**) Four piezoelectric modules connected in series; (**c**) Two piezoelectric modules connected in parallel; (**d**) Four piezoelectric modules connected in parallel

The pre-testing experiment excited the piezoelectric with direct force in order to study the circuit configuration that is suitable for installation in the subsequent investigations among two different speeds. The opened circuit voltage and the current when connected one kilohm resistance are measured, the output is calculated. From the experimental result, it can be seen that an average speed of 120 rpm had higher power output than 75 rpm because the piezoelectric module oscillated a greater vibration amplitude. At the same speed, the parallel configuration has a higher current and output power than the series configuration with constant voltage. Therefore, the parallel piezoelectric coupling was chosen to harvest energy from the exercise bike.

In the design of the rectifier circuit, the approximate input voltage from piezoelectric is set to 30 V; from this, the single-phase bridge rectifier KBP202G, capacitor (C1, C2, C3 = 0.1 μF) values, and positive voltage regulator L7815CV are selected to perform the energy harvesting circuit; the output voltage is equal to 15 V. The two cells are lithium-ion batteries, operating at 3.7 V, with a capacity of 2850 mAh. The energy harvesting circuit and schematic is shown in Figure 4.

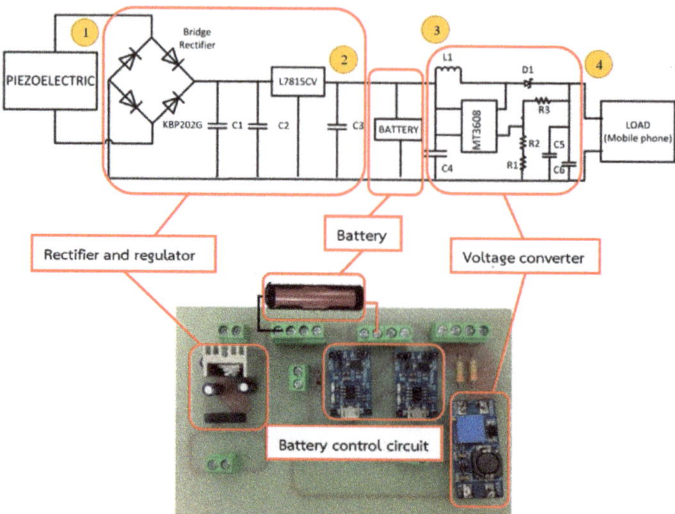

Figure 4. Piezoelectric energy harvesting circuit and measurement position.

Two cells, batteries A and B, are collecting energy and supplying the load alternatively, controlled by an open-source electronics platform, Arduino, set to interlocking mode. When the two cells are charging at the same time, the load consumes more current, resulting in rising heat loss and reduced battery life. The controller performs the following actions:

1. If battery A has less than or equal to 50% capacity, and battery B has a more than 50% capacity, Arduino commands battery A to store energy from the piezoelectric and battery B serves the load;
2. If the battery A has more than 50% capacity, and the battery B has less than or equal to 50% capacity, Arduino orders the battery A to supply the load and battery B harvest energy from the piezoelectric;
3. If the batteries A and B have less than or equal to 50% capacity, Arduino commands battery A to store energy from the piezoelectric;
4. If batteries A and B have more than 50% capacity, Arduino orders battery A to supply load.

The speed measuring device uses a magnetic switch as a signal detector, with the switch consisting of two magnets positioned on the steel beam near the rear wheels, and the spokes of the rear wheels, as shown in Figure 5. When cycling, the magnet attached to the wheel spokes moves through the other magnets mounted on the steel beam. Arduino calculates and shows the speed on the LCD display mounted on the bars.

The electrical quantities were measured at three cycling speeds: 100, 120, and 140 rpm, with the measurement points shown in Figure 6. Point 1 is the AC voltage from the eight piezoelectric energy harvesters, connected in parallel. Point 2 is the DC voltage after the voltage regulation circuit. Point 3 is the DC voltage behind the battery. Point 4 is the DC voltage after the converter. The magnetic distance between the bike wheels and the piezoelectric devices was varied with different spacings, i.e., 6, 9 and 12 mm, and the results are shown in Table 2.

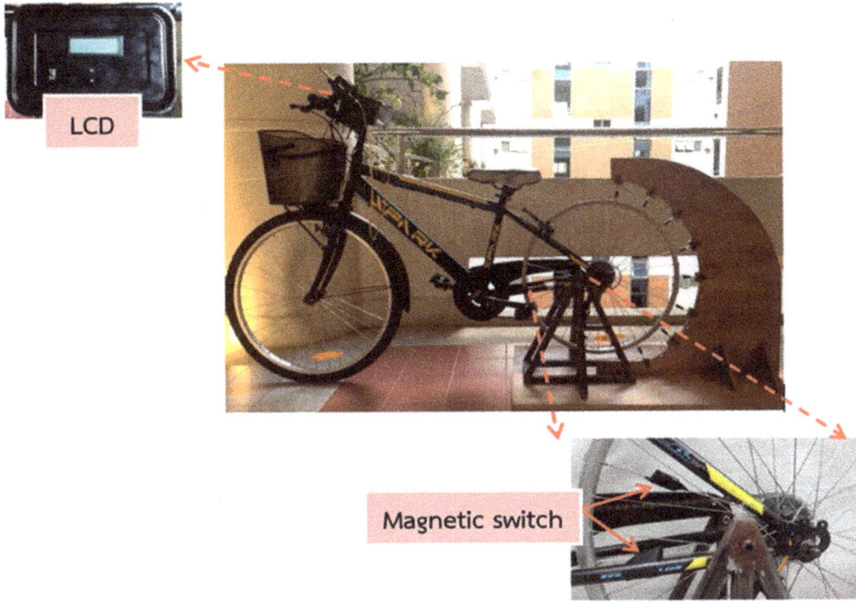

Figure 5. The speed measuring devices mounted on energy harvesting exercise bicycle.

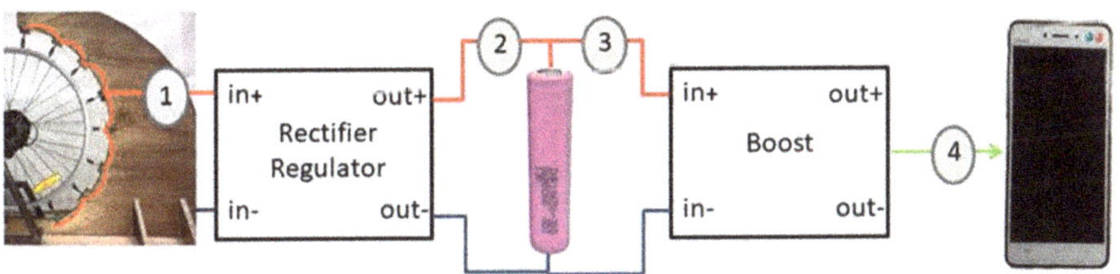

Figure 6. Measurement points of electrical quantities within the energy harvesting circuit.

Table 2. Experimentally measured electrical quantities at various cycling speeds at magnetic distances of 6, 9 and 12 mm.

Measurement Points		Magnetic Distance 6 mm			Magnetic Distance 9 mm			Magnetic Distance 12 mm		
		100 rpm	120 rpm	140 rpm	100 rpm	120 rpm	140 rpm	100 rpm	120 rpm	140 rpm
Point 1	Voltage (V)	5.67	6.03	5.11	5.20	6.09	5.15	5.25	5.69	5.14
	Current (mA)	18	20	14	14	20	15	14	18	15
	Power (mW)	102.1	120.6	71.54	72.8	121.8	77.25	73.5	102.4	77.1
Point 2	Voltage (V)	3.76	4.16	3.92	3.52	4.20	3.20	3.36	4.16	3.20
	Current (mA)	8	9	8	8	9	8	8	9	8
	Power (mW)	30.08	37.44	31.36	28.16	37.8	25.6	26.88	37.44	25.6
Point 3	Voltage (V)	3.92	3.92	3.92	3.92	3.92	3.92	3.92	3.92	3.92
	Current (mA)	88	88	88	88	88	88	88	88	88
	Power (mW)	345	345	345	345	345	345	345	345	345
Point 4	Voltage (V)	5.68	5.68	5.68	5.68	5.68	5.68	5.68	5.68	5.68
	Current (mA)	88	88	88	88	88	88	88	88	88
	Power (mW)	500	500	500	500	500	500	500	500	500

From the results it can be seen that at each magnetic distance, the piezoelectric apparatus generated comparable voltage output. A magnet spacing of 12 mm is chosen, as this length produced the highest magnetic force and provided clearance to protect the piezoelectric material from collision during operation.

Furthermore, at speeds of 100 and 140 rpm it was observed that the harvested energy was not being stored efficiently as battery charge. However, at a speed of 120 rpm the maximum AC voltage was observed, and the energy was stored in the battery efficiently. This is due to the optimized magnetizing force, which allows the piezoelectric modules to have the appropriate vibrating amplitude and frequency corresponding to the natural frequency of the piezoelectric material, resulting in more efficient power transfer.

3. Micro-Mobility Based on Piezoelectric Energy Harvesting

In this section, energy from the vibration force while cycling is used as the mechanical energy to generate AC in the piezoelectric material and harvested; this is achieved through an exerted electromotive force by a magnet configuration mounted to the wheels. A bicycle has been modified and developed to harvest energy; the apparatus consists of a piezoelectric circuit that is mounted on the front wheel hub and the back-wheel mudguard, a 3D-printed magnet apparatus, detection sensors and a control box in the bicycle basket. The harvesting circuit was controlled by Arduino ESP 32 Wi-Fi module and controlled via mobile application. The energy is stored in a 3200 mAh, 5 V battery in order to supply power to sensors mounted on the bicycle including a humidity and temperature sensor, ultrasonic sensor, hazardous gas detection sensor and a relay to control a 12 V light emitting diode (LED) at the front-wheel hub.

Two piezoelectric installation areas were selected on the bike: the front wheel hub, and the bike fender. The circuit control box is attached to the basket area, and contains the circuitry used to harvest energy from the piezoelectric, this includes: a charger circuit, Arduino Model ESP 32, two batteries, and a boost converter to supply LED load. The front wheel hub of the bicycle consists of eight devices, while the rear wheel mudguard was mounted with seven devices on each side. They are fastened by a bioplastic material, and a magnet attached near to the piezoelectric module imparts momentum on the piezoelectric material when the wheel spins.

The front wheel hub of the bicycle is designed to house eight piezoelectric devices on each side. At the tip of each piezoelectric device is a magnetic bar to generate an electromotive force on the other four magnets, placed on the opposite side at the inner wheel, causing the piezoelectric material to vibrate during cycling. The energy harvesting apparatus on the front wheel uses sixteen piezoelectric devices and twenty-four magnets in total. Moreover, the bike's rear-wheel fender area is designed to attach magnets to seven piezoelectric devices on each side, with a distance of 1.3 cm between the opposite two magnets on each side of the moving wheel. In total, eighteen magnets and fourteen piezoelectric devices are mounted on the rear wheel area, as seen in Figure 7.

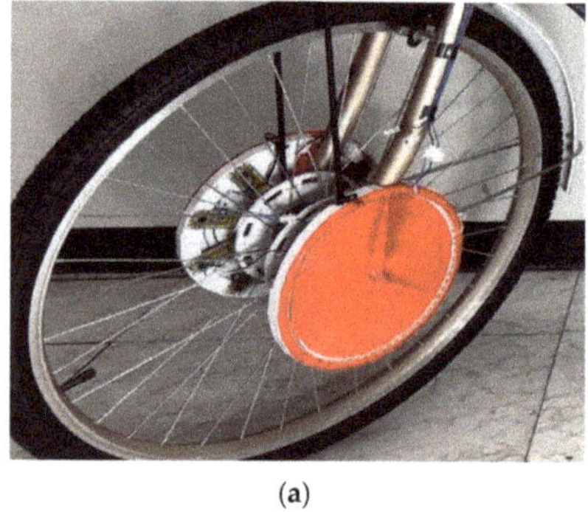

(**a**)

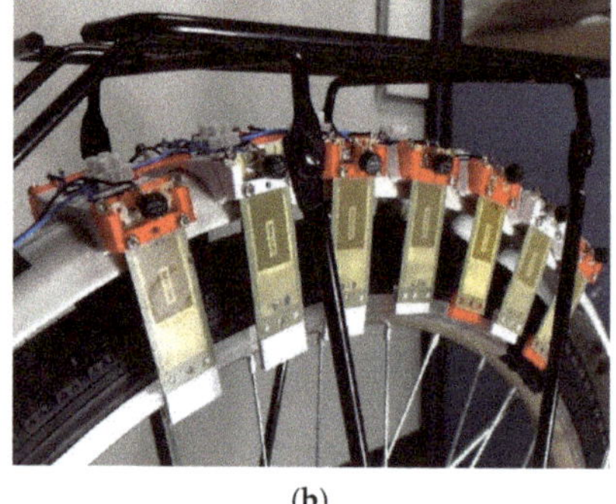

(**b**)

Figure 7. The actual installation of piezoelectric modules on a bicycle: (**a**) Piezoelectric modules installation distance at front wheel; (**b**) Piezoelectric modules installation distance at rear wheel fenders.

Experimental voltages were measured for a distance between the magnet and the piezo pad at the front wheel hub of 2.2 cm, and a rear wheel center area of 1.3 cm, by continuously cycling for 3 min and measuring the instantaneous voltage. The spin speed was varied between 30, 60, and 90 rpm, and the average voltage, current, and power are measured at four locations in the circuit: DC voltage after the rectifier circuit, behind the capacitor, behind the battery, and after the boost converter. The measurement point is different from the initial experiment and is shown in Figure 8, and observed electrical signals are displayed in Figure 9.

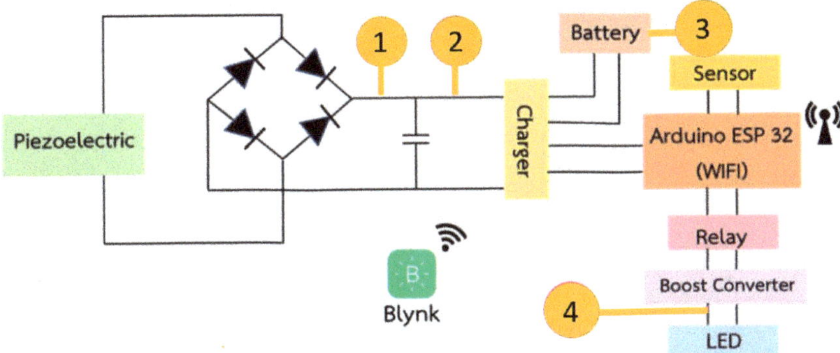

Figure 8. The measurement points in bicycle experiment.

When measuring instantaneous voltage, the maximum voltage range at each speed is the result of the repulsion of the magnets, and it is seen to gradually decrease until the next magnet exerts an additional force. From the acquired signal, the voltage is continuously increasing and decreasing, with each range being unstable due to the ringing effect of the second-order system behavior, and the damping of the system in this case. When measuring at the capacitor, a steadier signal is obtained. Although the average voltage is higher, there is still an unstable range, possibly the result of unstable cycling force and magnetic contact range. The experimental results of each speed in all measurement areas are summarised in Table 3.

Table 3. Electrical quantities obtained from cycling at various speeds

Measurement Points		Cycling Speed (rpm)		
		30	60	90
Point 1	Voltage (V)	1.93	2.81	2.62
	Current (mA)	0.37	1.72	1.79
	Power (mW)	0.71	4.83	4.69
Point 2	Voltage (V)	2.07	7.59	7.04
	Current (mA)	0.33	0.63	0.63
	Power (mW)	0.68	4.78	4.56
Point 3	Voltage (V)	3.80	3.80	3.80
	Current (mA)	1.46	1.46	1.46
	Power (mW)	5.55	5.55	5.55
Point 4	Voltage (V)	10.00	10.00	10.00
	Current (mA)	0.50	0.50	0.50
	Power (mW)	5.00	5.00	5.00

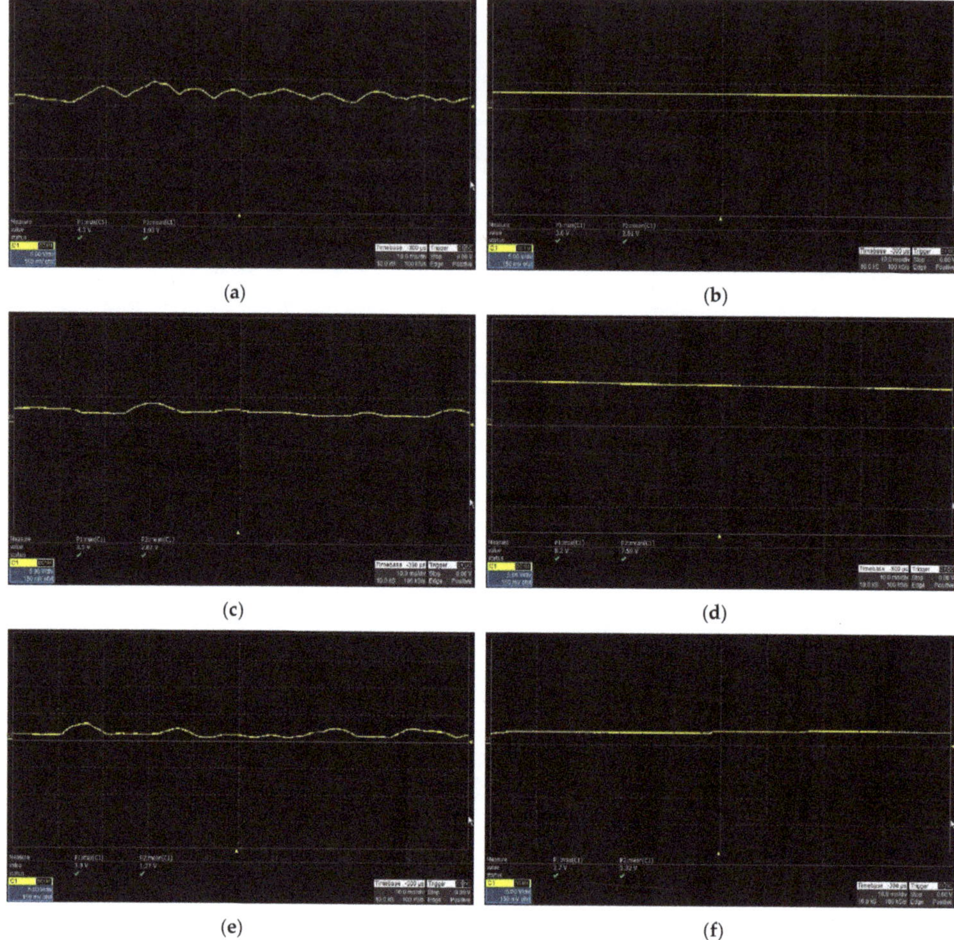

Figure 9. The DC voltage signal at position 1 and position 2 at various speeds: (**a**) Position 1 at the speed of 30 rpm; (**b**) Position 2 at the speed of 30 rpm; (**c**) Position 1 at the speed of 60 rpm; (**d**) Position 2 at the speed of 60 rpm; (**e**) Position 1 at the speed of 90 rpm; (**f**) Position 2 at the speed of 90 rpm.

The small power production in the 90 rpm because the magnetic force exerted on the piezoelectric material is over too short a duration for the material to produce electrical energy at the appropriate frequency range. At 30 rpm, there is less power generation due to there being fewer vibrations in the frequency range that can produce power. Point 3 and point 4 are the areas for connecting the battery and the voltage circuit. Therefore, the resulting value is constant at all revolutions.

The influence of road surface on power generation was examined. Experiments to obtain combined voltage values of the front and back wheels when cycling on three different road surfaces at 30, 60 and 90 rpm were conducted by measuring the electrical quantities behind the capacitor under three different road surfaces using a cyclist load weight of 70 kg; the results are shown in Table 4. The three different road surfaces are: road surface type 1—smooth road surface; road surface type 2—moderately rough road surface featuring gravel; and road surface type 3—rough road surface featuring rocks and gravel.

Table 4. The electrical quantities obtained from cycling at various speeds on various road surfaces.

Measurement Conditions		Cycling Speed (rpm)		
		30	60	90
Smooth road surface	Voltage (V)	8.54	11.44	9.73
	Current (mA)	0.88	1.15	0.95
	Power (mW)	7.53	13.11	9.20
Moderately rough road surface	Voltage (V)	10.69	11.49	11.49
	Current (mA)	0.86	1.18	0.98
	Power (mW)	9.21	13.59	11.21
Rough road surface	Voltage (V)	10.46	12.92	10.74
	Current (mA)	0.92	1.17	1.25
	Power (mW)	9.67	15.08	13.38

From the results shown in the table, it can be seen that the road surface plays a role in the energy harvested during cycling. The maximum power was obtained when cycling in areas where the road surface is rough. As a result of the vibrating force of the magnets on the front wheels and rear wheel fenders, the front wheel area experiences greater vibrational force during the impact on rough surfaces, allowing more energy to be harvested. A rough road surface was seen to produce the most power compared to a moderately rough and smooth road surface, respectively.

The energy harvesting results obtained under different load weight are obtained and analysed. Values were obtained from both the front and back wheels, using different rider load weights to see the effect on electric power production. Cycling speeds of 30, 60, and 90 rpm are compared for three weight ranges 40–50 kg, 50–65 kg, and 65–80 kg, by measuring electrical quantities behind the capacitor, on a smooth road surface. The results are shown in Table 5.

Table 5. Electrical quantities obtained from cycling at various speeds on various weight ranges.

Measurement Conditions		Cycling Speed (rpm)		
		30	60	90
Weight 40–50 kg.	Voltage (V)	8.54	11.49	9.73
	Current (mA)	0.88	1.18	0.95
	Power (mW)	7.53	13.59	9.20
Weight 50–65 kg.	Voltage (V)	8.21	12.09	9.62
	Current (mA)	0.92	1.01	0.96
	Power (mW)	7.57	12.19	9.24
Weight 65–80 kg.	Voltage (V)	8.79	11.79	9.39
	Current (mA)	0.79	1.06	0.73
	Power (mW)	6.91	12.51	6.92

From the results in Table 5, it can be that, in the case of all three rider load weights, with the same cycling speed, the power produced in each range is similar. Therefore, it can be concluded that the weight of the load has a minimal effect on the power that can be produced. The differing values may be due to spinning at uneven speeds. It was seen that the speed range that can produce the most power is 60 rpm, consistent with earlier results.

4. Conclusions

In this paper, an exercise bike has been modified by mounting a piezoelectric material apparatus on the frame, to be used as an electromechanical converter, converting the energy generated by exercise by-products into electrical energy. The installation and configuration of the piezoelectric harvesting apparatus were examined, and the location and distance of permanent magnets were investigated, to generate the maximum output energy.

From the results of the experiment adapting the energy harvesting system with piezoelectric in the first part of the experiment, a system with a moving exercise bike was developed. The efficiency of energy harvesting through the harvesting circuit was examined, with electrical measurements taken at four different positions in the circuit: the DC voltage after converter circuit, behind the capacitor, behind the battery, and after the boost converter circuit. The distance between the magnet attached to the piezoelectric plate and the magnet attached to the wheel frame was 2.2 cm for the front wheel, and 1.3 cm between the rear wheel fenders. The effect of the road surface on power generation showed that cycling on rough surfaces harvested the most energy, with more power being generated because of the additional shaking of the piezoelectric device. The influence of cyclist load weight on power generation was examined and was seen to not affect the energy harvested.

There is scope to improve the energy harvesting of this piezoelectric apparatus, through the addition of more piezoelectric modules or by increasing the magnet distance to the bicycle structure, or moving the magnets closer to each other; improving the energy harvested through the further optimization of physical parameters and density of piezoelectric modules is an area of future investigation. The results have shown promise for Internet of Things applications, in a more extensive system, that is able to support wireless device connection, self-powered micro-mobility devices are capable of generating significant data, and are an exciting prospect for Smart City design.

Author Contributions: C.J. conceptualization, formal analysis and design the experiment. P.S. performed the experiments, investigation and analyzed the data; C.J. funding acquisition, and A.N. contributed resource; C.J. wrote the paper, and A.N review and editing. All authors have read and agreed to the published version of the manuscript.

Funding: This research was funded by KING MONGKUT'S INSTITUTE OF TECHNOLOGY LADKRABANG RESEARCH FUND, THAILAND, grant number No. KREF046106.

Acknowledgments: The authors wish to gratefully acknowledge financial support for this research (No. KREF046106) from King Mongkut's Institute of Technology Ladkrabang Research fund, Thailand.

Conflicts of Interest: The authors declare no conflict of interest.

References

1. Mamat, R.; Sani, M.S.; Khoerunnisa, F.; Kadarohman, A. Target and demand for renewable energy across 10 ASEAN countries by 2040. *Electr. J.* **2019**, *32*, 1066–1070. [CrossRef]
2. Hongtao, L.; Wenjia, L. The analysis of effects of clean energy power generation. *Energy Procedia* **2018**, *152*, 947–952. [CrossRef]
3. Thailand Power Development Plan 2015–2036. Available online: https://www.egat.co.th/en/about-egat/strategy-policy/policy (accessed on 5 November 2019).
4. Sarker, M.R.; Julai, S.; Sabri, M.F.; Said, S.M.; Islam, M.M.; Tahir, M. Review of piezoelectric energy harvesting system and application of optimization techniques to enhance the performance of the harvesting system. *Sen. Actuators A Phys.* **2019**, *300*, 1116–1134. [CrossRef]
5. Chalioris, C.E.; Karayannis, C.G.; Angeli, G.M.; Papadopoulos, N.A.; Favvata, M.J.; Providakis, C.P. Applications of smart piezoelectric materials in a wireless admittance monitoring system (WiAMS) to Structures—Tests in RC elements. *Case Stud. Constr. Mater.* **2016**, *5*, 1–18. [CrossRef]
6. Micek, P.; Grzybek, D. Wireless stress sensor based on piezoelectric energy harvesting for a rotating shaft. *Sen. Actuators A Phys.* **2019**, *301*, 1117–1144. [CrossRef]
7. Bhuyan, S.; Sivanand, K.; Panda, S.K.; Kumar, R.; Hu, J. Resonance-Based Wireless Energizing of Piezoelectric Components. *IEEE Magn. Lett.* **2011**, *2*, 6000204. [CrossRef]
8. Dong, H.; Wu, J.; Zhang, H.; Zhang, G. Measurement of a Piezoelectric Transducer's Mechanical Resonant Frequency Based on Residual Vibration Signals. In Proceedings of the 2010 IEEE International Conference on Information and Automation, Harbin, China, 20–23 June 2010; pp. 1872–1876.
9. Lau, B.P.L.; Marakkalage, S.H.; Zhou, Y.; Hassan, N.U.; Yuen, C.; Zhang, M.; Tan, U.X. A survey of data fusion in smart city applications. *Inf. Fusion* **2019**, *52*, 357–374. [CrossRef]

10. Anjomshoaa, A.; Duarte, F.; Rennings, D.; Matarazzo, T.J.; de Souza, P.; Ratti, C. City Scanner: Building and Scheduling a Mobile Sensing Platform for Smart City Services. *IEEE Internet Things J.* **2018**, *5*, 4567–4579. [CrossRef]
11. Nižetić, S.; Djilali, N.; Papadopoulos, A.; Rodrigues, J.J. Smart technologies for promotion of energy efficiency, utilization of sustainable resources and waste management. *J. Clean. Prod.* **2019**, *231*, 565–591. [CrossRef]
12. Ismagilova, E.; Hughes, L.; Dwivedi, Y.K.; Raman, K.R. Smart cities: Advances in research—An information systems perspective. *Int. J. Inf. Manag.* **2019**, *47*, 88–100. [CrossRef]
13. Korczak, J.; Kijewska, K. Smart Logistics in the development of Smart Cities. *Transp. Res. Procedia* **2019**, *39*, 201–211. [CrossRef]
14. Haarstad, H.; Wathne, M.W. Are smart city projects catalyzing urban energy sustainability? *Energy Policy* **2019**, *129*, 918–925. [CrossRef]
15. Cellina, F.; Castri, R.; Simão, J.V.; Granato, P. Co-creating app-based policy measures for mobility behavior change: A trigger for novel governance practices at the urban level. *Sustain. Cities Soc.* **2020**, *53*, 1011–1019. [CrossRef]
16. Battarra, R.; Gargiulo, C.; Tremiterra, M.R.; Zucaro, F. Smart mobility in Italian metropolitan cities: A comparative analysis through indicators and actions. *Sustain. Cities Soc.* **2018**, *41*, 556–567. [CrossRef]
17. Peprah, C.; Amponsah, O.; Oduro, C. A system view of smart mobility and its implications for Ghanaian cities. *Sustain. Cities Soc.* **2019**, *44*, 739–747. [CrossRef]
18. Carreiro, I.L.; Monzon, A. Evaluating sustainability and innovation of mobility patterns in Spanish cities. Analysis by size and urban typology. *Sustain. Cities Soc.* **2018**, *38*, 684–696. [CrossRef]
19. Din, S.; Paul, A.; Hong, W.H.; Seo, H. Constrained application for mobility management using embedded devices in the Internet of Things based urban planning in smart cities. *Sustain. Cities Soc.* **2019**, *44*, 144–151. [CrossRef]
20. Sun, J.; Li, M.; Zhang, Z.; Xu, T.; He, J.; Wang, H.; Li, G. Renewable energy transmission by HVDC across the continent: System challenges and opportunities. *CSEE J. Power Energy Syst.* **2017**, *3*, 353–364. [CrossRef]
21. Rahbar, K.; Chai, C.C.; Zhang, R. Energy Cooperation Optimization in Microgrids with Renewable Energy Integration. *IEEE Trans. Smart Grid* **2018**, *9*, 1482–1493. [CrossRef]
22. Du, E.; Zhang, N.; Hodge, B.M.; Wang, Q.; Kang, C.; Kroposki, B.; Xia, Q. The Role of Concentrating Solar Power Toward High Renewable Energy Penetrated Power Systems. *IEEE Trans. Power Syst.* **2018**, *33*, 6630–6641. [CrossRef]
23. Zhou, B.; Xu, D.; Li, C.; Chung, C.Y.; Cao, Y.; Chan, K.W.; Wu, Q. Optimal Scheduling of Biogas–Solar–Wind Renewable Portfolio for Multicarrier Energy Supplies. *IEEE Trans. Power Syst.* **2018**, *33*, 6229–6239. [CrossRef]
24. Qazi, A.; Hussain, F.; Rahim, N.A.; Hardaker, G.; Alghazzawi, D.; Shaban, K.; Haruna, K. Towards Sustainable Energy: A Systematic Review of Renewable Energy Sources, Technologies, and Public Opinions. *IEEE Access* **2019**, *7*, 63837–63851. [CrossRef]
25. Huang, W.; Zhang, N.; Yang, J.; Wang, Y.; Kang, C. Optimal Configuration Planning of Multi-Energy Systems Considering Distributed Renewable Energy. *IEEE Trans. Smart Grid* **2019**, *10*, 1452–1464. [CrossRef]
26. Cao, Y.; Figueroa, J.; Pastrana, J.J.; Li, W.; Chen, Z.; Wang, Z.L.; Sepúlveda, N. Flexible ferroelectret polymer for self-powering devices and energy storage systems. *ACS Appl. Mater. Interfaces* **2019**, *11*, 17400–17409. [CrossRef]
27. Li, W.; Torres, D.; Wang, T.; Wang, C.; Sepúlveda, N. Flexible and biocompatible polypropylene ferroelectret nanogenerator (FENG): On the path toward wearable devices powered by human motion. *Nano Energy* **2016**, *30*, 649–657. [CrossRef]
28. Wan, H.; Cao, Y.; Lo, L.W.; Xu, Z.; Sepúlveda, N.; Wang, C. Screen-printed soft triboelectric nanogenerator with porous PDMS and stretchable PEDOT: PSS electrode. *J. Semicond.* **2019**, *40*, 112601. [CrossRef]
29. Wang, Y.; Zhang, Q.; Zhao, L.; Kim, E.S. Non-Resonant, Electromagnetic Broad-Band Vibration-Energy Harvester Based on Self-Assembled Ferrofluid Liquid Bearing. *IEEE/ASME J. Microelectromech. Syst.* **2017**, *26*, 809–819. [CrossRef]
30. Zhang, Q.; Wang, Y.; Zhao, L.; Kim, E.S. Integration of microfabricated low resistance and thousand-turn coils for vibration energy harvesting. *J. Micromech. Microeng.* **2016**, *26*. [CrossRef]
31. Ali, F.; Raza, W.; Li, X.; Gul, H.; Kim, K.H. Piezoelectric energy harvesters for biomedical applications. *Nano Energy* **2019**, *57*, 879–902. [CrossRef]

32. Kulkarni, H.; Zohaib, K.; Khusru, A.; Aiyappa, K.S. Application of piezoelectric technology in automotive systems. *Mater. Today Proc.* **2018**, *5*, 21299–21304. [CrossRef]
33. Chen, J.; Qiu, Q.; Han, Y.; Lau, D. Piezoelectric materials for sustainable building structures: Fundamentals and applications. *Renew. Sustain. Energy Rev.* **2019**, *101*, 14–25. [CrossRef]
34. Tang, B.; Akbari, H.; Pouya, M.; Pashaki, P.V. Application of piezoelectric patches for chatter suppression in machining processes. *Measurement* **2019**, *138*, 225–231. [CrossRef]
35. Xu, X.; Cao, D.; Yang, H.; He, M. Application of piezoelectric transducer in energy harvesting in pavement. *Int. J. Pavement Res. Technol.* **2018**, *11*, 388–395. [CrossRef]
36. Yang, Z.; Zhou, S.; Zu, J.; Inman, D. High-Performance Piezoelectric Energy Harvesters and Their Applications. *Joule* **2018**, *2*, 642–697. [CrossRef]
37. Elhalwagy, A.M.; Ghoneem, M.Y.M.; Elhadidi, M. Feasibility Study for Using Piezoelectric Energy Harvesting Floor in Buildings' Interior Spaces. *Energy Procedia* **2017**, *115*, 114–126. [CrossRef]
38. Garimella, R.C.; Sastry, V.R.; Mohiuddin, M.S. Piezo-Gen—An Approach to Generate Electricity from Vibrations. *Procedia Earth Planet. Sci.* **2015**, *11*, 445–456. [CrossRef]
39. Laumann, F.; Sørensen, M.M.; Lindemann, R.F.J.; Hansen, T.M.; Tambo, T. Energy harvesting through piezoelectricity—Technology foresight. *Energy Procedia* **2017**, *142*, 3062–3068. [CrossRef]
40. Yang, Z.; Erturk, A.; Zu, J. On the efficiency of piezoelectric energy harvesters. *Extrem. Mecha. Lett.* **2017**, *15*, 26–37. [CrossRef]
41. Wei, C.; Jing, X. A comprehensive review on vibration energy harvesting: Modelling and realization. *Renew. Sustain. Energy Rev.* **2017**, *74*, 1–18. [CrossRef]
42. Zhang, H.; Corr, L.R.; Ma, T. Issues in vibration energy harvesting. *J. Sound Vib.* **2018**, *421*, 79–90. [CrossRef]
43. Tran, N.; Ghayesh, M.H.; Arjomandi, M. Ambient vibration energy harvesters: A review on nonlinear techniques for performance enhancement. *Int. J. Eng. Sci.* **2018**, *127*, 162–185. [CrossRef]
44. Yildirim, T.; Ghayesh, M.H.; Li, W.; Alici, G. A review on performance enhancement techniques for ambient vibration energy harvesters. *Renew. Sustain. Energy Rev.* **2017**, *71*, 435–449. [CrossRef]
45. Cao, Y.; Figueroa, J.; Li, W.; Chen, Z.; Wang, Z.L.; Sepúlveda, N. Understanding the dynamic response in ferroelectret nanogenerators to enable self-powered tactile systems and human-controlled micro-robots. *Nano Energy* **2019**, *63*. [CrossRef]
46. Cao, Y.; Li, W.; Sepúlveda, N. Performance of Self-Powered, Water-Resistant Bending Sensor Using Transverse Piezoelectric Effect of Polypropylene Ferroelectret Polymer. *IEEE Sen. J.* **2019**, *19*, 10327–10335. [CrossRef]

Review

Reviewing Truck Logistics: Solutions for Achieving Low Emission Road Freight Transport

Tommi Inkinen * and Esa Hämäläinen

Brahea Centre, University of Turku, 20014 Turku, Finland; esa.hamalainen@utu.fi
* Correspondence: tommi.inkinen@utu.fi; Tel.: +358-50-313-0689

Received: 31 July 2020; Accepted: 18 August 2020; Published: 19 August 2020

Abstract: Low emission logistics have become an expected and desired goal in all fields of transportation, particularly in the European Union. Heavy-duty trucks (HDTs) are significant producers of emissions and pollution in inland transports. Their role is significant, as in multimodal transport chains truck transportation is, in most cases, the only viable solution to connect hinterlands with ports. Diesel engines are the main power source of trucks and their emission efficiency is the key challenge in environmentally sound freight transportation. This review paper addresses the academic literature focusing on truck emissions. The paper relies on the preliminary hypothesis that simple single solutions are nonexistent and that there will be a collection of suggestions and solutions for improving the emission efficiency in trucks. The paper focuses on the technical properties, emission types, and fuel solutions used in freight logistics. Truck manufacturing, maintenance, and other indirect emissions like construction of road infrastructure have been excluded from this review.

Keywords: research review; trucks; logistics; emission; regulations

1. Introduction

Global pressures to decrease airborne emissions in all industrial operations are increasing. This includes freight logistics that are heavily reliant on trucks. Heavy-duty trucks (HDTs) are used in long distance hinterland transportation and, respectively, light-duty trucks (LDTs) are dominant in intra-urban transports, as they are used for the last-mile customer door-to-door deliveries. According to Mahesh et al. [1] diesel trucks contribute significantly to the total volume of traffic emissions. This leads to a deterioration of urban air quality, and several emission studies have tried to measure the actual airborne emissions from trucks. These studies often focus on cities where the air quality problems are the most tangible, e.g., [2].

Road freight transport produces increasing volumes of pollutants and emissions (e.g., dioxides; polycyclic aromatic hydrocarbons (PAH); and particulate matter (PM0.5 and PM2.5)). All of these pollutants (separately and together) have severe health impacts especially in inner cities with high density populations. They also have an increasing impact on global warming. In freight transport, the main challenge is that current truck fleets produce a high share of road traffic emissions even though their total numbers in traffic are relatively small. A number of 3.8 million heavy-duty trucks have been sold every year (globally) between the years 2011 and 2018. Almost all of them are in commercial use and fitted with diesel engines [3]. Diesel-powered engines have been the dominant choice for heavy-duty trucks for decades.

Torrey and Murray [4] referenced American Transportation Research Institute results indicating that more than 90% of road transport traffic use fossil fuels. Similarly, the U.S. Department of Transportation [5] state that trucks comprise only around 1% of on-road vehicles in 2013. The distribution of energy use by mode has remained relatively stable. In total, trucks are handling approximately 70% of all annual freight transports in the U.S. Trucks are the backbone of regional and local logistics networks in all countries, highlighting their importance in emission studies. According to Truckers report [6], one commercial truck uses up to 20,000 gallons of fuel in one year (estimated with a driving distance of 230,000 km) and a standard family car uses 500 gallons (20,000 km). These consumption numbers indicate the extent of a fuel economy that is based on current engine technologies. On one hand, they have a strong and scalable impact on all traffic emissions, and on the other, this major reason for emission growth has been well known for decades, but is complicated to tackle.

Diesel engine technology has developed extensively during the last decades. Manufacturers also use fuels that are based on renewable and recycled materials, decreasing emissions. However, emission levels have remained relatively high, and replacing engine and fuel technologies are still under development, e.g., [7–9]. One solution could be electric engine solutions, but at the moment, electric vehicles are expensive to manufacture and their purchasing prices for customers are high. Combined with a shorter operating range and a limited charging station network, they are not yet economically attractive. In the case of e-HDTs (and electric-drive cars in general) the key drivers for cost-efficiency are battery prices and developments in battery technologies. They should provide an economically feasible alternative for current diesel engines [2]. There are also production challenges (particularly batteries): Are there enough natural resources available to build such a number of battery-powered HDTs or LDTs as the future need would require? Additionally, battery-powered vehicles are not totally CO2-free over their full life-cycle due to production and maintenance requirements.

This study is a review of current academic literature on the topics of road freight-based truck emissions, pollution prevention, and mitigation solutions. The analyzed articles are limited to truck logistics that address technical engine solutions, and existing (currently used) and new fuels. It is expected that the papers provide a scattered array of solutions as has been the case in maritime studies (for corresponding literature review, see [10]). Based on these considerations, the purpose for this selective literature review is to understand: How does the selected academic literature approach and propose solutions for the mitigation of emissions from trucks in freight transport, and what are the most relevant methods for reducing them in this article set?

2. Data Collection and Sources

The data is collected from ScienceDirect (Elsevier) using tools of the integrated Scopus database. The articles were queried with a filter phrase 'lower truck emissions.' Only peer-reviewed research papers were selected. The search phrase was deliberately defined to be general and inclusive in order to gain a diversified sample of journal articles from different disciplines. The selection query was based on abstracts, not just keywords. One goal was to look at how this topic has been researched during the last 25 years: the first included article was published in 1995, after which the interest in researching emissions in freight transportation gained increasing popularity. In total, there were 161 papers found in the search. After the initial search, articles were grouped into content categories and their bibliographic information was used in content classification. Web of Science (WoS) impact factor (IF-index) was added to indicate how journals are ranked by citations.

The original 161 articles were further refined by using the keywords 'pollution' and 'emissions' in order to divide the large original set into a smaller and more clearly defined set. This procedure resulted in 21 articles for the detailed reviewing. The variety of listed keywords indicates that LDT and HDT emissions have been studied practically in all fields of science. This variety highlights the need for interdisciplinary research required in solving 'wicked' problems. The solutions evidently require analyses of disruptive environmental innovations, green investment strategies, non-voluntary legislation, and strategic global agreements.

Time-wise, the number of published articles has clearly increased after 2011. The trend has continued all the way until the final year 2019. This is reinforced by international regulation and legislation that have also experienced difficulties, e.g., in major international environmental summits in the commitment of major pollution producing countries. In addition, public awareness concerning discharges in traffic and transportation has increased, and the Intergovernmental Panel on Climate Change (IPCC) [11] report has stated that significant agreements should be reached quickly if the effects of climate change are to be limited to sustainable levels.

Table 1 indicates that, unsurprisingly, research articles (145) were the dominant form of papers. The most dedicated journal for the topic is Atmospheric Environment with 33 published research articles. As the journal title indicates, it focuses on all types of airborne emissions affecting the environment. Other significant journals focus on clean-tech in production, transportation studies, and general environmental issues.

Table 1. Article types and journal names. N refers to the number of studied articles in each venue.

Article Type (N)	Journal Name (N)
Review article (4)	Atmospheric Environment (33)
Research article (145)	Journal of Cleaner Production (16)
Book chapters (6)	Science of the Total Environment (14)
Conference abstracts (2)	Transportation Research Part D: Transport and Environment (12)
Short communications (4)	Applied Energy (9)
-	Energy (4)
-	Energy Procedia (4)
-	Energy Conversion Management (3)
-	International Journal of Hydrogen Energy (3)
-	Energy Policy (2)

Table 1 shows variety for a reasonable sample enabling the examination of how to achieve or take steps towards zero emission land transport. Similar research has been done earlier in the maritime context. For example, Hämäläinen and Inkinen [10] summarized that low emissions development will (or has to) cover several (clean) technology domains, starting from bio-fuels up to the physical transport routing optimization and waste water handling. In practice, this means that in hybrid ship designs there are no single solutions to convert an environmentally hazardous vessel into a responsible clean one. Also, pollution can be understood very widely, because especially trucks with diesel engines produce extensive amounts of various types of harmful emissions (e.g., PAH, CO2, NOx). Finally, scalability is significant as it refers to decreasing marginal costs where the increasing use of certain solution leads to decreasing costs.

3. Perspectives on Truck Freight Emissions and Pollution

The global efforts to accomplish zero emission logistics have been studied for several decades now. Changes are however slow, as Siskos and Moysoglou [12] remind us that the European Commission introduced, as late as 2018, CO_2 emission standards for truck manufacturers aimed at achieving reductions in air emissions caused by trucks. This is significant, as trucks are the second largest source of CO_2 emissions in the EU's road traffic.

There are numerous regional studies from around the world (e.g., from Latin America [13]; from Scandinavia [14]) indicating volumes of truck emissions in national contexts. Similarly, Shamayleh et al. [15] point out the significance of civic concerns and currently developing legislation on environmental issues. This results into the fact that enterprises should take up a more rigid view on responsibility. In discharge mitigation, a direct solution would be subsidization for clean-tech and bio-based fuels. However, the current availability of alternative fuel sources is still limited and it would be a challenge if demand increased significantly. There are also indirect impacts that bio-fuel production may entail, e.g., impacts on food markets and prices. For example, bio-gases, among others, are an alternative option for consideration. Additional solutions come from fiscal tools, e.g., [16].

Heavy taxation on transport would, however, also decrease total economic activity and thus lead into problems in all economic sectors.

In terms of engines, battery electric trucks (BETs) have been nominated as modern alternatives in the mitigation of the CO2 emissions of trucks [17]. There are several global examples of this, and Tanco et al. [13] carried out a study in Latin-America (Argentina, Brazil, Chile, Colombia, and Uruguay). They analyzed the main barriers for BET adoption. The initial investment requirement in particular is significant in total cost (current price difference favors diesel trucks) between the alternatives. Thus, the difference between the costs of electricity and fuel is highly important in achieving cost neutrality.

Handler et al. [18] analyzed the air emitted pollution from the forestry sector. Their empirical case focused on roundwood and was conducted in the USA (Michigan). The study fruitfully illustrates the needs of combining harvesting equipment adjustment and different operational scenarios in forestry. In addition, transport emissions are significant as majority of wood collection takes place in remote locations requiring specific truck types. Generic results prove that fuel types are essential in aiming to lower emission levels in freights, also as in [19].

Accordingly, Nanaki and Koroneos [20] noted that alternative fuel solutions (such as blended gasoline versions and biodiesels) have made a breakthrough on the fuel markets, for example, in the EU where fuel properties are strongly regulated. These considerations also remind that when using 'cleaner' fuels, they may have non-desired side effects even though they lower the worst pollution emissions. National differences are also significant as the study indicates. For example, Le and Leung [21] studied particle matter 2.5 volumes in Ho Chi Minh City (Vietnam). The measured particle levels indicate clear health risks as their measurements provide extensive coverage of the significant exceeding in particle concentrations in the air (measured for an hour per day out of the acceptable maximum). In these situations, spatial planning is one way to reduce engine emissions in densely populated areas.

Mitigating emissions is not only an engine technology or fuel matter. Traffic flow regulation and management provides a feasible platform for reducing emissions with responsible planning. These environmentally sound decisions may improve local air quality and diminish total emission volumes. Efficient regulation manifests itself e.g., in speed limitations. Panis et al. [22] point out the importance of simple speed limit regulation as they are the most common tools in improving road accident figures. For example, 30 kph zones have become commonplace in several EU countries. Slower maximum speeds also decrease air pollution—it is environmentally advisably due to the associated reduced fuel consumption leading to lower emissions. This is verified by a study where two different base speeds were used: normal private vehicles (small roads 30 to 50 kph) and trucks on motorways (80 to 90 kph). The results confirmed significant impact of speed in emission control and cost savings (fuel). The presented outcomes are examples of causalities relevant for decision making and implementing speed management policies.

4. Findings from the Articles

The selected 21 articles (19 of them with keywords, presented in Table 2) cover numerous aspects which could help in building zero emission freight transportation including both long and short delivery distances. These articles cover the years 1995–2019. The results are presented in Tables 2 and 3. They should be considered as a collective description of the articles. The selected articles are focusing on technology, fuel types, urban pollution, standards, fleet optimization (battery vs. diesel) and emission inventories, and electric vehicles.

Table 3 lists solutions and purposes of the studied articles. In order to support the classification, Appendix 1 presents a more detailed content list of the papers. In the case of single articles, the first one (number 1 in Table 3) analyzed how the continuous ethanol transportation by HDTs could be replaced by a pipeline where physical construction costs are extensive, but after the initial investments, pipelines do not produce as much HDTs emission and CO2 gases to the atmosphere.

Studies on regulation effects are widely present (e.g., numbers 2–8 in Table 3). In general, they have a unified message: more regulation results in lower emissions. This is verified with measurements before and after a specific environmental law or regulation comes to force. The papers include cases from around the world. For example, during the Beijing Olympic Games 2008 strict traffic regulations were put in force, which significantly improved local air quality. The time period is of the essence. Over short time periods (e.g., mega-events such as Olympics) quick improvements are achieved, but in the longer term, restrictions start to have a negative effect on economic performance. Thus, the balancing between legislative regulation and economic goals is evident.

Some collective interpretations may be drawn from Table 3. First, emissions control programs should include tools to remove high pollution emitters from the traffic or alternatively improve emission efficiency in existing vehicles. Wang et al. [23] remark that the median black carbon (BC) concentration after the control was significantly reduced, if compared to those days that were not under traffic control. Similarly, Wang et al. [24] studied black carbon through 'heavy emitters' (old diesel engine trucks) on the roads. Their study found out that the most polluting trucks caused often more than half of the total emissions (all particle matter) ranging from 41% up to 70%. Similar results were obtained also in the cases of other pollutants such as particle matter and carbon dioxides.

Table 2. Article keywords (19 out of 21, as two articles are without keywords).

2019 emission factors; euro VI hdvs; CO_2 emissions; NOx emissions; solid PN emissions
2019 fleet mix optimization; heavy-duty truck robust optimization; hybrid life cycle assessment; alternative fuel adoption; battery-electric heavy-duty truck; robust pareto optimal solutions
2019 urban air quality; ultrafine particles; active transport; mobile measurements; cycling infrastructure
2019 warehousing receiving process; detention fee; traffic congestion; environmental pollution; discrete-event simulation; truck check-in
2018 urban air pollution; mobile combustion sources; biofuels; emission policies; bus rapid transit
2016 externalities; transportation infrastructure; occupational safety; life-cycle assessment (LCA); economic input–output (EIO) analysis; greenhouse gas emissions
2016 vehicle; emissions; control; air pollution; China
2016 emission inventory; vehicular pollutants; COPERT model; the PYRD
2016 economic input–output-based hybrid LCA; electric delivery truck; multi-objective linear programming; conventional air pollution externalities
2013 externalities; freight transport; trends; fundamental factors; vehicle technology
2012 electric vehicles; economic replacement model; urban freight
2012 low emission zones; trucks; local traffic policies; traffic; air pollution
2012 emission factor; diesel; climate change; air pollution; nitrogen oxides; black carbon
2011 climate change; air quality; diesel; size distribution
2010 diesel particles; emission factor; composition; composite diesel PM2.5 profile; Bangkok
2009 olympics; air pollution; black carbon; climate change; health effects
2003 particulate-bound PAH; urban air pollution; Zaragoza; seasonal trend; emission sources
2003 fuel/propulsion system; greenhouse gas; global warming; life cycle analysis; life cycle assessment; alternative fuels
1995 exhaust emissions; motor vehicles; emission requirements; tax incentives; air pollution control

Table 3. First author, journal, year, impact factor (IF) index, title, and main purpose of the selected 21 papers. Texts in "main purpose" are direct quotes from the article descriptions.

Authors, Journal, Year, IF-Index, Title	Main Purpose
1 Strogen et al. (2016). Applied Energy. 7.9. Environmental. public health. and safety assessment of fuel pipelines and other freight transportation modes.	The construction of an ethanol pipeline from the Midwest to Northeast United States.
2 Wang et al. (2012). Atmospheric Environment. 4.012. On-road diesel vehicle emission factors for nitrogen oxides and black carbon in two Chinese cities.	Multi-pollutant control strategies and in-use compliance programs are imperative to reduce emissions from the transportation sector.
3 Wang et. al. (2009). Atmospheric Environment. 4.012. Evaluating the air quality impacts of the 2008 Beijing Olympic Games: On-road emission factors and black carbon profiles.	The emission control measures implemented to improve air quality during 2008.
4 Wang et. al. (2011). Atmospheric Environment. 4.012. On-road emission factor distributions of individual diesel vehicles in and around Beijing.	A field study of on-road emissions of diesel vehicles in and around Beijing, during November and December of 2009.
5 Grigoratos et al. (2019). Atmospheric Environment. 4.012. Real world emissions performance of h-d Euro VI diesel vehicles.	Real-world diesel Euro VI HDVs emissions of both gaseous pollutants and solid PN. For that reason, five HDVs, tested on-road under typical driving conditions.
6 Oanh et al. (2019). Atmospheric Environment. 4.012. Compositional characterization of PM2.5 emitted from in-use diesel vehicles.	PM2.5 emissions from diesel vehicles in Bangkok, providing Emission Factors appropriate for developing countries.
7 Wu et a. (2016) Environmental Pollution. 5.714. Assessment of vehicle emission programs in China during 1998–2013: Achievement. challenges and implications.	In China, vehicles are major sources of air pollution problems and it has adopted control measures to mitigate vehicle emissions. A local emission model (EMBEV) to assess China's first fifteen-year (1998–2013) efforts in controlling vehicle emissions.
8 Song et al. (2016). Journal of Cleaner Production. 6.395. Vehicular emission trends in the Pan-Yangtze River Delta in China between 1999 and 2013.	Emission factors from the COPERT IV model were used to determine emission inventories of CO, NMVOCs, NOx, BC, OC, PM2.5 and PM10 between 1999 and 2013.
9 Kuo et al. (2015). Journal of Transport & Health. 2.583. A06 San Pedro Bay Ports Clean Air Analysis.	Study analyzes the co-benefits of these policies on the reduction of greenhouse gases and regional pollutants, particularly as expressed through positive impacts on human health.
10 Feng et al. (2012). Procedia - Social and Behavioral Sciences. 0.78. Conventional vs Electric Commercial Vehicle Fleets: A Case Study of Economic and Technological Factors Affecting the Competitiveness of Electric Commercial Vehicles in the USA.	Competitiveness of commercial electric vehicles and trucks have the potential to substantially reduce greenhouse gas emissions and pollution and lower per-mile operating and maintenance costs.
11 Pérez-Martínez & Vassallo-Magro (2013). Research in Transportation Economics. 1.798. Changes in the external costs of freight surface transport In Spain.	Analyses the external costs of surface freight transport in Spain and finds that a reduction occurred over the past15 years.
12 MacLean & Lave (2003). Progress in Energy and Combustion Science. 26.467. Evaluating automobile fuel/propulsion system technologies.	Fuel emissions technologies, customers require rethinking of regulations, design of vehicles and appeal to consumers over the next decades. Vehicles more than 35mpg make up less than 1% of new car sales.
13 Sen et al. (2019). Resources. Conservation and Recycling. 7.044. Robust Pareto optimal approach to sustainable heavy-duty truck fleet composition.	Sustainable trucking, objectives are considered, minimizing the life-cycle costs (LCCs), life-cycle GHGs (LCGHGs), and life-cycle air pollution externality costs (LCAPECs).
14 Boogaard et al. (2012). Science of The Total Environment. 5.589. Impact of low emission zones and local traffic policies on ambient air pollution concentrations.	Air pollution at street level before and after low emission zones (LEZ) directed at heavy-duty vehicles (trucks) in five Dutch cities in different background locations 2008 and 2010.
15 Olsson (1994). Science of The Total Environment. 5.589. Motor vehicle air pollution control in Sweden.	Light-duty and heavy-duty trucks and buses also need to be certified against stringent emission requirements. The equipment's ability to meet the use requirements.
16 Mastral et al. (2003). Science of The Total Environment. 5.589. Spatial and temporal PAH concentrations in Zaragoza. Spain.	The concentration of polycyclic aromatic hydrocarbons (PAH) was measured in the Zaragoza (North-East of Spain) atmosphere using fluorescence spectroscopy in the synchronous mode (FS).
17 Smith & Srinivas (2019). Simulation Modelling Practice and Theory. 2.426. A simulation-based evaluation of warehouse check-in strategies for improving inbound logistics operations.	Minimize the detention fees paid to the carrier by enhancing the check-in process of the inbound trucks, with the secondary goal of reducing the CO^2 emissions.
18 Walsh (1998). Studies in Surface Science and Catalysis. 1998 Global trends in motor vehicle pollution control: a 1997 update.	Air pollution is a common phenomenon necessitating aggressive motor vehicle pollution control efforts. Survey of what is presently known about transportation related air pollution problems.
19 Zhao et al. (2016). Sustainable Production and Consumption. 1.4. Life cycle based multi-criteria optimization for optimal allocation of commercial delivery truck fleet in the United States.	Alternative fuel trucks may mitigate environmental impacts. Cost of these el trucks is higher than those of diesel. Environmental, social, economic indicators are studied, a model provides solutions for a fleet of 30 commercial delivery trucks.
20 Le & Leung (2018). The Lancet Planetary Health. 2.736. Associations between urban road-traffic emissions. health risks. and socioeconomic status in Ho Chi Minh City. Vietnam: a cross-sectional study.	The public health associated with urban road-traffic emission in HCMC, and whether reducing air pollution will decrease hospital admissions, premature deaths, and years of life lost. The association between air pollution and socioeconomic status.
21 Policarpo et al. (2018). Transportation Research Part D: Transport and Environment. 2.34. Road vehicle emission inventory of a Brazilian metropolitan area and insights for other emerging economies.	Vehicle emissions of carbon monoxide (CO), non-methane hydrocarbons (NMHC), aldehydes (RCHO), nitrogen oxides (NOx), and particulate matter (PM) in a metropolitan area using a bottom-up method, between 2010 and 2015.

One solution is to develop engine technology. For example, investments in environmentally efficient fleet are a solid method for diminishing pollution from freights. Different regulative standards play a role here. For example, EU standardization levels II, III, and IV in buses are significantly different in terms of pollutant tolerance. This is exceedingly important in densely populated areas (e.g., large Chinese cities, see [25]). It is considered that, particularly in economically growing developing countries, instead of acquiring older diesel engine HDTs there should be a direct leap into clean-tech transport vehicles and low polluting engines following the strictest standards.

Policarpo et al. [26] studied the road emissions and fleet properties in Brazil. The number of vehicles in their study area of Ceará has grown significantly, almost doubled since 2008. They estimated trends from various harmful emission types caused by the road traffic in 2010–2015 with a macro-simulation. Their results showed that the implementation of environmental regulation and policies are efficient means to decrease emissions. They cause and motivate an accelerated phase in the adoption of clean-tech in transport business. Alternatively, an optimization model, designed by Zhao et al. [27], considers tailpipe emission constraints focusing on combustion emission after treatment. Their study indicated that the environmental performance may be categorized into three main impacts (classes of economy, environment, and health). They made a distinction to dichotomous categories and gave recommendations for selecting different engine types according to economic conditions. The end results clearly indicated that hybrid solutions should be preferred when transport demand (high utilization) is in progress. Diesel engines are viable options in the non-probable scenario where the oil economy is booming and transport demand is low.

In China, and generally in Asia, combustion engine vehicles are the major source of harmful emissions and therefore the mitigation of vehicle emissions are paramount. Oanh et al. [28] point out that especially particle matter (PM) EF levels are lower for new vehicles, indicating consistent development with progression in engine technologies and low emission engine standards. Accordingly, electronic vehicles are an interesting venue for research. However, they were studied the first time in 2012 among the selected 21 papers. In the broader original 161 paper data set, the first paper on electric vehicles was published in 2003. However, this topic is still considered 'experimental' as the maturity level of electronic trucking is still very low. They will remain a marginal means of transport in professional use in the decade-long time-span until 2030.

Walsh et al. [29] stated more than 20 years ago that four trends are the most important in road traffic market development, including clean-tech and emission control:

1. Global population growth;
2. Increasing wealth of countries with lesser development increases global traffic volume;
3. Cleaner environment results into healthier population;
4. Governmental responses and regulations. Continuous restrictions on newly produced model emissions levels.

In order to conclude, it is worthy to remember that China produced approximately double the global greenhouse gas (GHG) emissions in comparison to that of the second largest polluter (US) and three times that of the European Union [30]. The annual growth rate is also the highest in China. The total number of motorized vehicles also soon exceeds 700 million globally. Out of that number, approximately half a billion are small cars, and approximately 150 million are trucks and buses. The rest are motorcycles. The growth rates are also significant. Vehicle growth rates have slowed down in economically advanced countries where natural population growth is slow. The main growth areas for vehicles are in developing countries that are experiencing high economic growth together with increasing urbanization.

5. Summary and Conclusions

The summarizing results of this review are presented in Table 4. There are some overlapping topics in these categories, but it is considered that the table provides the best interpretive view of the data. The studied articles exposed several environmentally friendly solutions, which are reasonable and executable in the near future. However, some of the solutions might be easier and faster to implement while some need slower incremental implementation. The summary is presented according to these categories:

1. Fuels and engine innovations;
2. Other innovations and methods to lower emissions;
3. Infrastructure: Route, spatial planning, controls.

Table 4. Summary of the methods to lower and mitigate emissions in HDT transport.

1. Fuels and Engine Innovations	2. Other Innovations and Methods to Lower Emissions	3. Infrastructure: Route, Spatial Planning, Control
HDTs motor innovations: - Electric motor - Battery capacity, route planning - City deliveries	**HDTs and LDTs in fleet optimization:** - Valuable things, combination of electric and HDTs and LDTs depending of routes and places in use - Replacing whole or part of the fleet - Capacity, routes	**HDTs and traffic control:** - Emission control measures implemented to improve air quality, - HDTs can be responsible for 50% of total BC emissions, and 20% of trucks are responsible for 50% CO_2, PAH emissions control - Reducing black carbon
HDTs with new fuels solutions: - Fuel emissions technologies, flex fuel vehicles, noticeable reduction in NOx and PM emissions, emission standards - LNG-trucks, Methanol, Biofuels, Hydrogen	**HDTs replaced with other innovations:** - Pipelines, electric rails, electric ships	**HDTs and traffic planning:** - Shifting freight deliveries from peak to off-peak hours, lower local emissions
HDTs and exhaust gas types and treatment: - Large and growing vehicle pollution control market, especially with regard to exhaust after treatment systems. - Try to reveal exhaust gas types.	**Vehicle buyers, customers, consumer behavior is changing:** - Consumers now demand larger, more powerful personal vehicles, ignoring fuel economy and emissions of pollutants. - Legislation, tax-policy	**HDTs and spatial planning:** - Reducing motor emissions from motorcycles, trucks, and buses, produce health benefits based on better land-use and transport planning, - Low emission zones for urban road-traffic emissions

HDTs, heavy-duty trucks; LDTs, light-duty trucks; PAH, polycyclic aromatic hydrocarbons.

The articles verify that in traditional truck and freight transportation, the environmental focus has not been in airborne emissions and mitigation. There are also different opinions in the research regarding the urgency of the matter: Some articles stated that the transportation industry in developing countries should be allowed to use old trucks that are producing larger amounts of airborne emissions until their gross national products are comparable to developed countries. The economic growth would allow increasing investments in more sustainable vehicles. Generally, this view is considered invalid, e.g., by the IPCC. The best solution would be that all countries should have similar standards and global regulative directives should be used as is the case in the maritime sector following International Maritime Organization (IMO) regulations.

In the leftmost column (number 1) of Table 4, solutions are presented that require extensive and significant changes to current fleets. They address technological engine innovations and solutions such as renewable bio-fuels. They would impact the emission volume greatly but their costs are still exceedingly high and the maturity of these technologies young. However, their use would create a direct and clear change towards low (and even zero) emission goals in LDT and HDT freight transports. The most important management actions and key changes in order to implement these technological advancements include taxation legislation (regulation), after sales support systems and continuous product enhancement system based on the operational performance (malfunctions, fatigue, battery duration) indicators and feedback. An alternative in some geographical locations would also be to replace HDTs with other cleaner means of transport (e.g., direct pipelines, railways, or short sea shipping). However, these solutions are highly dependent on the existing infrastructure and land-water area properties. It will remain the responsibility of LDTs to manage the final short delivery distances. Therefore, innovative truck development will continue to be an essential part of emission reduction and control.

There is also a chance to mitigate emissions by using company-level optimization in daily operations. For example, on long and midrange routes biodiesel trucks should be preferred, and light-duty battery trucks (LDBT) are preferable instead for very short distance transports in urban environments. This type of fleet building is already a reality in economically advanced countries. However, fleet building takes time and therefore regulative exhaust limitations are needed to accelerate the process. A quicker and cheaper solution for the near future would the continuation of the use of existing HDTs and LDTs. They could be converted to use biodiesel, methanol, or liquefied natural gas (LNG). This would gradually replace old diesel trucks with zero or low emissions ones.

In the middle column (number 2), the final row addresses the role of truck markets, buyers, and end customers. The market-driven change towards cleaner freight transportation would be easier if all customers of transport companies would require only clean truck deliveries for their products. However, this not yet possible (on a large scale) and a reason for this is that prices are too high. A second simple reason is the limited availability of low emission trucks. In the case of an absolute need to mitigate emissions (e.g., in inner city deliveries), the best way would be to use electric (battery driven) HDTs. An alternative solution is to restrict the amount of traffic as has been done e.g., in Beijing and other megacities suffering from low air quality. One fuel-related solution is to use more than one fuel tank in trucks. In this case a truck may switch between different fuel types in accordance to local emission restrictions. A challenge is that the fuel types store energy with different capacities. This has a direct impact on the transportation distances. Diesel has the highest storage capability whereas hydrogen and electricity have the lowest energy storage characteristics.

In larger countries with long transport distances, diesel HDTs are now the most used vehicles in freights and this will remain so for several years into the future. In these countries, modal change could be made (in some scale) in order to use trains as feeders, for example in container deliveries, and final connections could be organized by diesel (to some extend even refined from renewable materials) or gradually in the future by battery-driven trucks. In this modal transition a common challenge is that train network connections cover only limited areas. Railway building is an expensive investment and is viable only in locations with enough transport volume now and especially in the long run. This is one of the main challenges in nationwide transport planning in several countries that experience unbalanced population concentrations (e.g., sparsely populated Nordic countries).

The rightmost (number 3) column presents solutions focusing on infrastructure. These include issues related to route and spatial planning, control technology, and emission measurement. The articles address the fact that pollution levels are lower in locations where regulations and restrictions have been established. Improvement of measurement is crucial action and the related studies among this third category papers indicate that HDTs may produce more than half of total BC emissions. There has also been continuing research in which different time intervals have been applied in order to tackle temporal change in air quality. In addition, the case locations are places where air quality is low or very low to begin with. It is evident that direct measurement studies are the best way to indicate emission levels and air quality. They should also be in the best position to be applied in emission regulation decision making. The final main action should be the recognition and understanding that road and spatial planning have the best potential to aid emission problems by distributing the HDT traffic during peak hours, particularly in cities. These efforts do not necessarily lower the overall emission amount, but distributing it to broader time-spans helps with the most severe pollution peaks.

The production of electric vehicles has an impact on environmental pollution mainly for two reasons that were restricted outside our paper: First, it is not well known what the environmental impact is on the production means and disposal of newly emerged energy sources, especially batteries. This is an important study subject for the future. Second, the production of energy that is used to charge and power electric vehicles comes mainly from the combustion of coal, which is of great importance, especially in countries and regions where this energy is generated. Therefore, it should be taken into account in the final assessments of the impact of car transport on environmental pollution. An additional future research task emerges from the development potential of hydrogen trucks and the use of digital applications, enabling e.g., "truck sharing," thus developments taken place already in the use of private cars.

To conclude, these specifically targeted 21 peer-reviewed research papers provided a large variety of approaches, measurements, and case studies from all around the world. The papers proved the diversity of options and obstacles in the pursuit for low- or zero-level emissions in freight transport. These findings are focusing on impact that the diverse set of actions may have in relation to emissions levels. Understandably, the goal of reaching zero-level emission levels is realistically challenging to achieve, perhaps just only lower if the whole manufacturing, maintenance, and other indirect processes are considered. For example, the manufacturing of batteries for electric vehicles requires significant amounts of valuable metals, of which some are rare. Mining processes, refining and transportation of (raw) materials to battery manufacturing and assembling sites require energy and thus produce emissions. Therefore, in actual road transportation business, all possible operations should be reconsidered from the viewpoint of engine and fuel technologies, which should produce clearly lower emission levels (i.e., green innovative disruptions). If these are not achieved, then modal changes from roads to train and maritime transportation could bring benefits in long time span with some logistics limitations such as how to reach markets outside train and ports. This study clearly addresses that re-designing future clean freight transport based on trucks system needs to be done in a broad way to include the latest fuel and engine technologies (and related solutions), parallel to driving road logistic towards the lowest possible emission levels and mitigation of harmful substances. This could be concluded as the only answer to reach an improved environmental situation bringing positive impacts on human health. At the same time, it is important to remember that freight transportation on roads is expected to increase globally in the coming decades.

Author Contributions: Conceptualization, T.I. and E.H.; methodology, E.H.; investigation, E.H.; writing—original draft preparation, T.I. and E.H.; writing—review and editing, T.I. All authors have read and agreed to the published version of the manuscript.

Funding: This research received no external funding.

Conflicts of Interest: The authors declare no conflict of interest.

References

1. Mahesh, S.; Ramadurai, G.; Nagendra, S.M.S. On-board measurement of emissions from freight trucks in urban arterials: T Effect of operating conditions, emission standards, and truck size. *Atmos. Environ.* **2019**, *212*, 75–82. [CrossRef]
2. Moultak, M.; Lutsey, N.; Hall, D. Transitioning to zero-Emission Heavy-Duty Freight Vehicles, White Paper. 2017. Available online: https://theicct.org/sites/default/files/publications/Zero-emission-freight-trucks_ICCT-white-paper_26092017_vF.pdf (accessed on 8 July 2020).
3. Statista. Heavy Truck Production Worldwide from 2011 to 2018, by Region (in 1000 Units). 2019. Available online: https://www.statista.com/statistics/261156/heavy-truck-production-worldwide-by-region/ (accessed on 1 July 2020).
4. Torrey, W.F.; Murray, D. An Analysis of the Operational Costs of Trucking: 2015 Update, Research Associate American Transportation Research Institute Atlanta, September 2015. Available online: http://truckingresearch.org/wp-content/uploads/2015/09/ATRI-Operational-Costs-of-Trucking-2015-FINAL-09-2015.pdf (accessed on 15 July 2020).
5. Department of Transportation, Bureau of Transportation Statistics. Transportation Statistics Annual Report 2018. Available online: https://www.bts.gov/sites/bts.dot.gov/files/docs/subdoc/3416/tsar2016r.pdf (accessed on 15 July 2020).
6. Truckers Report. The Real Cost of Trucking—Per Mile Operating Cost of A Commercial Truck. 2013. Available online: https://www.thetruckersreport.com/infographics/cost-of-trucking/ (accessed on 18 July 2020).
7. Lajevardi, S.M.; Axsen, J.; Crawford, C. Examining the role of natural gas and advanced vehicle technologies in mitigating CO2 emissions of heavy-duty trucks: Modeling prototypical British Columbia routes with road grades. *Transp. Res. Part. D Transp. Environ.* **2018**, *62*, 186–211. [CrossRef]
8. Kim, N.S.; Van Wee, B. Toward a better methodology for assessing CO_2 emissions for intermodal and truck-only freight systems: A European case study. *Int. J. Sustain. Transp.* **2014**, *8*, 177–201. [CrossRef]
9. Rogers, M.M.; Weber, W.L. Evaluating CO_2 emissions and fatalities tradeoffs in truck transport. *Int. J. Phys. Distrib. Logist. Manag.* **2011**, *41*, 750–767. [CrossRef]
10. Hämäläinen, E.; Inkinen, T. Big data in emission producing manufacturing industries—An explorative literature review. ISPRS Annals of the Photogrammetry. *Remote Sens. Spat. Inf. Sci.* **2019**, *4*, 57–64. [CrossRef]

11. IPCC Refinement to the 2006 IPCC Guidelines for National Greenhouse Gas Inventories, September, The Intergovernmental Panel on Climate Change. 2019. Available online: https://www.ipcc.ch/report/2019-refinement-to-the-2006-ipcc-guidelines-for-national-greenhouse-gas-inventories/ (accessed on 10 July 2020).
12. Siskos, P.; Moysoglou, Y. Assessing the impacts of setting CO2 emission targets on truck manufacturers: A model implementation and application for the EU. *Transp. Res. Part. A Policy Pract.* **2019**, *125*, 123–138. [CrossRef]
13. Tanco, M.; Cat, L.; Garat, S. A break-even analysis for battery electric trucks in Latin America. *J. Clean. Prod.* **2019**, *228*, 1354–1367. [CrossRef]
14. Pinchasik, D.R.; Hovi, I.B. A CO_2-fund for the transport industry: The case of Norway. *Transp. Policy* **2017**, *53*, 186–195. [CrossRef]
15. Shamayleh, A.; Hariga, M.; As'ad, R.; Diabat, A. Economic and environmental models for cold products with time varying demand. *J. Clean. Prod.* **2019**, *212*, 847–863. [CrossRef]
16. Carling, K.; Håkansson, J.; Meng, X.; Rudholm, N. The effect on CO_2 emissions of taxing truck distance in retail transports. *Transp. Res. Part. A Policy Pract.* **2017**, *97*, 47–54. [CrossRef]
17. Zhou, T.; Roorda, M.J.; MacLean, H.J.; Luk, J. Life cycle GHG emissions and lifetime costs of medium-duty diesel and battery electric trucks in Toronto, Canada. *Transp. Res. Part. D Transp. Environ.* **2017**, *55*, 91–98. [CrossRef]
18. Handler, R.M.; Shonnard, D.R.; Lautala, P.; Abbas, D.; Srivastava, A. Environmental impacts of roundwood supply chain options in Michigan: Life-cycle assessment of harvest and transport stages. *J. Clean. Prod.* **2014**, *76*, 64–73. [CrossRef]
19. Mazlan, N.A.; Yahya, W.J.; Ithnin, A.M.; Hasannuddin, A.K.; Ramlan, N.A.; Sugeng, D.A.; Adib, A.R.M.; Koga, T.; Mamat, R.; Sidik, N.A.C. Effects of different water percentages in non-surfactant emulsion fuel on performance and exhaust emissions of a light-duty truck. *J. Clean. Prod.* **2018**, *179*, 559–566. [CrossRef]
20. Nanaki, E.A.; Koroneos, C.J. Comparative LCA of the use of biodiesel, diesel and gasoline for transportation. *J. Clean. Prod.* **2012**, *20*, 14–19. [CrossRef]
21. Le, L.T.P.; Leung, A. Associations between urban road-traffic emissions, health risks, and socioeconomic status in Ho Chi Minh City. Vietnam: A cross-sectional study. *Lancet Planet. Health* **2018**, *2*, S5. [CrossRef]
22. Panis, L.I.; Beckx, C.; Broekx, S.; De Vlieger, I., Schrooten, L.; Degraeuwe, B.; Pelkmans, L. PM, NOx and CO_2 emission reductions from speed management policies in Europe. *Transp. Policy* **2011**, *18*, 32–37. [CrossRef]
23. Wang, X.; Westerdahl, D.; Chen, L.C.; Wu, Y.; Hao, J.; Pan, X.; Guo, X.; Zhang, K.M. Evaluating the air quality impacts of the 2008 Beijing Olympic Games: On-road emission factors and black carbon profiles. *Atmos. Environ.* **2009**, *43*, 4535–4543. [CrossRef]
24. Wang, X.; Westerdahl, D.; Hu, J.; Wu, Y.; Yin, H.; Pan, X.; Max Zhang, K. On-road diesel vehicle emission factors for nitrogen oxides and black carbon in two Chinese cities. *Atmos. Environ.* **2012**, *46*, 45–55. [CrossRef]
25. Song, X.; Hao, Y.; Zhang, C.; Peng, J.; Zhu, X. Vehicular emission trends in the Pan-Yangtze River Delta in China between 1999 and 2013. *J. Clean. Prod.* **2016**, *137*, 1045–1054. [CrossRef]
26. Policarpo, N.A.; Silva, C.; Lopes, T.F.A.; Araújo, R.D.S.; Cavalcante, F.S.; Pitombo, C.S.; Oliveira, M.L.M.D. Road vehicle emission inventory of a Brazilian metropolitan area and insights for other emerging economies. *Transp. Res. Part. D Transp. Environ.* **2018**, *58*, 172–185. [CrossRef]
27. Zhao, Y.; Ercan, T.; Tatari, O. Life cycle based multi-criteria optimization for optimal allocation of commercial delivery truck fleet in the United States. *Sustain. Prod. Consum.* **2016**, *8*, 18–31. [CrossRef]
28. Oanh, N.T.K.; Thiansathit, W.; Bond, T.C.; Subramanian, R.; Winijkul, E.; Paw-armart, I. Compositional characterization of PM2.5 emitted from in-use diesel vehicles. *Atmos. Environ.* **2019**, *44*, 15–22. [CrossRef]
29. Walsh, M.P. 1998 Global trends in motor vehicle pollution control: A 1997 update. *Stud. Surf. Sci. Catal.* **1998**, *116*, 3. [CrossRef]
30. Muntean, M.; Vignati, E.; Crippa, M.; Solazzo, E.; Schaaf, E.; Guizzardi, D.; Olivier, J.G.J. *Fossil CO_2 Emissions of All World Countries—2018 Report*; EUR 29433 EN; Publications Office of the European Union: Luxembourg, 2018; ISBN 978-92-79-97240-9. [CrossRef]

Perspective

Mobile-Energy-as-a-Service (MEaaS): Sustainable Electromobility via Integrated Energy–Transport–Urban Infrastructure

Mahinda Vilathgamuwa [1], Yateendra Mishra [1], Tan Yigitcanlar [2,*], Ashish Bhaskar [3] and Clevo Wilson [4]

1 School of Electrical Engineering and Robotics, Queensland University of Technology, 2 George Street, Brisbane, QLD 4000, Australia; mahinda.vilathgamuwa@qut.edu.au (M.V.); yateendra.mishra@qut.edu.au (Y.M.)
2 School of Architecture and Built Environment, Queensland University of Technology, 2 George Street, Brisbane, QLD 4000, Australia
3 School of Civil and Environmental Engineering, Queensland University of Technology, 2 George Street, Brisbane, QLD 4000, Australia; ashish.bhaskar@qut.edu.au
4 School of Economics and Finance, Queensland University of Technology, 2 George Street, Brisbane, QLD 4000, Australia; clevo.wilson@qut.edu.au
* Correspondence: tan.yigitcanlar@qut.edu.au

Citation: Vilathgamuwa, M.; Mishra, Y.; Yigitcanlar, T.; Bhaskar, A.; Wilson, C. Mobile-Energy-as-a-Service (MEaaS): Sustainable Electromobility via Integrated Energy–Transport–Urban Infrastructure. *Sustainability* **2022**, *14*, 2796. https://doi.org/10.3390/su14052796

Academic Editor: Luca D'Acierno

Received: 4 February 2022
Accepted: 25 February 2022
Published: 27 February 2022

Abstract: The transport sector is one of the leading contributors of anthropogenic climate change. Particularly, internal combustion engine (ICE) dominancy coupled with heavy private motor vehicle dependency are among the main issues that need to be addressed immediately to mitigate climate change and to avoid consequential catastrophes. As a potential solution to this issue, electric vehicle (EV) technology has been put forward and is expected to replace a sizable portion of ICE vehicles in the coming decades. Provided that the source of electricity is renewable energy resources, it is expected that the wider uptake of EVs will positively contribute to the efforts in climate change mitigation. Nonetheless, wider EV uptake also comes with important issues that could challenge urban power systems. This perspective paper advocates system-level thinking to pinpoint and address the undesired externalities of EVs on our power grids. Given that it is possible to mobilize EV batteries to act as a source of mobile-energy supporting the power grid and the paper coins, and conceptualize a novel concept of Mobile-Energy-as-a-Service (MEaaS) for system-wide integration of energy, transport, and urban infrastructures for sustainable electromobility in cities. The results of this perspective include a discussion around the issues of measuring optimal real-time power grid operability for MEaaS, transport, power, and urban engineering aspects of MEaaS, flexible incentive-based price mechanisms for MEaaS, gauging the public acceptability of MEaaS based on its desired attributes, and directions for prospective research.

Keywords: mobile-energy-as-a-service (MEaaS); mobile energy; urban electromobility; electric vehicle; renewable energy resource; bidirectional electric vehicle charging

1. Introduction

The climate change crisis arising from increasing greenhouse gas (GHG) emissions from all sectors of the economy, including energy and transport, demand decisive action towards sustainability and smooth transition to cleaner energy sources [1,2]. After the recent UN Climate Change Conference (COP-26), several governments and businesses in the industrialized countries are beginning to show a paradigm shift in addressing these challenges [3]. As road transport is a significant contributor to GHG emissions [4,5], the uptake of electric vehicles (EV) in large numbers has been touted as one of the pathways to lower the carbon footprint from the transport sector [6]. Accordingly, the world's major automotive industries have pledged to phase out internal combustion engines (ICEs) and replace them with the EV technology. For example, the designated timelines for this conversion are Volvo in 2030, Mazda in 2030, GM in 2035, and Nissan in the early 2030s [7].

EVs form the spine of electromobility [8], where the term electromobility is defined as "a set of activities related to the use of EVs, as well as technical and operational EV solutions, technologies and charging infrastructure, as well as social, economic and legal issues pertaining to the designing, manufacturing, purchasing and using EVs" [9] (p. 1). There are many challenges in establishing electromobility in cities. These range from EV costs and charging station availability and accessibility, from charging speed and battery capacity and cost to purchasing power and government incentives, from cost of electricity to vehicle to grid adoption and consumer awareness on green technologies [10–13].

In many countries, EVs are expected to proliferate over the coming decade [14]. Based on Bloomberg New Energy Financial predictions, after a tipping point in ICE/EV comparative costs in the coming years, 60% of the cars will be fully electric, and the rest of them will have the plug-in capacity by 2040. Globally, more than seven million EVs are sold in 2021 alone, indicating the serious market penetration [7]. Although EVs have significant environmental benefits, given the potential to use renewable power for charging purposes, widespread adoption is limited due to the lack of available charging infrastructure and the capability of the existing electricity distribution network to handle cumulative peak demand [15–17]. In developing a comprehensive EV charging regime, the locations, including customer premises, workplaces, public stations, online charging on-route and car parks, are critical [18,19].

In the absence of any smart charging facilities, uncoordinated and erratic bidirectional charging (charge and discharge) of EVs will hamper the stability of the electricity grid. Based on [20], most customers tend to connect their EVs to the grid immediately after returning home from work between 6:00 p.m. and 10:00 p.m. For instance, a study in the Netherlands reports uncoordinated charging with just 30% EVs penetration, leading to a 54% increase in peak demand [21]. A similar study in Western Australia reports that a 62% EV penetration rate will result in a 2.57 time increase in electricity loading [22]. Hence, large-scale uncoordinated EV penetration will lead to excessive feeder overloading and poor power quality [23,24]. The existing power quality issues, which are primarily over-voltage problems, owing to the large numbers of rooftop photovoltaics (PVs) in the distribution grids [25–28], will become worse due to the uncoordinated integration of EVs.

Therefore, there is a need to develop an intelligent scheduling and control mechanism that encourages EV owners to relieve the overloading issue in the electricity grid. In addition, optimal placements of EV charging infrastructures considering transport and urban real-estate constraints are paramount to alleviate the electricity loading constraints. Moreover, an incentive-based market mechanism where EV owners are rewarded for their bidirectional charging behavior between EVs and grid (vehicle-to-grid (V2G) and grid-to-vehicle (G2V)) helping the electricity grid is necessary. Additionally, the bidirectional power flow can also be used to provide high-quality ancillary services such as voltage and frequency regulation, peak power management and improvement to the load factor [29–32]. This should be combined with appropriate market research where the public acceptability of this mechanism is evaluated. Furthermore, treating EVs as mobile energy sources has created the concept of the mobile energy internet [33].

This perspective paper introduces a novel concept of Mobile-Energy-as-a-Service (MEaaS), a well-planned mechanism incorporating transport, power, and urban infrastructure aspects of mobile energy for EVs and the flexible incentive-based pricing schemes to handle challenges introduced by the upcoming wave of EVs. Provided that this mechanism is planned well, challenges due to widespread EV uptake on urban power systems could be handled via rapidly advancing mobile energy technology [34–37]. In other words, EV batteries could serve as mobile energy sources to compensate the pressures on the grid during peak times. Nevertheless, this requires a careful system design for large metropolitan cities that can accommodate an MEaaS system.

One of the attractive features is that MEaaS creates a platform for EV users to trade energy in an established market which can be operated via an app in smartphones. Using the arbitrage market (i.e., power price differential during peak and off-peak hours), EV owners could cover not only the costs of running their vehicles but even profit from it. The other advantage is convenience, timesaving, reliability of electricity supply to EVs and the opportunity to be connected to a market on a 24/7 basis. Therefore, this new technology adds extra value to the rapidly emerging process of prosumaging—Prosumage is a term used for PROroduction, conSUMption, and storAGE [38]. The rollout of MEaaS should further accelerate the uptake of EVs, increase the number of prosumagers, lead to an increase in renewable energy driven by market forces [39], lower prices and a reduction in GHG emissions from the transport sector. In this sense, MEaaS provides an opportunity where large numbers of EVs, with their batteries, form a giant battery when aggregated. The batteries not only will take the pressure off the grid, but their mobility provides an opportunity for energy to be delivered to consumers on-demand, both to households as well as EV users in a certain area where shortages exist.

In this perspective paper, we introduce a novel MEaaS system approach and offer a discussion around the issues of: (a) Measuring optimal real-time power grid operability; (b) Utilizing transport, power, and urban infrastructures; (c) Establishing a flexible incentive-based price mechanisms, and (d) Gauging the public acceptability of MEaaS based on its desired attributes. All statements in this perspective paper are based on a thorough review of the current literature, research, developments, trends, and applications.

It should be noted that this paper is written in the form of a perspective piece, where the authors express their personal experiences and opinions on a new perspective about emerging research and development (i.e., MEaaS) on a particular topic (i.e., urban electro-mobility). After this introduction, the rest of the paper is structured as follows. Section 2 introduces the state-of-the-art in mobile energy and transport. Section 3 presents the key aspects of MEaaS as the future of urban mobile-energy. Section 4 concludes the paper by presenting remarks on the future research directions.

2. Mobile Energy and Transport: The State-of-the-Art

Smart bidirectional EV chargers can regulate the grid frequency by charging and discharging the EV batteries [40]. Using the arbitrage market, EVs can be financially attractive to potential owners, covering EVs' running costs to make a profit. For example, one study reports that the EV-owner can gain between $3777 to $4000 per year for sharing an EV's power reserve with the power grid with a regulating power of 10–15 kW [41].

Load-levelling and peak-shaving are other potential benefits of G2V and V2G applications [42,43]. With the help of V2G, it is possible to discharge the extra power of EV batteries to the grid during daily peak demand (peak-shaving). On the other hand, with the help of G2V, EVs can be charged during off-peak hours, improving the load profile during the day (load-levelling). According to [44], if New York City's EV population is approximately 100,000, representing a 50% penetration level, up to 10% of the peak power can be provided by EV batteries—valued at $110 million per year.

In addition, the renewable energy sector can benefit from the presence of G2V and V2G charging. Due to the intermittent nature of RERs, it is possible to use EV batteries as storage units during periods of high peak generation and discharge them during peak demand. Such a market will lead to an inevitable further increase in the uptake of rooftop PVs, EVs and, at the same time, reduce the stress on the grid. Increasing amounts of renewable energy generated, through an increased number of PVs, will eventually lead to lower electricity prices even during peak hours, replacing fossil fuel-based generation, leading to an eventual decrease in GHG emissions.

On the other hand, smart bidirectional charging makes it possible to determine the charging time. This coordinated system can help decrease daily electricity costs, transformer and conductor current ratings and flatten the power profile of the grid. Authors in [45] report that a 50% peak load increase can be avoided at a 10% EV penetration rate with a coordinated bidirectional charging strategy for the US power grid. Adopting a coordinated mechanism requires specialized equipment such as sensors and communication devices and related policies that can encourage customers to adopt EVs and adhere to associated bi-directional charging protocols. Some policies incentivize customers to purchase EVs in tax credits/rebates and subsidize charging installation or discount in-building parking [46].

For instance, in the case of Australia, limited policies that exist at the state level look surprisingly more proactive than those at the federal level. For example, the states of Queensland and South Australia offer up-front financial incentive programs for EV buyers and subsidies for EV bidirectional charging stations. On the other hand, some policies involve bidirectional charging scheduling strategies. In the context of coordinated bidirectional charging algorithms, most policies are based on multi-level pricing that are established using power demand and RER generation to encourage customers to shift to off-peak power demand. Several countries are lagging other developed countries in terms of coordinated scheduling with no clear framework in place to set state and federal initiatives [7]. Policymakers need to consider numerous local and national factors, including the existing infrastructure; types of bidirectional charging stations and their locations; the EV penetration rate, customer mobility profiles, convenience, preferences and acceptability.

Urban planning is another aspect that can be significantly affected by integrating EVs to the power grid [47–49]. Public bidirectional charging stations that are optimally located and easily accessible are critical in boosting the adoption of EVs. Moreover, range anxiety is a prohibitive concern for the rapid growth of EVs. Many studies have been dedicated to maximizing the satisfaction level with respect to charging demand and limited budgets [50,51]. Nevertheless, optimization of EV bidirectional charging locations must be expanded to encompass traffic concerns, equitable distribution of stations, the capacity of roads/cities and existing infrastructure limitations. Furthermore, the current public charger locating strategies do not consider the impact of having household chargers, which may lead to the excessive location of the public charger in residential areas.

The increasing number of studies indicates that the number of EVs on our roads will rise as they are preferred as personalized transport given the environmentally friendly benefits and the acceptance of autonomous driving technology [52]. The uptake of autonomous driving will also be revolutionizing smart urban mobility, where electrification of such vehicles is also a desired outcome [53]. The levels of EVs on roads and their changing environmental roles globally in recent years indicate that EVs will be the future of personalized transport, especially when autonomous EV technology is concerned [54,55].

Although it is expected that 28% of total sold vehicles will be battery-powered by 2030, little work has been done to estimate the required infrastructure and the corresponding budget to meet this growing demand. The practicability of constructing a complex infrastructure across the country requires time, deep analysis of the issues and considerable financial expenditure. Since the rapid growth of EV numbers has already started, the analysis of the current grid system is essential to ensure that grid resilience is not jeopardized. Although extra headroom for facilities is considered when the power grid is constructed to account for future power demand increases, it is unlikely to be sufficient once the EV fleet is integrated into the transport system. Thus, system augmentation is inevitable, and potential flaws must be identified.

In addition, the increasing number of EVs as public transport necessitates the new bidirectional charging stations with appropriate technologies for public stations, which is different from the household stations. A substantial investment is required to develop the needed technology, and significant financing is necessary to install and operate these stations. Due to the complexity of the cost estimation and its dependency on the local and national parameters, so far, no comprehensive analysis has been carried out to estimate the expected investment and the gained profit for many countries. Internationally, only two scenarios have been widely investigated for charging infrastructure requirements [56].

The first one is based on bulky battery packages that can guarantee the daily driven distance of an ordinary individual. The EV is charged at night using a moderate power transmission rate in this strategy. From an investment perspective, this approach requires a low investment cost at the household level since most installed infrastructures in houses can tolerate the required power transmission based on existing ratings of the installed wires. However, the main concerns of this approach are the lack of providing the required ancillary services during the daytime and the increasing power demand across residential area feeders at night [57].

The second strategy is based on the constant charging and discharging of the EVs with a small battery-package volume that would be feasible with the help of new emerging charging technologies such as 'wireless power transfer systems' (WPTS), which can charge EVs on-road. Although the continuous drive option might be an ambitious and appealing target, it requires a significant modification in both electrical and roadway infrastructures and, therefore, an unjustifiable expense, particularly for low population density regions. In general, commercialized bidirectional chargers can be classified as Level 2 chargers (~5–10 kW) and Level 3 chargers (fast chargers, ~50 kW) [58]. The former suffers from slow bidirectional charging time, making it more suitable for household applications. Level 3 charges can substantially increase power flows during peak power usage periods, making it more convenient for public stations.

Ideally, the adopted bidirectional charging facilities should not be restricted to a single technology. Thus, analyses related to EV infrastructure and corresponding impacts need to be expanded to encompass various bidirectional charging options. For example, so far, the investigated business models of bidirectional charger stations have been focused only on a single technology across a system (i.e., only Level 2 or only Level 3 is adopted), while the cost optimization of hybrid bidirectional charging facilities (including Level 2, Level 3, and on-route bidirectional chargers) has not been investigated (Figure 1). Furthermore, how EVs as mobile energy carriers would affect the electricity grid power quality and stability and the way the electricity and transport networks are interwound have not been comprehensively investigated.

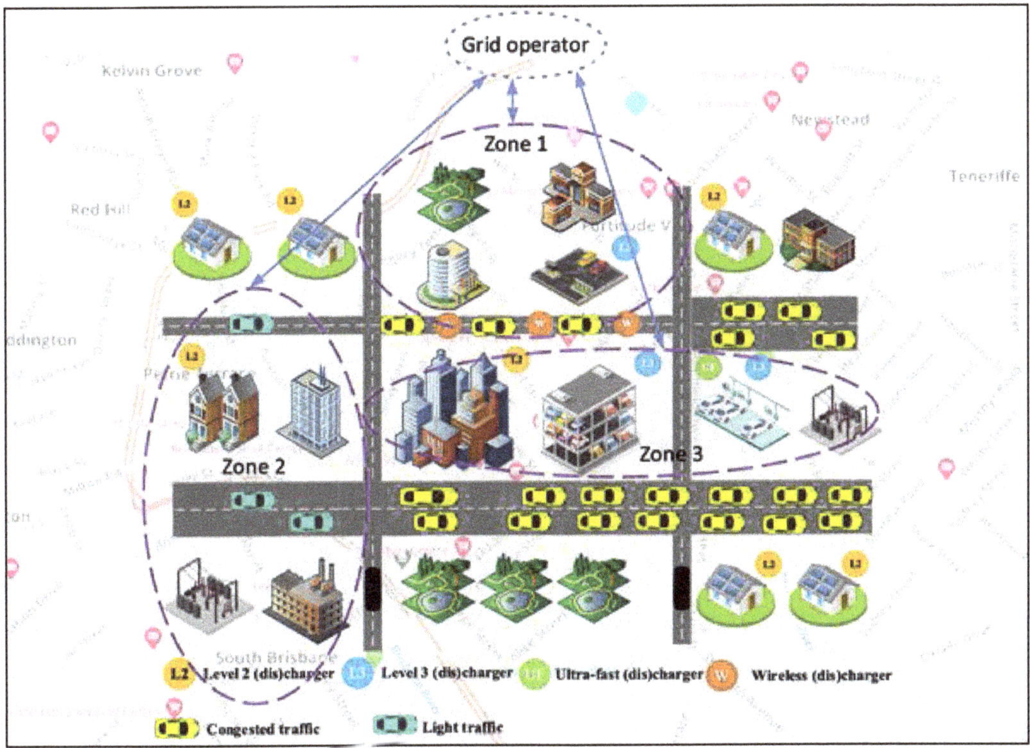

Figure 1. Bidirectional EV charging technologies and their interactions with power and transport grids.

Thus, this perspective paper is concerned with how the variability of electricity unit price from RERs will influence the state-of-the-art stationery and dynamic bidirectional charging costs and related operating expenses. In this way, a generic model for large metropolitan cities can be developed to offer the optimal combination of bidirectional chargers for the public (including on-route and stationary chargers) and household applications.

From an urban planning perspective, there are limited but growing studies concerning the location of EV charging stations [59]. In the urban planning literature, integration of land use and transport has been widely covered [60], whereas the inclusion of power infrastructure is rather new. The existing literature on the location of charging stations does not adequately factor in power, transport and urban planning aspects [61–63]. Moreover, while there are studies on wireless power transfer with one-directional charging consideration [64–66], there are, to our knowledge, no comprehensive studies conducted to determine the optimal location of bidirectional charging stations or infrastructure.

So far, accessibility has been the main factor in determining public EVs' location bidirectional chargers. As a result, parking lots are assumed to be a reasonable location for the installation of such stations [67–69]. Nevertheless, factors such as balancing the power generation from RER and EV charging power demand at each urban locality (e.g., suburbs or neighborhoods); system augmentation; the impact of household chargers; traffic-related concerns; the capacity of the civil infrastructure and the change in the behavior of drivers on the face of potential charging/discharging incentivizing schemes have not been covered systematically in the literature [70].

Despite a growing literature in this domain, it only covers limited aspects in identifying optimal locations for bidirectional charging infrastructure [71]. Additionally, while MEaaS provides a system approach to urban mobility including EVs [72], there is no comprehensive and system-level approach to EVs and bidirectional charging infrastructure in the context of metropolitan cities.

Against this backdrop, this perspective paper advocates a system thinking to pinpoint and address the negative externalities of EVs on the power grid of our cities. Hence, in the next section, the paper coins and conceptualizes a novel concept of MEaaS to operationalize energy, transport and urban infrastructures for establishing sustainable electromobility in large metropolitan cities.

3. Mobile-Energy-as-a-Service (MEaaS): The Future of Urban Mobile-Energy

Accordingly, this paper aims to provide a multidisciplinary perspective—of power and transport engineering, urban planning, and social science—for the smooth transition from the grid in its present form to a network that can support a dominance of EVs on roads via an MEaaS. Furthermore, this research intends to analyze the effect of the power system when EVs act as sources of mobile energy. Such mobile energy sources can be utilized to supply energy to specific areas that experience energy deprivation at any given time or absorb energy from certain areas that experience energy surplus. Thus, with proper forecasting of the mobility over the transport network, EVs can be used as dynamic energy sources to alleviate power system operational issues using bidirectional charging. This shows the symbiotic relationship between transport and energy networks that can be managed to be economically optimized.

In essence, MEaaS presents how an operational framework can manage such mobile energy sources to benefit EV owners and urban transport and energy networks, particularly in large metropolitan cities. As the geographical context of MEaaS, the primary implementation target is naturally the large metropolitan cities (cities offering increased urban socioeconomic activities, density and population, and diverse and advanced infrastructure and smart mobility options) [73–75]. Moreover, urban administrations of some global cities (also branded as knowledge or smart cities), particularly in developed nations, have the capability and interest to be the leader in urban innovation [76].

This perspective paper advocates the identification of appropriate charging systems for domestic requirements and for addressing some of the current shortcomings of systems.

Such broad scope and the multidisciplinary research perspective also include a public preference/acceptability study to assist researchers in comprehensively investigating and developing a charging system that tracks the influence of optimization in one aspect over the other (Figure 2).

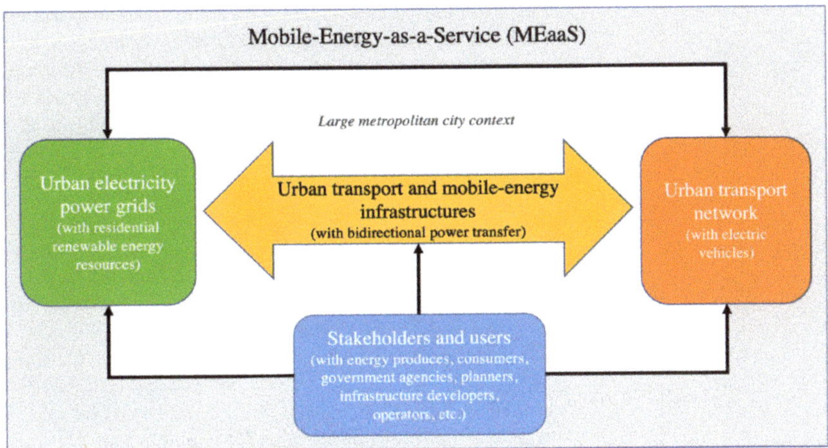

Figure 2. Conceptual framework of MEaaS.

This paper proposes MEaaS as a novel concept to address some of the undesired externalities that EVs will create on urban electricity power grids. Turning the MEaaS concept into reality will require thorough investigations despite this benefit. Four critical ones are listed below and elaborated in the following subsections:

- Identification of optimal real-time power grid operationality when incorporating MEaaS (where this involves, inter alia, a reliability analysis of the existing urban power grid with respect to the future uptake of EVs and RERs) (further details are presented in Section 3.1);
- Determination of the structure of MEaaS in the large metropolitan city context through smart urban infrastructure design guidelines encompassing transport, power and civil engineering aspects (where this involves analysis of the urban topography and urban form and designating the optimal locations for public charging stations) (further details are presented in Section 3.2);
- Development of the flexible incentive-based pricing mechanisms for MEaaS (where this involves optimal bidirectional charging through V2X/X2V of mobile/stationary EVs) (further details are presented in Section 3.3);
- Assessment of public acceptability of MEaaS and identifying its salient attributes under which the system could be widely deployed (where this involves public technology adoption surveys and interviews) (further details are presented in Section 3.4).

3.1. Measuring Optimal Real-Time Power Grid Operationality for MEaaS

The impact of EVs on the operations of electricity network utilities must be investigated to avoid imposing an unprecedented burden leading to the degradation of power quality, high-stress levels on local transformers or cables and poor system reliability. This involves identifying optimal real-time power grid power flows incorporating MEaaS—that is, a reliability analysis of the existing power grid with respect to the future uptake of EVs and RERs. Researchers have investigated the impact on electricity grid power flows from EV charging purely from a supply point of view [77]. However, the dynamic spatio-temporal electricity demand distribution and its interaction with RERs have been overlooked. A recent report from Energeia [78] indicates the likely significant variance in cumulative EV uptake over the next decade with respect to the government intervention level and how it can significantly impact the power grid performance in large metropolitan cities.

We suggest the following procedures to measure optimal real-time power grid operationality for MEaaS involving EVs and the advanced bidirectional charging infrastructures. First, the electricity network should be categorized into several modelling zones, which in turn are defined in terms of the electricity distribution system and the local statistical areas. Electricity grids are represented by several substation zones based on local statistical areas, whereas transport networks consider several statistical areas based on socioeconomic characteristics. This needs to be amalgamated to determine the impact of EVs on the power grid. The worst-case condition in each zone, in terms of the required power consumption, can be estimated as Equation (1):

$$P_{tot,EV} = \sum N_i \, P_{nom,i} \quad (1)$$

where the subscript "i" designates the adopted bidirectional charging technology, N_i is the number of users of the i-th technology in each zone, $P_{nom,€}$ is the nominal power of the bidirectional charger, which is a known value for different technologies (the influence of private and public charging stations is accounted here). N_i represents private users which can be approximated from the historical census records of each zone ($N_i = \alpha_i * N_h$, where N_h is the number of houses in each zone, and α_i is the penetration rate of EV in each zone). Poisson distribution can be used to estimate the worst-case scenario of the number of public station users at a specific timeslot [20].

With the help of Poisson distribution, the probability of the customers of public stations (including the stationary and dynamic bidirectional chargers) can be determined to estimate the corresponding power consumption/generation from Equation (1).

Customer behavior regarding the bidirectional charging operation is influenced by several factors: electricity price, mobility trip purpose (work/recreational/shopping), time (day/week/month) and customer convenience. The different electricity demand scenarios should be modelled considering the aforementioned factors, and Equation (1) should be solved for different times of the day. This analysis adjusts the current/projected daily load curve by incorporating each zone's RER generation distribution and demand scenarios from the available power grid database. It should determine the limitations of the existing power grid in each zone and the need for corresponding network augmentation.

3.2. Transport, Power and Urban Engineering Aspects of MEaaS

Integration of a large EV fleet into the power grid requires a placement strategy that can justify the financial outlay and consideration of urban-related issues such as congestion and social welfare. Moreover, the applications and implications of EVs are overshadowed by the uncertainties and potential consequences for the operation of the existing power grid and urban infrastructures. Regarding the placement strategies for bidirectional chargers, several critical factors such as the travel anxiety level, the required time to charge with respect to the adopted technology, and fair access to these facilities must be considered. The power grid infrastructure limitations such as the power system transformer ratings must also be taken into consideration.

In addition, the advent of modern technologies such as wireless powered bidirectional chargers provides a unique opportunity for on-the-go bidirectional charging to be employed. Although the required technology for on-route bidirectional charging is more expensive than stationary chargers [64], it offers valuable features such as reducing the waiting time in queues and the requirement for the EV battery size [79].

Furthermore, in locations with sufficient traffic, the revenue generated from cars using bidirectional charging on the route can meet the investment expenses. There is limited research on developing frameworks for identifying the most suitable locations for public chargers in locations such as shopping centers and parking lots [50,56,66]. Nonetheless, most drivers were shown to prefer mobility rather than profitability and indicated social welfare, grid augmentation and urban planning-related concerns. Noticeably, the simultaneous impact of more than one technology is not considered [80].

The grid augmentation and economic viability concerns should be conducted in parallel with the grid modelling. In terms of various available bidirectional charging options and topography of the city, there is a clear gap in the quantitative evaluation of the placement of these stations with respect to urban planning and transport operational constraints.

Access to the strategic transport demand model and relevant socioeconomic and land use data is necessary, as this provides an opportunity to demonstrate the applicability of the research on a real network and establish evidence-based findings. The research should consider spatio-temporal travel patterns and the recent advancements in artificial intelligence and machine learning techniques (such as non-negative matrix factorization) on the transport data can be explored for accurate and reliable modelling of base case transport patterns.

Different scenarios specified by MEaaS should be considered to model EV customer profiles with respect to the adoption of EVs. The profiling should consider different types of EV customers and their socioeconomic characteristics. Different modelling scenarios based on technology enthusiastic early adopters should be used to assess the price sensitivity where there is mass ownership of EVs. A structured review of the emerging international findings on EV and similar technology adoptions should provide a strong foundation for the design of scenarios.

Based on the MEaaS scenarios, the charging location of EVs should be optimized considering various factors such as spatio-temporal demand distribution, temporal vehicle trip distribution, driving behavior (on the face of incentivizing schemes), power grid impact and other technical factors related to charging type (e.g., fast/slow, wireless/plug-in, reverse).

Different multi-objective mathematical function formulations (*Z*) for the objective function should be considered. In general, *Z* is defined as a function of the traffic delay on the network (*Tg*) and impact on power grid (*Pg*) as represented by Equation (2). The sensitivity and stability of the different formulations should be thoroughly tested systematically. The values will be based on the simulation using the strategic transport model and the power grid simulator:

$$Z = f(Tg, Pg, \text{other constraints}) \quad (2)$$

Potential algorithms that can be considered for optimization include simulated annealing, advanced meta-heuristic algorithms, gradient descent and its variations, evolutionary computation, hybridized algorithms and reinforcement learning.

The optimization protocol should take the following factors (not limited to) into account: (a) Spatio-temporal distribution of mobile energy; (b) Travel patterns; (c) Strategic transport model; (d) Customer profiling with respect to the adoption of EVs; (e) Various urban scenarios; and (f) Optimization for charging infrastructure location with respect to different technology types.

To accomplish a public placement bidirectional charger plan, an MEaaS feasibility/optimization study—looking into transport, power, and urban engineering aspects of MEaaS—should utilize a zone-based geographic partitioning technique to predict electricity generation and demand at each zone. Given the available local data and the type of land used (e.g., residential/commercial/industrial/recreational), one can forecast the behavior and demand of users in each zone. The strategy should be based on optimal placement of the public bidirectional charging stations (Level 3) in areas with the low RER and high-power demand (such as apartment zones), where household bidirectional chargers (Level 2) are more effective. Given the traffic flow at each zone, the modelling should pinpoint those areas where there is potential for application of ultra-fast and wireless bidirectional chargers.

The interaction between the price of electricity for selling or buying, user decision to choose the travel route and the capacity of roads to accommodate the number of EVs generates a coupled equilibrium in power and transportation.

Such optimization study can learn from [50] the interaction between the price of electricity and destination choice of EVs users, the optimized profit for both the stations and users with respect to the optimal location of the stations. To this end, a combined distribution and assignment model should describe the user destination choices based on the price of charging and price paid for the purchase of excess electricity by charging stations. Research can be further extended to model the effect of vehicle discharging on traveler behavior and road capacity, not considered in [50].

3.3. Flexible Incentive-Based Pricing Mechanisms for MEaaS

The EV charging behavior of users can significantly impact the power grid performance [79]. Currently, the charging infrastructure is limited to specific locations such as workplaces, residential, commercial, and recreational premises. The existing charging pattern and its impact on the grid is primarily governed by user convenience and the cost associated with existing tariffs (pay-per-kWh and pay-for-time). Such patterns are not system optimal and significantly impact the power grid.

With the advancement of technologies, users would actively participate in vehicle-to-everything or everything-to-vehicle (V2X/X2V) mechanisms through bidirectional charging and associated Smart/IT/apps/technologically based solutions. Although consequential system augmentation is inevitable to some extent, adopting a reasonable policy or pricing mechanism to incentivize users should result in optimal capital investments in power, transport and urban infrastructures.

Furthermore, due to the high penetration of RERs in some nations, such as Australia, and the lack of synchronous generators (low inertia), as pointed out in the introduction, the grid is prone to instability, and the connection of EVs to the grid can be a beneficial alternative if managed competently. Thus, there is a unique opportunity for many national electricity utilities (grid and retail) and charging station operators to benefit from the optimal integration of EVs. Nevertheless, there are neither mechanisms nor incentives for EV owners at this stage. Several studies offer guidelines for the economic operation of charging stations, ignoring the multiple options for charging EVs. Furthermore, the literature has rarely modelled strategies regarding the discharging process and sharing stored electricity of EVs with the power grid [81].

Therefore, it is imperative to incorporate multiple bidirectional charging technologies connected to the electricity grid, charging stations and other energy storage facilities with a flexible incentive-based pricing scheme for the optimal operation of MEaaS. Such a mechanism should consider many priorities, including the RER generation profile, demand and generation of power (supply) within each geographic zone and capturing the sensitiveness of buyers and sellers of electricity to changes in price to make the market dynamic.

It should be noted here that this system should be highly flexible, incentive-based and entirely dependent on the existing demand and supply of electricity at any given time and location. Strategically, the operating system automatically, for example, increases (decreases) the sale price during peak (off-peak) demand to discourage (encourage) customers to charge (discharge) them to sell (buy) to (from) the grid, charging stations, EVs and other storage devices. In addition, it should be able to provide a travel plan for EV drivers depending on their battery capacity where possible economically beneficial pathways for them can be identified in terms of charging/discharging during travel.

Accordingly, the peak-load demands of the grid can be minimized, leading to more optimal usage of the infrastructure. One potential pricing mechanism that can be considered for the MEaaS is an improvement/variant of, for example, Australia's existing wholesale spot market pricing scheme, which is operated by the Australian Energy Market Operator (AEMO), where the electricity supply and demand is 'matched' simultaneously using real-time spot market dispatched every five minutes [82].

The AEMO operated spot market pricing scheme is not accessible to the public to directly purchase electricity from generators. Nonetheless, given the technology involved with MEaaS and because it operates within defined zones, it is argued here that the spot market for mobile energy should be instantaneous because such pricing could be arranged on a zone-by-zone basis as shown in Figure 1 to reach market equilibrium prices. Markets clear when quantity demanded (Q_d) equals quantity supplied (Q_s), as represented in Equation (3):

$$Q_d = Q_s \tag{3}$$

where Q_d and Q_s are quantities of electricity demanded and supplied, respectively. Using primary and secondary data, it is possible to solve for equilibrium prices and quantities. Once the price is settled in each zone, the price is instantaneously made available to all system users (buyers and sellers) via an app. The main merit of this strategy is that the price of electricity is instantaneously determined locally. Thus, with respect to RER generation, EV users can decide to charge/discharge (bidirectional charging) in a zone based on the prevailing price. Furthermore, since providing the ancillary services is crucial for grid operators, they can offer further incentives for EV users to participate in the market through V2X/X2V mechanisms. It is advocated that a fully flexible, highly incentivized pricing system that affords the opportunity to arbitrage is a key determinant of success for the uptake of the MEaaS system.

3.4. Public Acceptability of and Appropriate Business Models for MEaaS

Creating a dynamic pricing system for arbitrage is one of the key essentials for the success of MEaaS. Another key determinant for the uptake of this emerging technology hinges on the public acceptability of mobile energy as a realistic product with the desired attributes linked to its use. For public acceptability, it is imperative to showcase the key attributes of the technology. They include the costs, monetary gains from the use of the technology, convenience, reliability and the inclusion of renewable energy and uptake of EVs instead of ICEs.

For this purpose, a common approach is to conduct a consumer choice model, where the most desired attributes mentioned above could be tested, and consumer preferences ranked and highlighted. It is also possible to test whether consumers are willing to adopt the change, under what circumstances they would do so or whether they would prefer the status quo. In this case, ICEs or EVs minus the MEaaS system. It is also possible to rank consumers' acceptability for each attribute in monetary terms. The choice modelling is usually embedded in a survey of consumers where the technology is most likely to be adopted.

A consumer choice model assumes that a consumer reveals his/her preference. Here, the consumer selects the alternative that provides the highest utility [83], say between ICE and EV with MEaaS technology. That is, a consumer n selects choice I if $U_i > U_j$, "$_j$ Î $C_{n,}$ $i \neq j$, where U_i is decomposed into a A deterministic (observed), V_{nj}, and random (unobserved) part, ε_{nj}, as represented in Equation (4):

$$U_i = V_i + \varepsilon_i \tag{4}$$

where V_i is a deterministic component, and ε_i is a random error component that captures any influences on individual choices that are unobservable to the researcher. The deterministic component, V_i, is the function of the MEaaS attributes and socioeconomic characteristics of the consumer, and can be expressed as represented in Equation (5):

$$V_i = \beta_0 + \beta_1 X_1 + \beta_2 X_2 + \ldots.. + \beta_n X_n + \beta_a S_1 + \beta_b S_2 + \ldots.. \beta_m S_k \tag{5}$$

where β_0 is the alternative specific constant, X_1 to X_n are the attributes, S_1 to S_k are the social, economic and attitudinal characteristics of the consumer and β_1 to β_n and β_a to β_m are attached to the vectors of attributes and vectors of the consumer characteristics, both of which influence utility. Models such as multinomial logit (MNL), random parameter logit (RPL) and other models can be used to analyze the choice data [84,85].

Researchers have used choice experiments to test consumers' acceptability of autonomous vehicles and one of the studies that provide before and after provision of information about the emerging technologies include [86]. These experiments which gather socioeconomic (including income, education and gender) and attitudinal data on surveyed residents provide numerous indications, including which groups are likely to adopt such technology and those who are unlikely. That is to say, identifying key determinants of such technology is vital for technology developers, policy planners and investors. Such modelling can also indicate barriers and identify the most preferred attributes and provides an excellent basis to measure the strengths of these markets and to what extent and under what conditions consumers are likely to embrace such technology.

In summary, choice surveys, if well-executed, could elicit critical data that will enable the planning and execution of a business model that considers customer preferences, costs and benefits. This exercise will also provide valuable insights into the pricing mechanisms (including the development of relevant apps) that need to be put in place and obtain an understanding of the challenges and opportunities that MEaaS provides the various stakeholders.

4. Conclusions: Future Research Directions

Increasing urban population and their energy and mobility needs have created major energy and transport sustainability problems for particularly large metropolitan cities [87–89]. In this perspective paper, we argued for the need for MEaaS to operationalize energy, transport and urban infrastructures for establishing sustainable electromobility in large metropolitan cities to address the energy and transport sustainability problems. Prospective studies on MEaaS are needed as they provide numerous benefits, where some of them are elaborated as below.

Prospective studies on MEaaS will disclose new knowledge and scientific outcomes: The chief specific outcome of prospective research on MEaaS will be new scientific knowledge, as the MEaaS concept is the first comprehensive attempt to investigate future urban energy systems for sustainable electromobility in large metropolitan cities, where many of them today call themselves smart cities. These investigations will generate critical knowledge and an evidence base that will enable government agencies to follow pathways in adopting appropriate urban energy systems for sustainable electromobility.

Prospective studies on MEaaS will generate economic returns flowing from scientific outcomes: There are likely to be significant economic returns flowing from such scientific outcomes. These investigations will be of direct benefit to government agencies, and many others internationally—as these research studies will unveil new knowledge on adopting appropriate urban energy systems for sustainable electromobility. The large-scale commercialization of these MEaaS will generate an economic return.

Prospective studies on MEaaS will provide social and environmental returns flowing from scientific outcomes: The flow-on benefits will not just be economical; there will be significant societal and environmental benefits, as these studies will identify socio-spatial negative externalities of urban mobility and prescribe responsible solutions for government agencies to address the adverse effects on the communities and the environment. The MEaaS system with its financial and other benefits will accelerate the uptake of EVs, thus reducing the use of fossil-based fuels. These investigations will also reduce the strain imposed on the existing power grid.

Prospective studies on MEaaS will support informed urban, transport and energy policy and debate: These studies, throughout their investigation phase, will generate research outcomes, which will be communicated regularly to inform urban, transport and energy policy and debate, thereby raising the awareness of governments and the public regarding the importance of MEaaS solutions.

Future research will inform urban, transport and energy policy circles and the research community by leading a public discourse on urban energy systems that foster sustainable electromobility in large metropolitan cities. Our research team will also continue to embark on different facets of MEaaS and complimentary aspects of future urban, transport and power technologies.

We conclude this perspective piece by quoting [90], "the electricity grid with a high penetration of renewable energy can enable travelers to travel free of emissions using state-of-the-art EVs. Extensive EV demands at the peak-times, and an increase in electricity consumption due to population growth, have led to higher utility infrastructure investments. Mobile energy systems can be used as an innovative demand-side management solution to reduce long-term utility infrastructure investments. They can store and release electricity to the grid based on consumer demand. However, a scientific planning approach for grid integration has been overlooked (p. 1)", and our study in this paper offers a new conceptualization of such a system integration with MEaaS.

Author Contributions: Conceptualization, methodology, investigation, writing—original draft preparation, and writing—review and editing, M.V., Y.M., T.Y., A.B. and C.W. All authors have read and agreed to the published version of the manuscript.

Funding: This research did not receive any specific grant from funding agencies in the public, commercial or not-for-profit sectors.

Institutional Review Board Statement: Not applicable.

Informed Consent Statement: Not applicable.

Data Availability Statement: The paper did not use any data beyond the literature, where cited and quoted literature works are listed in the references.

Acknowledgments: The authors thank the editor and anonymous referees for their constructive comments.

Conflicts of Interest: The authors declare no conflict of interest.

References

1. Nuttall, W.J.; Manz, D.L. A new energy security paradigm for the twenty-first century. *Technol. Forecast. Soc. Chang.* **2008**, *75*, 1247–1259. [CrossRef]
2. Webb, J.; de Silva, H.N.; Wilson, C. The future of coal and renewable power generation in Australia: A review of market trends. *Econ. Anal. Policy* **2020**, *68*, 363–378. [CrossRef]
3. Yilmaz, M.; Krein, P.T. Review of the Impact of Vehicle-to-Grid Technologies on Distribution Systems and Utility Interfaces. *IEEE Trans. Power Electron.* **2013**, *28*, 5673–5689. [CrossRef]
4. Holz-Rau, C.; Scheiner, J. Land-use and transport planning—A field of complex cause-impact relationships. Thoughts on transport growth, greenhouse gas emissions and the built environment. *Transp. Policy* **2019**, *74*, 127–137. [CrossRef]
5. Mahbub, P.; Goonetilleke, A.; Ayoko, G.; Egodawatta, P.; Yigitcanlar, T. Analysis of build-up of heavy metals and volatile organics on urban roads in gold coast, Australia. *Water Sci. Technol.* **2011**, *63*, 2077–2085. [CrossRef]
6. Li, F.; Ou, R.; Xiao, X.; Zhou, K.; Xie, W.; Ma, D.; Liu, K.; Song, Z. Regional comparison of electric vehicle adoption and emission reduction effects in China. *Resour. Conserv. Recycl.* **2019**, *149*, 714–726. [CrossRef]
7. Global EV Sales Set to Smash Records with 7 Million Cars in 2021 While Crossing 10% Annual Threshold. Ecogeneration. Available online: https://www.ecogeneration.com.au/global-ev-sales-set-to-smash-records-with-7-million-cars-in-2021-while-crossing-10-annual-threshold/#:~{}:text=annual%20threshold%20%7C%20EcoGeneration-,Global%20EV%20sales%20set%20to%20smash%20records%20with%207%20million,while%20crossing%2010%25%20annual%20threshold&text=As%20Australia%20putters%20along%2C%20unsure,the%20world%20is%20accelerating%20ahead (accessed on 13 February 2022).
8. Yigitcanlar, T. Towards Smart and Sustainable Urban Electromobility: An Editorial Commentary. *Sustainability* **2022**, *14*, 2264. [CrossRef]
9. Macioszek, E. E-mobility Infrastructure in the Górnośląsko—Zagłębiowska Metropolis, Poland, and Potential for Develop-ment. In Proceedings of the 5th World Congress on New Technologies (NewTech'19), Lisbon, Portugal, 18–20 August 2019.
10. Szczuraszek, T.; Chmielewski, J. Planning Spatial Development of a City from the Perspective of Its Residents' Mobility Needs. In *Advances in Human Error, Reliability, Resilience, and Performance*; Springer Science and Business Media LLC: Cham, Switzerland, 2019; pp. 3–12.
11. Ling, Z.; Cherry, C.R.; Wen, Y. Determining the Factors That Influence Electric Vehicle Adoption: A Stated Preference Survey Study in Beijing, China. *Sustainability* **2021**, *13*, 11719. [CrossRef]
12. Coffman, M.; Bernstein, P.; Wee, S. Electric vehicles revisited: A review of factors that affect adoption. *Transp. Rev.* **2017**, *37*, 79–93. [CrossRef]

13. Chen, C.-F.; de Rubens, G.Z.; Noel, L.; Kester, J.; Sovacool, B.K. Assessing the socio-demographic, technical, economic and behavioral factors of Nordic electric vehicle adoption and the influence of vehicle-to-grid preferences. *Renew. Sustain. Energy Rev.* **2020**, *121*, 109692. [CrossRef]
14. Rietmann, N.; Hügler, B.; Lieven, T. Forecasting the trajectory of electric vehicle sales and the consequences for worldwide CO2 emissions. *J. Clean. Prod.* **2020**, *261*, 121038. [CrossRef]
15. Buekers, J.; Van Holderbeke, M.; Bierkens, J.; Panis, L.I. Health and environmental benefits related to electric vehicle introduction in EU countries. *Transp. Res. Part D: Transp. Environ.* **2014**, *33*, 26–38. [CrossRef]
16. Hannan, M.; Hoque, M.; Mohamed, A.; Ayob, A. Review of energy storage systems for electric vehicle applications: Issues and challenges. *Renew. Sustain. Energy Rev.* **2017**, *69*, 771–789. [CrossRef]
17. Un-Noor, F.; Padmanaban, S.; Mihet-Popa, L.; Mollah, M.N.; Hossain, E. A Comprehensive Study of Key Electric Vehicle (EV) Components, Technologies, Challenges, Impacts, and Future Direction of Development. *Energies* **2017**, *10*, 1217. [CrossRef]
18. Hardman, S.; Jenn, A.; Tal, G.; Axsen, J.; Beard, G.; Daina, N.; Figenbaum, E.; Jakobsson, N.; Jochem, P.; Kinnear, N.; et al. A review of consumer preferences of and interactions with electric vehicle charging infrastructure. *Transp. Res. Part D Transp. Environ.* **2018**, *62*, 508–523. [CrossRef]
19. Pagany, R.; Camargo, L.R.; Dorner, W. A review of spatial localization methodologies for the electric vehicle charging infrastructure. *Int. J. Sustain. Transp.* **2019**, *13*, 433–449. [CrossRef]
20. Jarvis, R.; Moses, P. Smart Grid Congestion Caused by Plug-in Electric Vehicle Charging. In Proceedings of the 2019 IEEE Texas Power and Energy Conference (TPEC), College Station, TX, USA, 7–8 February 2019; pp. 1–5.
21. Van Vliet, O.; Brouwer, A.S.; Kuramochi, T.; van den Broek, M.; Faaij, A. Energy use, cost and CO_2 emissions of electric cars. *J. Power Sources* **2011**, *196*, 2298–2310. [CrossRef]
22. Moses, P.S.; Masoum, M.A.S.; Hajforoosh, S. Overloading of distribution transformers in smart grid due to uncoordinated charging of plug-in electric vehicles. In Proceedings of the 2012 IEEE PES Innovative Smart Grid Technologies (ISGT), Washington, DC, USA, 16–20 January 2012; pp. 1–6.
23. Khan, W.; Ahmad, A.; Ahmad, F.; Alam, M.S. A Comprehensive Review of Fast Charging Infrastructure for Electric Vehicles. *Smart Sci.* **2018**, *6*, 1–15. [CrossRef]
24. Dubarry, M.; Devie, A.; McKenzie, K. Durability and reliability of electric vehicle batteries under electric utility grid operations: Bidirectional charging impact analysis. *J. Power Sources* **2017**, *358*, 39–49. [CrossRef]
25. Kharrazi, A.; Sreeram, V.; Mishra, Y. Assessment techniques of the impact of grid-tied rooftop photovoltaic generation on the power quality of low voltage distribution network—A review. *Renew. Sustain. Energy Rev.* **2020**, *120*, 109643. [CrossRef]
26. Hung, D.Q.; Mishra, Y. Impacts of Single-Phase PV Injection on Voltage Quality in 3-Phase 4-Wire Distribution Systems. In Proceedings of the 2018 IEEE Power & Energy Society General Meeting (PESGM), Portland, OR, USA, 5–10 August 2018; pp. 1–5.
27. Mishra, S.; Mishra, Y. Decoupled controller for single-phase grid connected rooftop PV systems to improve voltage profile in residential distribution systems. *IET Renew. Power Gener.* **2017**, *11*, 370–377. [CrossRef]
28. Esplin, R.; Nelson, T. Redirecting solar feed in tariffs to residential battery storage: Would it be worth it? *Econ. Anal. Policy* **2021**, *73*, 373–389. [CrossRef]
29. Ahmadian, A.; Sedghi, M.; Mohammadi-Ivatloo, B.; Elkamel, A.; Golkar, M.A.; Fowler, M. Cost-Benefit Analysis of V2G Implementation in Distribution Networks Considering PEVs Battery Degradation. *IEEE Trans. Sustain. Energy* **2018**, *9*, 961–970. [CrossRef]
30. Lam, A.Y.S.; Leung, K.-C.; Li, V.O.K. Capacity Estimation for Vehicle-to-Grid Frequency Regulation Services qith Smart Charging Mechanism. *IEEE Trans. Smart Grid* **2016**, *7*, 156–166. [CrossRef]
31. Zhang, Z.; Mishra, Y.; Yue, D.; Dou, C.-X.; Zhang, B.; Tian, Y.-C. Delay-Tolerant Predictive Power Compensation Control for Photovoltaic Voltage Regulation. *IEEE Trans. Ind. Inform.* **2021**, *17*, 4545–4554. [CrossRef]
32. Zhang, Z.; Mishra, Y.; Dou, C.; Yue, D.; Zhang, B.; Tian, Y.-C. Steady-State Voltage Regulation with Reduced Photovoltaic Power Curtailment. *IEEE J. Photovolt.* **2020**, *10*, 1853–1863. [CrossRef]
33. Jurdak, R.; Dorri, A.; Vilathgamuwa, M. A Trusted and Privacy-Preserving Internet of Mobile Energy. *IEEE Commun. Mag.* **2021**, *59*, 89–95. [CrossRef]
34. Yu, R.; Zhong, W.; Xie, S.; Yuen, C.; Gjessing, S.; Zhang, Y. Balancing Power Demand Through EV Mobility in Vehicle-to-Grid Mobile Energy Networks. *IEEE Trans. Ind. Inform.* **2015**, *12*, 79–90. [CrossRef]
35. Zhong, W.; Yu, R.; Xie, S.; Zhang, Y.; Yau, D.K.Y. On Stability and Robustness of Demand Response in V2G Mobile Energy Networks. *IEEE Trans. Smart Grid* **2018**, *9*, 3203–3212. [CrossRef]
36. Abdeltawab, H.; Mohamed, Y. Mobile energy storage scheduling and operation in active distribution systems. *IEEE Trans. Ind. Electron.* **2017**, *64*, 6828–6840. [CrossRef]
37. Bozchalui, M.C.; Sharma, R. Analysis of Electric Vehicles as Mobile Energy Storage in commercial buildings: Economic and environmental impacts. In Proceedings of the 2012 IEEE Power and Energy Society General Meeting, San Diego, CA, USA, 22–26 July 2012; pp. 1–8.
38. Webb, J.; Whitehead, J.; Wilson, C. Chapter 18—Who Will Fuel Your Electric Vehicle in the Future? You or Your Utility? In *Consumer, Prosumer, Prosumager*; Sioshansi, F., Ed.; Academic Press: Cambridge, MA, USA, 2019; pp. 407–429.
39. Webb, J.; Wilson, C.; Steinberg, T.; Stein, W. Chapter 19—Solar Grid Parity and its Impact on the Grid. In *Innovation and Disruption at the Grid's Edge*; Sioshansi, F.P., Ed.; Academic Press: Cambridge, MA, USA, 2017; pp. 389–408.

0. De Melo, H.N.; Trovao, J.P.F.; Pereirinha, P.G.; Jorge, H.; Antunes, C.H. A Controllable Bidirectional Battery Charger for Electric Vehicles with Vehicle-to-Grid Capability. *IEEE Trans. Veh. Technol.* **2017**, *67*, 114–123. [CrossRef]
1. Tomić, J.; Kempton, W. Using fleets of electric-drive vehicles for grid support. *J. Power Sources* **2007**, *168*, 459–468. [CrossRef]
2. Li, S.; Mi, C. Wireless Power Transfer for Electric Vehicle Applications. *IEEE J. Emerg. Sel. Top. Power Electron.* **2015**, *3*, 4–17.
3. Vilathgamuwa, D.M.; Sampath, J.P.K. Wireless Power Transfer (WPT) for Electric Vehicles (EVs)—Present and Future Trends. In *Plug in Electric Vehicles in Smart Grids: Integration Techniques*; Rajakaruna, S., Shahnia, F., Ghosh, A., Eds.; Springer: Singapore, 2015; pp. 33–60.
4. Chakraborty, S.V.; Shukla, S.K.; Thorp, J. A detailed analysis of the effective-load-carrying-capacity behavior of plug-in electric vehicles in the power grid. In Proceedings of the 2012 IEEE PES Innovative Smart Grid Technologies (ISGT), Washington, DC, USA, 16–20 January 2012; pp. 1–8.
5. Schneider, K.; Gerkensmeyer, C.; Kintner-Meyer, M.; Fletcher, R. Impact assessment of plug-in hybrid vehicles on pacific northwest distribution systems. In Proceedings of the 2008 IEEE Power and Energy Society General Meeting—Conversion and Delivery of Electrical Energy in the 21st Century, Pittsburgh, PA, USA, 20–24 July 2008; pp. 1–6.
46. Zhou, Y.; Wang, M.; Hao, H.; A Johnson, L.; Wang, H. Plug-in electric vehicle market penetration and incentives: A global review. *Mitig. Adapt. Strat. Glob. Chang.* **2015**, *20*, 777–795. [CrossRef]
47. Yin, X.; Zhao, X. Planning of Electric Vehicle Charging Station Based on Real Time Traffic Flow. In Proceedings of the 2016 IEEE Vehicle Power and Propulsion Conference (VPPC), Hangzhou, China, 17–20 October 2016; pp. 1–4.
48. Zhu, J.; Li, Y.; Yang, J.; Li, X.; Zeng, S.; Chen, Y. Planning of electric vehicle charging station based on queuing theory. *J. Eng.* **2017**, *2017*, 1867–1871. [CrossRef]
49. Cui, Q.; Weng, Y.; Tan, C.-W. Electric Vehicle Charging Station Placement Method for Urban Areas. *IEEE Trans. Smart Grid* **2019**, *10*, 6552–6565. [CrossRef]
50. Lin, H.; Bian, C.; Wang, Y.; Li, H.; Sun, Q.; Wallin, F. Optimal planning of intra-city public charging stations. *Energy* **2021**, *238*, 121948. [CrossRef]
51. Shaikh, P.W.; Mouftah, H.T. Intelligent Charging Infrastructure Design for Connected and Autonomous Electric Vehicles in Smart Cities. In Proceedings of the 2021 IFIP/IEEE International Symposium on Integrated Network Management (IM), Bordeaux, France, 17–21 May 2021; pp. 992–997.
52. Faisal, A.; Yigitcanlar, T.; Kamruzzaman; Paz, A. Mapping Two Decades of Autonomous Vehicle Research: A Systematic Scientometric Analysis. *J. Urban Technol.* **2020**, *28*, 45–74. [CrossRef]
53. Wu, J.; Liao, H.; Wang, J.-W.; Chen, T. The role of environmental concern in the public acceptance of autonomous electric vehicles: A survey from China. *Transp. Res. Part F Traffic Psychol. Behav.* **2019**, *60*, 37–46. [CrossRef]
54. Golbabaei, F.; Yigitcanlar, T.; Bunker, J. The role of shared autonomous vehicle systems in delivering smart urban mobility: A systematic review of the literature. *Int. J. Sustain. Transp.* **2021**, *15*, 731–748. [CrossRef]
55. Butler, L.; Yigitcanlar, T.; Paz, A. Smart Urban Mobility Innovations: A Comprehensive Review and Evaluation. *IEEE Access* **2020**, *8*, 196034–196049. [CrossRef]
56. Márquez-Fernández, F.J.; Bischoff, J.; Domingues-Olavarría, G.; Alaküla, M. Assessment of future EV charging infra-structure scenarios for long-distance transport in Sweden. *IEEE Trans. Transp. Electrif.* **2021**. [CrossRef]
57. Nykvist, B.; Sprei, F.; Nilsson, M. Assessing the progress toward lower priced long range battery electric vehicles. *Energy Policy* **2019**, *124*, 144–155. [CrossRef]
58. Lee, H.; Alex, C. *Charging the Future: Challenges and Opportunities for Electric Vehicle Adoption*; Harvard University: Cambridge, MA, USA, 2018.
59. Villeneuve, D.; Füllemann, Y.; Drevon, G.; Moreau, V.; Vuille, F.; Kaufmann, V. Future Urban Charging Solutions for Electric Vehicles. *Eur. J. Transp. Infrastruct. Res.* **2020**, *20*, 78–102.
60. Dur, F.; Yigitcanlar, T. Assessing land-use and transport integration via a spatial composite indexing model. *Int. J. Environ. Sci. Technol.* **2015**, *12*, 803–816. [CrossRef]
61. He, J.; Yang, H.; Tang, T.-Q.; Huang, H.-J. An optimal charging station location model with the consideration of electric vehicle's driving range. *Transp. Res. Part C Emerg. Technol.* **2018**, *86*, 641–654. [CrossRef]
62. Yang, J.; Dong, J.; Hu, L. A data-driven optimization-based approach for siting and sizing of electric taxi charging stations. *Transp. Res. Part C Emerg. Technol.* **2017**, *77*, 462–477. [CrossRef]
63. Chen, Z.; He, F.; Yin, Y. Optimal deployment of charging lanes for electric vehicles in transportation networks. *Transp. Res. Part B Methodol.* **2016**, *91*, 344–365. [CrossRef]
64. Ahmad, A.; Alam, M.S.; Chabaan, R. A Comprehensive Review of Wireless Charging Technologies for Electric Vehicles. *IEEE Trans. Transp. Electrif.* **2018**, *4*, 38–63. [CrossRef]
65. Kandasamy, K.; Vilathgamuwa, D.M.; Madawala, U.K.; Tseng, K. Inductively coupled modular battery system for electric vehicles. *IET Power Electron.* **2016**, *9*, 600–609. [CrossRef]
66. Zhang, S.; Yu, J.J. Electric Vehicle Dynamic Wireless Charging System: Optimal Placement and Vehicle-to-Grid Scheduling. *IEEE Internet Things J.* **2021**. [CrossRef]
67. Wu, H.; Niu, D. Study on Influence Factors of Electric Vehicles Charging Station Location Based on ISM and FMICMAC. *Sustainability.* **2017**, *9*, 484. [CrossRef]

68. Csiszár, C.; Csonka, B.; Földes, D.; Wirth, E.; Lovas, T. Urban public charging station locating method for electric vehicles based on land use approach. *J. Transp. Geogr.* **2019**, *74*, 173–180. [CrossRef]
69. Mirzaei, M.J.; Kazemi, A.; Homaee, O. A Probabilistic Approach to Determine Optimal Capacity and Location of Electric Vehicles Parking Lots in Distribution Networks. *IEEE Trans. Ind. Inform.* **2015**, *12*, 1963–1972. [CrossRef]
70. Foley, B.; Degirmenci, K.; Yigitcanlar, T. Factors Affecting Electric Vehicle Uptake: Insights from a Descriptive Analysis in Australia. *Urban Sci.* **2020**, *4*, 57. [CrossRef]
71. Habib, S.; Khan, M.M.; Abbas, F.; Tang, H. Assessment of electric vehicles concerning impacts, charging infrastructure with unidirectional and bidirectional chargers, and power flow comparisons. *Int. J. Energy Res.* **2018**, *42*, 3416–3441. [CrossRef]
72. Butler, L.; Yigitcanlar, T.; Paz, A. Barriers and risks of Mobility-as-a-Service (MaaS) adoption in cities: A systematic review of the literature. *Cities* **2021**, *109*, 103036. [CrossRef]
73. Yigitcanlar, T.; Bulu, M. Dubaization of Istanbul: Insights from the Knowledge-Based Urban Development Journey of an Emerging Local Economy. *Environ. Plan. A Econ. Space* **2015**, *47*, 89–107. [CrossRef]
74. Sarimin, M.; Yigitcanlar, T. Towards a comprehensive and integrated knowledge-based urban development model: Status quo and directions. *Int. J. Knowl. Based Dev.* **2012**, *3*, 175. [CrossRef]
75. Tan, Y.; Koray, V.; Scott, B. (Eds.) *Creative Urban Regions: Harnessing Urban Technologies to Support Knowledge City Initiatives*; IGI Global: Hershey, PA, USA, 2008.
76. Esmaeilpoorarabi, N.; Yigitcanlar, T.; Guaralda, M. Place quality in innovation clusters: An empirical analysis of global best practices from Singapore, Helsinki, New York, and Sydney. *Cities* **2018**, *74*, 156–168. [CrossRef]
77. Wang, X.; Shahidehpour, M.; Jiang, C.; Li, Z. Coordinated Planning Strategy for Electric Vehicle Charging Stations and Coupled Traffic-Electric Networks. *IEEE Trans. Power Syst.* **2019**, *34*, 268–279. [CrossRef]
78. Energeia. *Australian Electric Vehicle Market Study*; AREA: Canberra, Australia, 2018; pp. 1–103.
79. Li, X.; Xiang, Y.; Lyu, L.; Ji, C.; Zhang, Q.; Teng, F.; Liu, Y. Price Incentive-Based Charging Navigation Strategy for Electric Vehicles. *IEEE Trans. Ind. Appl.* **2020**, *56*, 5762–5774. [CrossRef]
80. Dorcec, L.; Pevec, D.; Vdovic, H.; Babic, J.; Podobnik, V. How do people value electric vehicle charging service? A gamified survey approach. *J. Clean. Prod.* **2019**, *210*, 887–897. [CrossRef]
81. Solanke, T.U.; Ramachandaramurthy, V.K.; Yong, J.Y.; Pasupuleti, J.; Kasinathan, P.; Rajagopalan, A. A review of strategic charging–discharging control of grid-connected electric vehicles. *J. Energy Storage* **2020**, *28*, 101193. [CrossRef]
82. Zhang, S.; Mishra, Y.; Shahidehpour, M. Utilizing distributed energy resources to support frequency regulation services. *Appl. Energy* **2017**, *206*, 1484–1494. [CrossRef]
83. Hanley, N.; Mourato, S.; Wright, R.E. Choice Modelling Approaches: A Superior Alternative for Environmental Valuatioin? *J. Econ. Surv.* **2001**, *15*, 435–462. [CrossRef]
84. Lin, Z.; Greene, D. A Plug-in Hybrid Consumer Choice Model with Detailed Market Segmentation. In Proceedings of the Transportation Research Board 89th Annual Meeting, Washington, DC, USA, 10–14 January 2010.
85. Hensher, D.A.; Johnson, L.W. *Applied Discrete-Choice Modelling*, 1st ed.; Routledge: London, UK, 2018.
86. Webb, J.; Wilson, C.; Kularatne, T. Will people accept shared autonomous electric vehicles? A survey before and after receipt of the costs and benefits. *Econ. Anal. Policy* **2019**, *61*, 118–135. [CrossRef]
87. Yigitcanlar, T.; Dodson, J.; Gleeson, B.J.; Sipe, N.G. Travel Self-Containment in Master Planned Estates: Analysis of Recent Australian Trends. *Urban Policy Res.* **2007**, *25*, 129–149. [CrossRef]
88. Ingrao, C.; Messineo, A.; Beltramo, R.; Yigitcanlar, T.; Ioppolo, G. How can life cycle thinking support sustainability of buildings? Investigating life cycle assessment applications for energy efficiency and environmental performance. *J. Clean. Prod.* **2018**, *201*, 556–569. [CrossRef]
89. Lantz, T.; Ioppolo, G.; Yigitcanlar, T.; Arbolino, R. Understanding the correlation between energy transition and urbanization. *Environ. Innov. Soc. Transitions* **2021**, *40*, 73–86. [CrossRef]
90. Khardenavis, A.; Hewage, K.; Perera, P.; Shotorbani, A.M.; Sadiq, R. Mobile energy hub planning for complex urban networks: A robust optimization approach. *Energy* **2021**, *235*, 121424. [CrossRef]

MDPI
St. Alban-Anlage 66
4052 Basel
Switzerland
Tel. +41 61 683 77 34
Fax +41 61 302 89 18
www.mdpi.com

Sustainability Editorial Office
E-mail: sustainability@mdpi.com
www.mdpi.com/journal/sustainability

www.ingramcontent.com/pod-product-compliance
Lightning Source LLC
LaVergne TN
LVHW070132170726
843515LV00007B/1780

* 9 7 8 3 0 3 6 5 4 2 7 4 4 *